# Lecture Notes in Engineering

The Springer-Verlag Lecture Notes provide rapid (approximately six months), refereed publication of topical items, longer than ordinary journal articles but shorter and less formal than most monographs and textbooks. They are published in an attractive yet economical format; authors or editors provide manuscripts typed to specifications, ready for photo-reproduction.

# Lecture Notes in Engineering

Edited by C. A. Brebbia and S. A. Orszag

10

C. A. Brebbia, H. Tottenham,
G. B. Warburton, J. M. Wilson,
R. R. Wilson

# Vibrations of Engineering Structures

Springer-Verlag
Berlin Heidelberg New York Tokyo

**Authors**
C. A. Brebbia
Civil Engineering Dept.
University of Southampton
Southampton
U. K.

H. Tottenham
11 Middlebridge Street
Romsey, Hampshire
SO 5 8 HJ
U.K.

G. B. Warburton
Dept. of Mechanical Engg.
The University of Nottingham
University Park
Nottingham NG 7 2 RD
U.K.

J. M. Wilson
Dept. of Engineering
University of Durham
Science Laboratories
South Road
Durham DH 1 3 LE
U.K.

R. R. Wilson
James Howden & Company Limited
195, Scotland Street
Glasgow, G 5 8 PJ
Scotland

ISBN-13: 978-3-540-13959-1 e-ISBN-13: 978-3-642-82390-9
DOI: 10.1007/978-3-642-82390-9

Library of Congress Cataloging in Publication Data
Main entry under title:
Vibrations of engineering structures.
(Lecture notes in engineering; 10)
1. Structural dynamics. 2. Vibration.
I. Brebbia, C. A. II. Series.
TA654.V49 1985 624.1'76 84-24800



2061/3020-543210

# FOREWORD

The increasing size and complexity of new structural forces in engineering have made it necessary for designers to be aware of their dynamic behaviour. Dynamics is a subject which has traditionally been poorly taught in most engineering courses. This book was conceived as a way of providing engineers with a deeper knowledge of dynamic analysis and of indicating to them how some of the new vibrations problems can be solved. The authors start from basic principles to end up with the latest random vibration applications. The book originated in a week course given annually by the authors at the Computational Mechanics Centre, Ashurst Lodge, Southampton, England. Special care was taken to ensure continuity in the text and notations.

Southampton 1984

CONTENTS

# CHAPTER 1

# INTRODUCTION TO VIBRATION

by

G.B. Warburton

## 1. Introductory Remarks

In recent years the number of structures, for which the dynamic forces, likely to be encountered in service, have required investigation at the design stage, has increased. Several factors have contributed to this increase: growth in size of structures of various types; consequential increased importance of wind forces; efforts to reduce the effects of earthquakes on structures and to prevent total collapse; design of off-shore structures. Two important questions are: why is it essential to include dynamic effects in structural analysis and why is this a more difficult task than conventional (static) structural analysis?

Suppose that the stresses in a structure are known for: (a) a static force P at a particular location; (b) a force at the same location that varies in magnitude with time and has a maximum value of P. Then the dynamic magnification factor is the maximum stress at a point for (b) / the stress at the same point for (a). This factor depends upon how the force varies with time, the distribution of stiffness and mass in the structure and the damping present. In certain circumstances it will be very large; in others very small. Obviously, if there is any possibility of the dynamic magnification factor being significantly greater than unity, a dynamic analysis of the structure is necessary. This book is primarily concerned with methods of determining dynamic magnification factors for various types of loads and structures. However, no simple rules exist for these factors. Thus there are greater conceptual difficulties for dynamic problems than for comparable static problems, as the intuition and experience, which help an engineer to form a reasonable view of the safety of a structure under static forces, do not lead to an estimate of the relevant dynamic magnification factors. Also the time dependence of stresses, displacements etc. and the necessity to include mass and damping effects make dynamic analysis more complex than its static counterpart. There are also practical difficulties; some dynamic loads, e.g. wind forces, and most damping forces can only be estimated.

In addition to the possibility of elastic failure of a structure if dynamic effects are neglected, long-time repetition of dynamic stresses, whose magnitudes would be considered to be safe from static considerations, may lead to cumulative fatigue failures.

In this chapter the concepts that are relevant to vibration analysis of structures will be discussed briefly. Emphasis is on the response of structures to dynamic forces and how different types of force time variation influence the choice of method. Many of the concepts are introduced by considering the simplest vibrating structure; then, as this simple structure has limited practical applications, general structures are discussed. For these the normal mode method of determining response is given particular attention, because it illustrates the physical behaviour of structures better than other methods. Lastly dynamic interaction problems are discussed; here interaction exists between the vibrations of a structure and those of the underlying soil or the surrounding fluid. Many current practical problems, and also much current research effort, involve interaction effects.

Naturally in a single chapter the major topics of structural vibration can only be mentioned. Most of these topics will be studied in depth in subsequent chapters. It is hoped that their introduction here will illustrate their interrelationship and show how they contribute to the determination of stresses in complex structures caused by various types of dynamic excitation.

## 2. Single Degree of Freedom Systems: Equation of Motion and Types of Problem

Although the dynamic response of a practical structure will be complex, it is necessary to begin our study by considering the fundamentals of vibration of simple systems. A rough guide to the complexity of a dynamical system is the number of *degrees of freedom* possessed by the system. This number is equal to the number of independent coordinates required to specify completely the displacement of the system. For instance, a rigid body constrained to move in the X Y plane requires three coordinates to specify its position completely - namely the linear displacements in the X- and Y-directions and the angular rotation about the Z-axis (perpendicular to the plane X Y); thus this body has three degrees of freedom. The displacement of an elastic body, e.g. a beam, has to be specified at each point by using a continuous equation so that an elastic body has an infinite number of degrees of freedom. In a dynamical problem the number of modes of vibration in which a structure can respond is equal to the number of degrees of freedom, thus the simplest structure has only one degree of freedom.

Figure 1 shows the conventional representation of a system with one degree of freedom; it consists of a mass m constrained to move in the X-direction by frictionless guides and restrained by the spring of stiffness k. It is assumed that the mass of the spring is negligible compared to m. Thus the displacement of the system is specified completely by x, the displacement of the mass, and the system has one degree of freedom. For the purpose of analysing their dynamic response it is possible to treat some simple structures as systems with one degree of freedom.

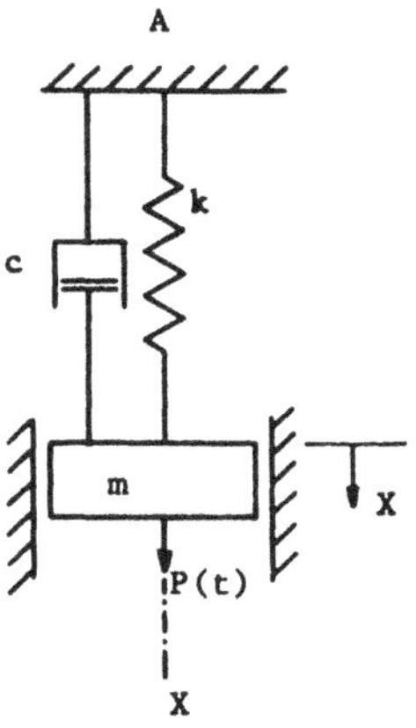


Figure 1 Single degree of freedom system

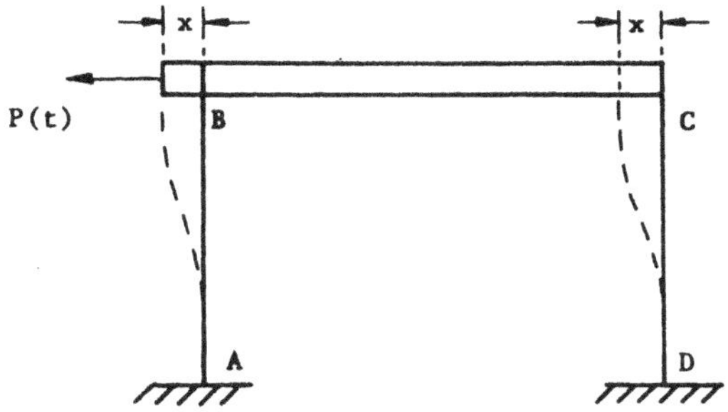


Figure 2 Simple frame with one degree of freedom

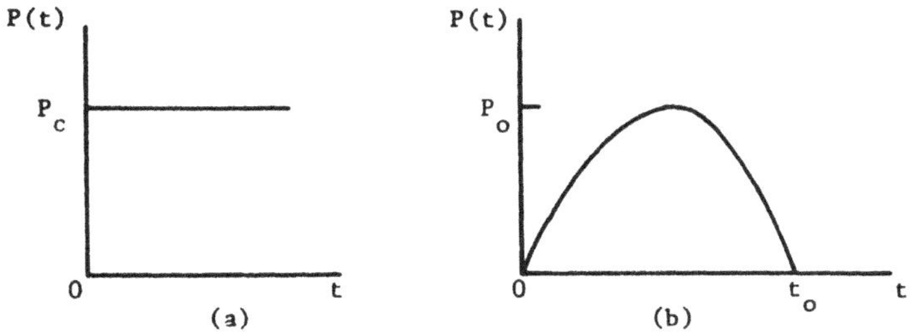


Figure 3 Examples of transient force excitation

In the simple frame of Figure 2 it is assumed that the horizontal member BC is rigid, that the vertical members AB and CD have negligible mass compared to that of BC, and that in any swaying motion BC remains horizontal. Then the motion of the system is given by the horizontal displacement of BC, x, and the frame can be treated as a system with one degree of freedom. Equations and results derived for the system of Figure 1 will be applicable to that of Figure 2.

The general equation of motion is derived by considering the forces acting on the mass m of Figure 1 at any time t. If the displacement of the mass, x, is measured from the position of static equilibrium, the gravity force mg need not be included in the equation as it is balanced by the restoring force in the spring $kx_s$ where $x_s$ is the static deflection of m and k is the stiffness of the spring or the force required to produce unit deflection in the spring; it is assumed that the spring is linear, i.e. k is a constant.

In any real system there will be some damping; this may take various forms, but here *viscous damping* will be assumed, and thus the damping force is proportional to the velocity $\dot{x}$ and opposes the motion. (A dot over a symbol indicates differentiation with respect to time; thus velocity $dx/dt \equiv \dot{x}$ and acceleration $d^2x/dt^2 \equiv \ddot{x}$.) Conventionally viscous damping is represented by the dashpot, shown in parallel with the spring in Figure 1; in practice the damping force is caused by internal friction in the spring etc. and thus is collinear with the spring force.

Newton's second law of motion is applied to the system; this can be expressed as the product of the mass and the resulting acceleration in the X-direction is equal to the net applied force in the X-direction. For this system the latter has three components, namely the applied force P(t), the restoring or spring force (-kx) and the damping force $(-c\dot{x})$. Thus the equation of motion is

$$m\ddot{x} = P(t) - kx - c\dot{x}$$

or

$$m\ddot{x} + c\dot{x} + kx = P(t) \qquad (1)$$

The solution of equation (1) gives the response of the mass to the applied force P(t)

Equation (1) represents also the motion of the member BC of the frame of Figure 2, if m is the mass of BC, k is the combined stiffness of the stanchions AB and DC, and it is assumed that a viscous damping force $c\dot{x}$ opposes the motion of BC. The equations of motion for various single degree of freedom systems are derived in Chapter 2.

Vibrations may be excited by impressed motion at the support. Considering Figure 2, to represent a simple structure, its response to vibrations transmitted through the ground by earthquakes, traffic, pile-drivers, hammers, explosions etc. is important in practice. Suppose that the support A in Figure 1 is given a vertical displacement $x_o(t)$ or the foundation AD in Figure 2 is given a horizontal displacement $x_o(t)$. In both cases the restoring force on the mass due to deformation of the spring or stanchions is $k(x-x_o)$. The damping force is proportional to the relative velocity across the dashpot (Figure 1) and is $c(\dot{x} - \dot{x}_o)$. If the force P(t), shown in Figure 1 and 2, is no longer acting, the equation of motion is:

$$m\ddot{x} = -k(x - x_o) - c(\dot{x} - \dot{x}_o)$$

$$\text{i.e.} \quad m\ddot{x} + c\dot{x} + kx = kx_o + c\dot{x}_o \tag{2}$$

Writing equation (2) in terms of the displacement of the mass relative to the support (i.e. the deformation of the spring or stanchions), $x_r = x - x_o$,

$$m\ddot{x}_r + c\dot{x}_r + kx_r = -m\ddot{x}_o \tag{3}$$

The solution of equation (3) yields the relative displacement, which is proportional to the stress in the elastic member. This solution can be obtained when the base acceleration $\ddot{x}_o$ is specified. In practical problems relating to excitation due to imposed motion of the base, the acceleration is usually known, rather than the displacement and velocity, although the latter can be found by integration.

Equations (1), (2) and (3) are mathematically similar. Thus discussion of the different types of excitation, i.e. how the applied force or base motion varies with time, applies to all three equations. Solutions obtained from one equation can be used to infer solutions for either of the others. Only a change of nomenclature is required to interchange solutions between equations (1) and (3).

Considering the force P(t), shown in Figures 1 and 2, there are three main types of excitation: (i) Harmonic forces, such as $P(t) = P_o \sin \omega t$ or $P(t) = C\omega^2 \sin \omega t$, (the latter is typical of a component of the force produced by out of balance in a rotating machine). A force which is periodic but not harmonic can be expressed as a sum of harmonic terms, using Fourier series, and for a linear system the total response can be obtained by superposing the individual response from each harmonic component of the force. Thus forces which are periodic but not harmonic will not be considered further. (ii) Transient or aperiodic forces: usually these are forces which are applied suddenly or for a short interval of time; simple examples, illustrating the two types, are shown in Figure 3(a) and (b). (iii) Random forces:

the force P(t) cannot be specified as a known function of time, but can be described only in statistical terms; forces due to gusts of wind form an example of this type of excitation.

For (i) the steady-state response of the mass to the harmonic force is required. For (ii) the transient response is required, usually the maximum displacement of the mass or the maximum extension of the spring (the stress in the elastic member of the system is proportional to this extension), occurring during the period of application of the force or in the motion immediately following this period, will be of greatest interest. For (iii) the response can only be determined statistically.

Mathematically, the solution of equation (1) consists of two parts: the complementary function, which is obtained by solving the equation with the right hand side equal to zero, i.e. $P(t) = 0$, and the particular integral which depends on the form of P(t). Physically, the complementary function represents free damped vibrations, i.e. the vibrations that occur if the mass is given an initial displacement or velocity and released. The solution for free vibrations can be written

$$x = \exp(-\gamma\omega_n t)(A \sin \omega_d t + B \cos \omega_d t) \quad (4)$$

where $\omega_d = \omega_n(1 - \gamma^2)^{1/2}$ (5)

$$\omega_n^2 = k/m, \quad \gamma = c/c_c \quad \text{and} \quad c_c = 2(km)^{1/2} = 2k/\omega_n = 2m\omega_n \quad (6)$$

In equation (4) the constants A and B are chosen to satisfy the initial conditions, i.e. the values of $x$ and $\dot{x}$ at time $t = 0$. Equation (4) represents a damped oscillation: $x \to 0$ as $t \to \infty$. It has been assumed that the damping ratio $\gamma < 1$. In practice, $\gamma \ll 1$; thus from equation (5) $\omega_d \simeq \omega_n$. Now $\omega_n$ is the (circular or radian) natural frequency of the system and is of great importance in vibration analysis. If an initial displacement is given to the mass in Figure 1 or 2, the frequency of the ensuing vibrations is strictly $\omega_d$, but provided that $\gamma \ll 1$ it can be assumed that the natural frequency $\omega_n$ has been measured. The assumption that $\gamma \ll 1$ can be checked by determining $\gamma$ from the rate of decay of successive oscillations. (See Chapter 2 for further details).

## 3. Response

The response of systems with one degree of freedom (Figures 1 and 2) to the various types of excitation force will be summarised.

Considering a *harmonic applied force*, i.e. $P(t) = P_o \cos \omega t$, where $P_o$ is a constant and $\omega$ is the (radian) frequency of the force, equation (1) becomes

$$m\ddot{x} + c\dot{x} + kx = P_o \cos \omega t \tag{7}$$

The complete solution consists of free damped vibrations [equation (4)] and a particular integral. However, the former dies out and thus the steady state solution is given by the particular integral, which can be shown to be

$$x = \frac{P_o \cos(\omega t - \beta)}{[(k - m\omega^2)^2 + c^2\omega^2)]^{\frac{1}{2}}} \tag{8}$$

with $\tan \beta = \dfrac{c\omega}{k - m\omega^2}$ .

Using definitions (6) and putting $r = \omega/\omega_n$, i.e. r is the ratio of the excitation frequency to the natural frequency, the steady state amplitude X from equation (8) is

$$\frac{kX}{P_o} = \frac{1}{[(1 - r^2)^2 + (2\gamma r)^2]^{\frac{1}{2}}} \tag{9}$$

Now $P_o/k$ is the static deflection of the mass due to a static force $P_o$, so $kX/P_o$ is the dynamic magnification factor. Equation (9) introduces the phenomenon of resonance. The dynamic magnification factor is a function of the frequency ratio r and the damping ratio $\gamma$. For $\gamma$ small it has a sharp peak when $r = 1$ and this peak value, obtained by putting $r = 1$ in equation (9), is $1/2\gamma$. Thus for practical systems with low damping the dynamic magnification factor is very large when the excitation and natural frequencies coincide. However, well away from resonance the dynamic magnification factor is not large. (See Chapter 2 for further details).

Looking ahead to more complex structures, the viscous damping mechanism, shown in Figure 1 and used in the above equations, causes the response at higher frequencies (strictly higher resonances) to be underestimated. To overcome this difficulty viscous damping is replaced by *hysteretic damping*, i.e. the damping term $c\dot{x}$ in equation (1) is replaced by $h\dot{x}/\omega$, where h is the hysteretic damping constant and $\omega$ is the excitation frequency. With the viscous damper the energy dissipated per cycle increases linearly with the frequency, although the amplitude of vibration is kept constant. For a hysteretic damper the energy dissipated per cycle is independent of the frequency. For hysteretic damping equation (7) is replaced by

$$m\ddot{x} + h\dot{x}/\omega + kx = P_o \cos \omega t \tag{10}$$

If $h/k = \mu$, the steady state amplitude is

$$\frac{kX}{P_o} = \frac{1}{[(1 - r^2)^2 + \mu^2]^{1/2}} \tag{11}$$

The maximum value of the dynamic magnification factor is $1/\mu$ and occurs when $r = 1$.

For a *general transient force* $P(t)$ the solution of equation (1) is given by the Duhamel integral, which is derived in Chapter 2, or by the convolution integral using Laplace transforms, and is

$$x = \frac{1}{m\omega_d} \int_o^t P(\tau) \exp[-\gamma\omega_n (t - \tau)] \sin \omega_d(t - \tau)d\tau \tag{12}$$

In equation (12) it is assumed that at $t = 0$ the displacement and velocity of the mass are zero. If these conditions are not satisfied, free vibrations, equation (4) must be added with A and B determined from the non-zero conditions. [Equation (12) could be used to determine the complete response to a harmonic applied force, but other methods of solution are simpler.] Considering the step function force of Figure 3a, $P(t) = P_o$, $t > 0$, equation (12) is integrated and the response

$$\frac{kx}{P_o} = 1 - \exp(-\gamma\omega_n t) \left[\cos \omega_d t + \frac{\gamma}{(1 - \gamma^2)^{1/2}} \sin \omega_d t\right] \tag{13}$$

This gives damped oscillations about the new mean position, given by $kx/P_o = 1$. The maximum response occurs when $\omega_d t = \pi$ and is given by

$$\left(\frac{kx}{P_o}\right)_{max} = 1 + \exp\left[\frac{-\pi\gamma}{(1 - \gamma^2)^{1/2}}\right] \tag{14}$$

The variation of the dynamic magnification factor from equation (14) with the damping factor $\gamma$ is shown in Table 1. For small damping the factor is relatively insensitive to $\gamma$. For comparison the maximum dynamic magnification factor associated with a harmonic force, namely $1/2\gamma$ for $r = 1$, is also given in the table.

Next consider the response of the system of Figure 1 to a sinusoidal force of finite duration, i.e.

$$P(t) = P_o \sin \frac{\pi t}{t_o}, \qquad 0 \leq t \leq Nt_o \tag{15}$$

$$P(t) = 0, \qquad t > Nt_o$$

where N is an integer. [ The force $P(t)$ is shown in Figure 3(b) for $N = 1$ ].

The response for $t \leq Nt_o$ is obtained by substituting equation (15) in (12). For $t > Nt_o$ free damped vibrations occur and are given by equation (4) with A and B chosen to give appropriate continuity conditions at $t = Nt_o$. Figure 4 shows the dynamic magnification factor, i.e. the maximum value of $kx/P_o$ with respect to time, plotted against $t_o/T$ for zero damping ($\gamma = 0$) and n = 1, 2 and 4; $T(=2\pi/\omega_n)$ is the period of the system. Outside the range of values of $t_o/T$ for which $(kx/P_o)_{max}$ has been plotted for N = 2 and 4, its behaviour is more complicated but values are significantly less than the peak values shown. When $t_o/T = 0.5$, $(kx/P_o)_{max} = N\pi/2$. Thus for an excitation force of two complete waves (N = 4) the dynamic magnification factor can be as large as 6.3, For the plotted ranges the maximum displacement occurs in the residual or free vibration era (i.e. $t > Nt_o$) if $t_o/T < 0.5$ and occurs in the forced vibration era (i.e. $t < Nt_o$) if $t_o/T > 0.5$.

For a *random variable* the spectral density shows the distribution of the harmonic content of the variable over the frequency range from zero to infinity. If the spectral density is specified, the mean value of the square of the variable can be obtained. For stationary ergodic random processes with Gaussian or normal probability distributions (these standard assumptions for random vibration theory are described in Chapter 14), if $S_p(\omega)$ and $S_x(\omega)$ are the spectral densities of the input force and response respectively for the system of Fig. 1, the mean square values are given by

$$< P^2(t) > \;=\; \frac{1}{2\pi}\int_0^\infty S_p(\omega)\, d\omega \tag{16}$$

and

$$< x^2(t) > \;=\; \frac{1}{2\pi}\int_0^\infty S_x(\omega)\, d\omega \tag{17}$$

It can be shown (Chapter 14) that the spectral densities are related by

$$S_x(\omega) \;=\; \frac{S_p(\omega)}{k^2[(1 - r^2)^2 + (2\gamma r)^2]} \tag{18}$$

[If hysteretic damping replaces viscous damping in Fig. 1, $2\gamma r$ in equation (18) is replaced by $\mu$]. If the spectral density of the force is known, the mean square value of the response is obtained from equations (17) and (18). The simplest force spectrum is: $S_p(\omega) = S_o$, a constant; i.e. the spectrum is uniform over the complete frequency range and is called white noise. The corresponding mean response is, from the calculus of residues,

$$< x^2(t) > \;=\; \frac{S_o\,\omega_n}{8\gamma\, k^2} \tag{19}$$

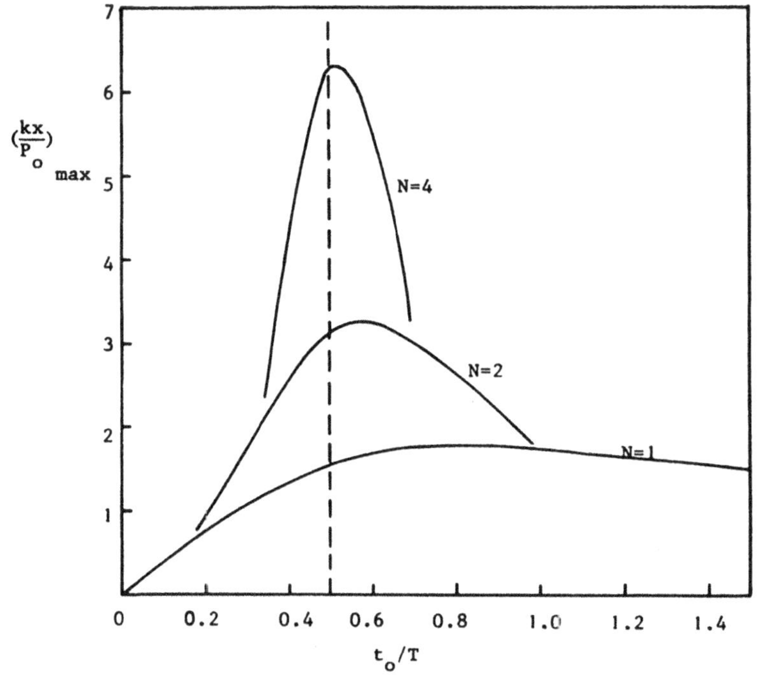


**Figure 4** Dynamic magnification factor for a single degree of freedom system, subjected to a transient force:

$$P(t) = P_o \sin \pi t/t_o, \quad 0 \le t \le Nt_o$$
$$= 0 \ , \quad t > Nt_o$$

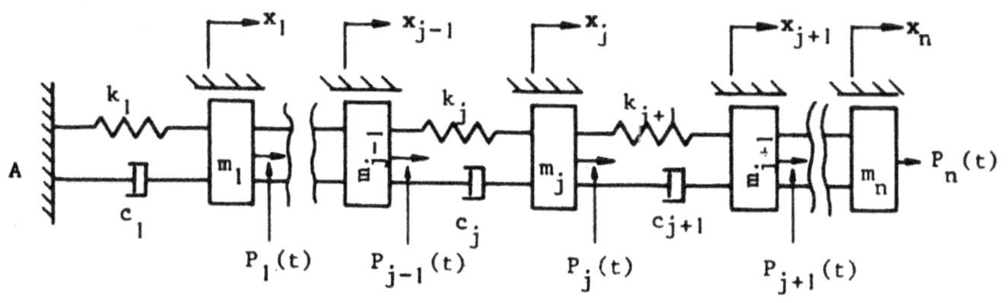


**Figure 5** Multi degree of freedom system

Although useful for analytical purposes, white noise predicts an infinite mean square value for P(t) [see equation (16)]. A more practical assumption is that $S_p(\omega)$ is uniform and equal to $S_o$ up to a cut-off frequency $\omega_c$, which is well above the natural frequency $\omega_n$, and $S_p(\omega) = 0$ for $\omega > \omega_c$. Provided that $\omega_c/\omega_n > 2$ and $\gamma < 0.1$ the mean square response is given to 1 per cent accuracy by equation (19). For comparison with the dynamic magnification factors for harmonic and transient excitation, values of $3k\ \sigma_x/\sigma_p$ are given in Table 1 for a spectrum $S_p(\omega)$ that is flat up to $\omega = \omega_c$, and $\omega_c/\omega_n = 2$; $\sigma_x$ and $\sigma_p$ are the root mean square values of response and force from equations (16) and (19), i.e. $\sigma_x = [ < x^2(t) > ]^{1/2}$. An absolute maximum value for x for a random process cannot be specified; the factor 3 has been introduced because for a Gaussian distribution the probability of $|x|$ exceeding $3\sigma_x$ is only 0.3 per cent.

In this section the dynamic response of single degree of freedom systems to the three major types of excitation, harmonic, transient and random, has been outlined. Although equations have been given in terms of force excitation, corresponding equations for excitation by support or foundation motion can easily be formulated. The possible existence of large dynamic magnification factors in unfavourable circumstances has been demonstrated by the simple examples for which results are given in Table 1 and Figure 4.

TABLE 1.

Dynamic Magnification Factors for Single Degree of Freedom Systems with Different Types of Excitation

| Excitation | Tabulated value | Damping Factor $\gamma$ 0 | 0.01 | 0.02 | 0.05 | 0.1 |
|---|---|---|---|---|---|---|
| Step function of Fig. 3a | $(kx/P_o)_{max}$ | 2 | 1.969 | 1.939 | 1.854 | 1.729 |
| Harmonic | $(kX/P_o)$ for r=1 | | 50 | 25 | 10 | 5 |
| Random, $S_p(\omega) = S_o$ with cut-off at $\omega_c = 2\omega_n$ | $3k\ \sigma_x/\sigma_p$ | | 18.8 | 13.3 | 8.41 | 5.95 |

## 4. General Structures: Equations of Motion

The concepts, developed for single degree of freedom systems, assist in the understanding of the dynamic behaviour of general structures. Figure 5 shows the conventional representation of a system with several degrees of freedom. Each mass is constrained to move only in the X-direction. Thus the displacement $x_j$ (measured from the position of static equilibrium) defines the instantaneous position of mass $m_j$. Figure 6 represents a shear building with n storeys; the assumptions are the same as those used to describe the structure of Figure 2. If $k_j$ is the combined stiffness in flexure of the pair of stanchions below the $j^{th}$ floor, $x_j$ is the horizontal displacement of the $j^{th}$ floor of mass $m_j$, and it is assumed that relative motion between adjacent floors is resisted by viscous damping forces in the stanchions between these floors, the equation of motion for mass $m_j$ in either Figures 5 or 6 is

$$m_j \ddot{x}_j + c_j(\dot{x}_j - \dot{x}_{j-1}) + c_{j+1}(\dot{x}_j - \dot{x}_{j+1}) + k_j(x_j - x_{j-1}) + k_{j+1}(x_j - x_{j+1}) = P_j(t) \tag{20}$$

There are n equations similar to equation (20). (See Chapter 3 for futher details). Obviously other multi degree of freedom systems can be defined. In order to accommodate different types of structure the following general matrix equation will be considered.

$$\underset{\sim}{M}\,\ddot{\underset{\sim}{x}} + \underset{\sim}{C}\,\dot{\underset{\sim}{x}} + \underset{\sim}{K}\,\underset{\sim}{x} = \underset{\sim}{P}(t) \tag{21}$$

In equation (21) $\underset{\sim}{x}$ is a vector (or column matrix) of the independent coordinates,

$$\text{i.e.} \quad \underset{\sim}{x} = \begin{bmatrix} x_1 \\ x_2 \\ \cdot \\ \cdot \\ \cdot \\ \cdot \\ \cdot \\ \cdot \\ x_n \end{bmatrix} \tag{22}$$

$\dot{\underset{\sim}{x}}$ and $\ddot{\underset{\sim}{x}}$ are the corresponding velocity and acceleration vectors. The vector $\underset{\sim}{x}$ may contain linear and angular displacements. The vector $\underset{\sim}{P}(t)$ lists the applied forces.

$$\text{i.e.} \quad \underset{\sim}{P}(t) = \begin{bmatrix} P_1(t) \\ P_2(t) \\ \vdots \\ \vdots \\ \vdots \\ P_n(t) \end{bmatrix} \tag{23}$$

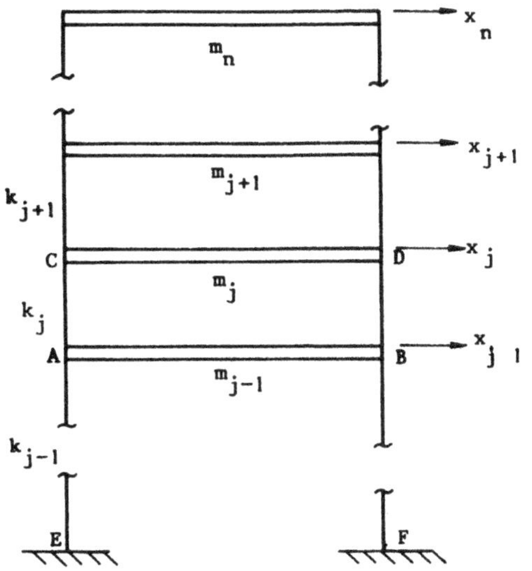


Figure 6 Frame with n degrees of freedom, the only permitted deformation is flexure of the light vertical members in the plane of the frame.

A component of the force vector, $P_j(t)$, is the resultant applied force on a mass and acts in the direction of the displacement $x_j$ of that mass. The mass matrix $\underset{\sim}{M}$, the damping matrix $\underset{\sim}{C}$ and the stiffness matrix $\underset{\sim}{K}$ are square and symmetric; $\underset{\sim}{K}$ is defined as

$$\underset{\sim}{K} = \begin{bmatrix} k_{11} & k_{12} & k_{13} & \cdots\cdots & k_{1n} \\ k_{21} & k_{22} & k_{23} & \cdots\cdots & k_{2n} \\ \cdots & \cdots & \cdots & \cdots & \cdots \\ k_{n1} & k_{n2} & k_{n3} & \cdots\cdots & k_{nn} \end{bmatrix} \tag{24}$$

where $k_{js} = k_{sj}$. This symmetry follows from the reciprocal theorem. For the systems and coordinates defined in Figures 5 and 6 the mass matrices are diagonal (i.e. $m_{js} = 0$ for $j \neq s$ and there is a single inertia term, typically $m_{jj}\ddot{x}_j$, in each equation).

The significance of equation (21) must be stressed. Engineering structures are usually complex and their response to specified inputs can be determined only by approximate methods. Three general approximate methods are Rayleigh-Ritz, finite elements and finite differences.

In these methods the actual structures are replaced by approximate mathematical models, which are represented by equation (21). The mass, stiffness and damping matrices are symmetric for models: from the Rayleigh-Ritz method; from the finite element displacement method; and from the finite difference method, if the recent energy formulation is used instead of conventional differences based on the equations of motion. Thus solutions of equation (21) are applicable to all linear, elastic structures.

Putting $\underset{\sim}{C} = 0$ and $\underset{\sim}{P} = 0$ in equation (21), the general equation for *free undamped vibration* is:

$$\underset{\sim}{M}\,\ddot{\underset{\sim}{x}} + \underset{\sim}{K}\,\underset{\sim}{x} = 0 \tag{25}$$

For harmonic motion, $x_j = \phi_j \sin(\omega t + \beta)$, $j = 1,2, \ldots n$ where $\omega$ is a natural frequency and $\phi_j$ is the amplitude of vibration at the point where the displacement $x_j$ is measured. This leads to the frequency determinant

$$\det\left[\, \underset{\sim}{K} - \omega^2 \underset{\sim}{M} \,\right] = 0 \tag{26}$$

In general, equation (26) gives n positive real roots for $\omega^2$, say $\omega_1^2, \omega_2^2, \omega_3^2, \ldots \omega_n^2$ with $\omega_1^2 < \omega_2^2 < \omega_3^2 \ldots < \omega_n^2$. Then $\omega_1, \omega_2, \omega_3 \ldots \omega_n$ are the natural frequencies of the system. Corresponding to a frequency $\omega_r$ there is a set of values of the amplitudes $\phi$. If $\phi_{rj}$ is the amplitude at the $j^{th}$ coordinate in the $r^{th}$ mode, substitution of $\omega_r{}^2$ in the equations yields only relative, rather than absolute, values of $\phi_{rj}$; e.g. we can find $\phi_{r2}/\phi_{r1}$, $\phi_{r3}/\phi_{r1}$, ... , $\phi_{rn}/\phi_{r1}$. This indeterminacy can be removed by introducing the vector

$$\underset{\sim}{z}_r = \left(\frac{1}{a_r}\right) \underset{\sim}{\phi}_r$$

where $a_r$ is a scalar, such that

$$\underset{\sim}{z}_r^T \underset{\sim}{M} \underset{\sim}{z}_r = 1 \tag{27}$$

Then the vectors $\underset{\sim}{z}_r$ are normalized.

Standard computer programs give the eigenvalues $\omega_r$ from equation (26) and also the corresponding eigenvectors, $\underset{\sim}{\phi}_r$ or $\underset{\sim}{z}_r$. If equation (21) represents an approximate mathematical model of a complex structure, then only the lower natural frequencies from equation (26) will approximate reasonably the natural frequencies of the structure. If the size of the matrix equation (21) is too large, the technique of eigenvalue economisers or reduction can be used to reduce the order of the equations without introducing serious errors into the values of the lower natural frequencies determined from the reduced equations [1][2]. (See Chapters 3 and 4 for further details).

5. Response

In this section we discuss some methods of determining the response of general structures, defined by equation (21), to excitation. This discussion is in terms of exciting forces, i.e. the vector $\underset{\sim}{P}(t)$, but the methods can be adapted to yield the response when support or foundation motion provides the excitation. The types of excitation, i.e. harmonic, transient or random, to which the methods are applicable, and the limitations of the methods will be summarised.

First we consider the *normal mode method;* this requires the natural frequencies $\omega_r$ and the normalized eigenvectors $\underset{\sim}{z}_r$ to be known. The original matrix equation (21) is transformed into a set of uncoupled equations by the change of variables

$$\underset{\sim}{x} = \underset{\sim}{Z} \underset{\sim}{q} \tag{28}$$

where $\underset{\sim}{q}^T$, $= [\, q_1 q_2 \ldots q_n]$ , are principal coordinates and the normalized vector

$\underset{\sim}{z}_r$, defined by equation (27), forms the $r^{th}$ column of $\underset{\sim}{Z}$. Provided that the damping matrix $\underset{\sim}{C}$ satisfies the relation

$$\underset{\sim}{C} = \lambda_m \underset{\sim}{M} + \lambda_k \underset{\sim}{K} \tag{29}$$

where $\lambda_m$ and $\lambda_k$ are constants, substituting equation (28) into (21) leads to uncoupled equations in $q_r$ (Chapter 9) of the form:

$$\ddot{q}_r + (\lambda_m + \lambda_k \omega_r^2) \dot{q}_r + \omega_r^2 q_r = \sum_{j=1}^{n} z_{rj} P_j(t) \tag{30}$$

The sclution of equation (30) can be found by analogy with the Duhamel integral (12) for a general transient excitation and by simpler means for harmonic excitation. In the general case

$$q_r = \frac{1}{\omega_{rd}} \int_0^t f_r(\tau) \exp [- \gamma_r \omega_r (t - \tau)] \sin \omega_{rd}(t - \tau) d\tau \tag{31}$$

where $\omega_{rd} = \omega_r (1 - \gamma_r^2)^{½}$, $\quad 2\omega_r \gamma_r = \lambda_m + \lambda_k \omega_r^2$

and $f_r(t) = \sum_{j=1}^{n} z_{rj} P_j(t)$ ,

provided that the initial conditions are zero. Evaluation of the principal coordinate $q_r$ from equation (31) and use of the transformation (28) gives the response $x_s$ at any coordinate. In practice, numerical integration of equation (31) will usually be necessary. In general, the response from the higher modes (larger values of r) is insignificant, and it is necessary to compute $q_r$ only for a limited range of values of r, $r = 1, 2, 3, \ldots n_1$ with $n_1 < n$. Unfortunately, no general rules can be given for deciding upon $n_1$; engineering judgement and trial computer runs are required to establish $n_1$ for a specific problem. Improved convergence can be obtained, i.e. the number of significant terms $n_1$ reduced, by the *modal acceleration technique*, originally devised by Williams for structures such as beams. Applied to equations (28) and (31), the response at coordinate s is

$$x_s = \frac{\sum_{j=1}^{n} a_{sj} P_j(t)}{\det |\underset{\sim}{K}|} - \sum_{r=1}^{n} \frac{z_{rs}}{\omega_r^2} \int_0^t \dot{f}_r(\tau) \exp [- \gamma_r \omega_r (t - \tau)]$$

$$[ \cos \omega_{rd}(t - \tau) + \gamma_r (1 - \gamma_r^2)^{-½} \sin \omega_{rd}(t - \tau)] \, d\tau \tag{32}$$

where $a_{sj}$ is a term in $\underset{\sim}{A}$, the adjoint matrix of $\underset{\sim}{K}$, $\dot{f}_r(t) = \frac{df_r}{dt}$ and it has been assumed that $P_j(0) = 0$, $j = 1,2, \ldots n$.

If the applied forces are *harmonic*, typically $P_j(t) = P_j \cos(\omega t + \beta_j)$ at coordinate j where $P_j$ is a constant, the steady state response at coordinate s is found from equation (28) and the direct solution of equation (30) as

$$x_s = \underline{\underline{\text{Re}}} \sum_{r=1}^{n} \left[ \frac{z_{rs} \left[ \sum_{j=1}^{n} z_{rj} P_j \exp(i\beta_j) \right] (\omega_r^2 - \omega^2 - 2i\gamma_r\omega_r\omega) \exp(i\omega t)}{(\omega_r^2 - \omega^2)^2 + 4\gamma_r^2\omega_r^2\omega^2} \right] \qquad (33)$$

where $i = (-1)^{\frac{1}{2}}$ and $\underline{\underline{\text{Re}}}$ signifies that the real part of the complete complex solution is taken. For light damping there will be successive resonances when the excitation frequency $\omega$ equals $\omega_1, \omega_2, \omega_3, \ldots$ ; in these circumstances the resonant response for the lower modes is usually determined by the resonant term in series (33) and depends upon frequency and damping as $1/2\gamma_r\omega_r^2$. (It depends also upon the terms in $z_r$ etc.)

For harmonic excitation hysteretic damping can be considered instead of viscous damping. $\underset{\sim}{H}\dot{\underset{\sim}{x}}/\omega$ replaces $\underset{\sim}{C}\dot{\underset{\sim}{x}}$ in equation (21) where $\underset{\sim}{H}$ is a symmetric matrix of hysteretic damping constants and must satisfy the condition

$$\underset{\sim}{H} = a_m \underset{\sim}{M} + a_k \underset{\sim}{K} \qquad (34)$$

where $a_m$ and $a_k$ are constants, in order that transformation (28) yields uncoupled equations. The response can be obtained from equation (33), if $2\gamma_r\omega$ is replaced by $\mu_r\omega_r$, where $\mu_r = a_k + a_m/\omega_r^2$. The resonant response now depends upon frequency and damping as $1/\mu_r\omega_r^2$. In mathematical models of elastic structures internal damping is often approximated by $\underset{\sim}{C} = \lambda_m \underset{\sim}{K}$ for viscous damping or $\underset{\sim}{H} = a_k \underset{\sim}{K}$ for hysteretic damping. Then, as $2\gamma_r = \lambda_k\omega_r$ and $\mu_r = a_k$, and $\lambda_k$ and $a_k$ are constants, the frequency dependence of the resonant response varies as $\omega_r^{-3}$ for viscous damping and $\omega_r^{-2}$ for hysteretic damping. Experimental evidence suggests that the latter behaviour is a better approximation to real structures.

The uncoupling conditions, equations (29) and (34) for viscous and hysteretic damping respectively, restrict the applications of the normal mode method. As discussed later, methods exist which avoid this restriction. Some recent numerical evidence [3] suggests that the restriction can be ignored. Mathematically, the equations in $\underset{\sim}{q}$ are uncoupled if $\underset{\sim}{Z}^T \underset{\sim}{C}\underset{\sim}{Z}$ (or $\underset{\sim}{Z}^T \underset{\sim}{H}\underset{\sim}{Z}$) is a diagonal matrix. Response curves for systems which do not satisfy condition (29), i.e. systems for which $\underset{\sim}{Z}^T \underset{\sim}{C}\underset{\sim}{Z}$ is not diagonal, agree very closely with approximate curves, obtained by the normal mode method with the non-diagonal terms in $\underset{\sim}{Z}^T\underset{\sim}{C}\underset{\sim}{Z}$ neglected.

The normal mode method may be used to determine the response to *random excitation*.

The spectral densities of response $S_x(\omega)$ and force $S_F(\omega)$ are related by

$$S_x(\omega) = | H_{sj}(\omega) |^2 S_p(\omega) \quad (35)$$

where $H_{sj}(\omega)$ is the complex frequency response or receptance, i.e. $H_{sj}(\omega) \exp(i\omega t)$ is the response at coordinate s to a force $\exp(i\omega t)$ at coordinate j. [Equation (18) for a single degree of freedom system is a special case of equation (35).] From equation (33)

$$H_{sj}(\omega) = \sum_{r=1}^{n} \left[ \frac{z_{rs} z_{rj}(\omega_r^2 - \omega^2 - 2i\gamma_r\omega_r\omega)}{(\omega_r^2 - \omega^2)^2 + 4\gamma_r^2\omega_r^2 \omega^2} \right] \quad (36)$$

for viscous damping and $\mu_r\omega_r$ replaces $2\gamma_r\omega$ to obtain $H_{sj}(\omega)$ for hysteretic damping. From equations (17) and (35) the mean square value of the response is given by

$$< x_s^2(t) > = \frac{1}{2\pi} \int_0^\infty |H_{sj}(\omega)|^2 S_p(\omega) \, d\omega \quad (37)$$

Approximations are often made when evaluating equation (37). If $(A_r + iB_r)$ is a typical complex term in the series (36) for $H_{sj}(\omega)$, products $A_r A_q$ and $B_r B_q$, $r \neq q$, are neglected in comparison with $A_r^2$ and $B_r^2$ respectively. Then

$$| H_{sj}(\omega) |^2 = \sum_{r=1}^{n} \left[ \frac{z_{rs}^2 z_{rj}^2}{(\omega_r^2 - \omega^2)^2 + 4\gamma_r^2\omega_r^2\omega^2} \right] \quad (38)$$

If $S_p(\omega)$ varies slowly, it is replaced by the value $S_p(\omega_r)$ when the $r^{th}$ term is considered. Then, from equations (37) and (38), and by analogy with equations (17) to (19),

$$< x_s^2(t) > = \sum_{r=1}^{n} \left[ \frac{S_p(\omega_r) z_{rs}^2 z_{rj}^2}{8 \gamma_r \omega_r^3} \right] \quad (39)$$

[For hysteretic damping $\mu_r$ replaces $2\gamma_r$.] Equation (39) is a reasonable approximation, provided that damping is light, $S_p(\omega)$ varies slowly in the vicinity of each resonant frequency and the natural frequencies $\omega_r$ are not too close together. For large structures the last condition will usually be the most difficult to satisfy.

The essence of the normal mode method is that it gives the response in terms of contributions from the undamped normal modes. There exist also methods, which avoid restriction (29) on the damping matrix and give the response in terms of damped modes. These methods are more complicated, algebraically and in concept, than the standard normal mode method and will not be considered here.

The *frequency response method* can be used to determine the steady state response to harmonic excitation; it imposes no restriction on the damping matrix, gives the response in closed form, instead of a series, but requires the inversion of a complex matrix of order n x n at each frequency for which the response is required. If the force at j is $P_j \cos(\omega t + \beta_j)$, the excitation vector, $\underset{\sim}{P}(t)$ in equation (21), is written $\underset{\sim}{P} \exp(i\omega t)$, where a typical term in $\underset{\sim}{P}$ is $P_j \exp(i\beta_j)$, $\dot{\underset{\sim}{x}} = i\omega \underset{\sim}{x}$ and $\ddot{\underset{\sim}{x}} = -\omega^2 \underset{\sim}{x}$ are used in equation (21), and the response is given by

$$\underset{\sim}{x} = \underline{\underline{Re}} \; [\, \underset{\sim}{J}^{-1} \underset{\sim}{P} \exp(i\omega t)] \qquad (40)$$

where $\underset{\sim}{J} = \underset{\sim}{K} - \omega^2 \underset{\sim}{M} + i\omega \underset{\sim}{C}$ for viscous damping
and $\underset{\sim}{J} = \underset{\sim}{K} - \omega^2 \underset{\sim}{M} + i \underset{\sim}{H}$ for hysteretic damping.

Many methods exist for the *numerical integration* of the general dynamic equation (21) (Chapter 9). In these, assumptions are made about the variation of either the displacements or accelerations during small time intervals; e.g. it may be assumed that during a small interval the displacement is a cubic function of time or that the acceleration varies linearly with time. With these assumptions the set of n second order differential equations (21) is replaced, in general by n simultaneous equations. The solution of the latter gives the displacements at the end of the short time step for known conditions at the beginning. Successive applications of this procedure give the response of the structure. Numerical stability of the computations and accuracy have to be considered. For some of the methods the time interval must be less than a certain value, given in terms of the period of the $n^{th}$ or highest mode of the system, if numerical stability is to be achieved. A criterion for accuracy is usually formulated in terms of $\Delta t/T_j$, where $\Delta t$ is the time step and $T_j$ is the period of the highest mode making a significant contribution to the response For example, from the survey of numerical integration methods by Bathe and Wilson 4] engineering accuracy is achieved with the Newmark $\beta$ method if $\Delta t/T_j < 0.1$. The methods can be used with any prescribed excitation vector $\underset{\sim}{P}(t)$ and there is no restriction on the form of the damping matrix.

Equation (21) is limited to elastic structures, where linear relations between forces and displacements exist. The normal mode method uses the principle of superposition and is applicable only to linear systems. In practice, the response of structures, when some components are stressed beyond the elastic range, is of interest, e.g. the response to earthquakes. Methods of numerical integration can be used to predict the response of non-linear systems, although the criteria for numerical stability and accuracy become more stringent.

5. Dynamic Interaction Problems

In the previous section methods of determining the response of structures have been outlined. These methods are satisfactory, if the exciting forces are known, and can be extended to deal with the excitation by known displacements or accelerations of the foundation. The excitation may be transient or aperiodic, harmonic or random. However, there exists a growing class of important dynamic problems, where the excitation mechanism is affected by the properties of the surrounding or underlying medium. Thus a proper determination of structural response requires some dynamic analysis of the medium. Broadly there are two types, interaction between ground and structure and between fluid and structure. The former occurs when determining the response of structures to earthquakes, as the known input may be the acceleration of the underlying bedrock or of the free surface of the ground, i.e. without the structure present. It occurs also when the source of excitation is located in one structure and the response of neighbouring structures is required. The latter occurs when the coupled vibrations of water and dams, piers or off-shore structures are considered. Indeed, both types of interaction may have to be considered, if, for example, the response of a dam to an earthquake is required. When aerodynamic effects are included, wind-induced oscillations of structures exhibit interaction effects.

Considering an elastic structure mounted on a rigid foundation, which is supported by soil, we require the structural response when the free-field acceleration at the ground surface is prescribed. Let $\underset{\sim}{x}_s$ be the vector of displacements of the constrained structure, i.e. the structure with the foundation clamped. [ For example, for the shear building of Figure 6 $\underset{\sim}{x}_s$ consists of the horizontal displacements $x_1, x_2, \ldots x_n$ of the masses $m_1, m_2, \ldots m_n$.] Let $\underset{\sim}{M}_s$, $\underset{\sim}{C}_s$ and $\underset{\sim}{K}_s$ be the mass, damping, and stiffness matrices of the constrained structure. Vector $\underset{\sim}{x}_F$ lists the displacements of the foundation; [if the foundation EF in Figure 6 undergoes horizontal translation $x_o$ and rocking $\theta$, $\underset{\sim}{x}_F$ consists of $x_o$ and $\theta$]; vector $\underset{\sim}{x}$ lists the displacements of the structure foundation system,

$$\text{i.e.} \qquad \underset{\sim}{x} = \begin{bmatrix} \underset{\sim}{x}_s \\ \hline \underset{\sim}{x}_F \end{bmatrix} \tag{41}$$

Using matrices with subscripts F for the foundation and with no subscripts for the complete system, equation (21) represents the complete system and can be partitioned as

$$\begin{bmatrix} \underset{\sim}{M}_s & \underset{\sim}{M}_{sF} \\ \underset{\sim}{M}_{sF}^T & \underset{\sim}{M}_F \end{bmatrix} \begin{bmatrix} \ddot{\underset{\sim}{x}}_s \\ \ddot{\underset{\sim}{x}}_F \end{bmatrix} + \begin{bmatrix} \underset{\sim}{C}_s & \underset{\sim}{C}_{sF} \\ \underset{\sim}{C}_{sF}^T & \underset{\sim}{C}_F \end{bmatrix} \begin{bmatrix} \dot{\underset{\sim}{x}}_s \\ \dot{\underset{\sim}{x}}_F \end{bmatrix}$$

$$\begin{bmatrix} \mathbf{K}_s & \mathbf{K}_{sF} \\ \mathbf{K}_{sF}^T & \mathbf{K}_F \end{bmatrix} \begin{bmatrix} \mathbf{x}_s \\ \mathbf{x}_F \end{bmatrix} = \begin{bmatrix} \mathbf{0} \\ \mathbf{P}_F \end{bmatrix} \tag{42}$$

assuming that there are no applied forces on the structure and excitation is caused by the forces $\mathbf{P}_F$ on the foundation due to the ground. If the free field displacement of the ground is $\mathbf{y}$, the interaction (or dynamic magnification) vector for the foundation, $(\mathbf{x}_F - \mathbf{y})$, is related to the interaction forces, or forces on the ground due to the foundation, $-\mathbf{P}_F$ by

$$-\mathbf{P}_F = \mathbf{K}_G [\mathbf{x}_F - \mathbf{y}] + \mathbf{C}_G [\dot{\mathbf{x}}_F - \dot{\mathbf{y}}] / \omega \tag{43}$$

The matrices $\mathbf{K}_G$ and $\mathbf{C}_G$ can be found analytically for simple geometries, or by the finite element method or experimentally. Equation (43) refers to steady state vibrations of frequency $\omega$; in general $\mathbf{K}_G$ and $\mathbf{C}_G$ are dependent upon $\omega$. [The analytical solutions and approximate forms of $\mathbf{K}_G$ and $\mathbf{C}_G/\omega$, which are independent of $\omega$, are discussed by Richart, Hall and Woods [5].] Substituting from equation (43) in (42) and rearranging terms,

$$\mathbf{M}\ddot{\mathbf{x}} + \mathbf{C}^*\dot{\mathbf{x}} + \mathbf{K}^*\mathbf{x} = \mathbf{P}^* \tag{44}$$

where

$$\mathbf{C}^* = \begin{bmatrix} \mathbf{C}_s & \mathbf{C}_{sF} \\ \mathbf{C}_{sF}^T & \mathbf{C}_F + \mathbf{C}_G/\omega \end{bmatrix} \qquad \mathbf{K}^* = \begin{bmatrix} \mathbf{K}_s & \mathbf{K}_{sF} \\ \mathbf{K}_{sF}^T & \mathbf{K}_F + \mathbf{K}_G \end{bmatrix}$$

$$\mathbf{P}^* = \begin{bmatrix} \mathbf{0} \\ \mathbf{K}_G y + \mathbf{C}_G \dot{y} / \omega \end{bmatrix}$$

If $\mathbf{K}_G$ and $\mathbf{C}_G/\omega$ can be treated as independent of frequency, the response $\mathbf{x}$ can be found by solving equation (44) by standard methods for any freee-field motion. If $\mathbf{K}_G$ and $\mathbf{C}_G/\omega$ depend upon $\omega$, Fourier transforms are used to obtain a solution. For $\mathbf{P}^*(t)$ known, its Fourier transform $\bar{\mathbf{P}}^*(\omega)$ is given by

$$\bar{\mathbf{P}}^*(\omega) = \int_0^\infty \mathbf{P}^*(t) \exp(-i\omega t)\, dt \tag{45}$$

Similarly, the Fourier transform of $\mathbf{x}(t)$ is defined as

$$\bar{x}(\omega) = \int_{-\infty}^{\infty} x(t) \exp(-i\omega t)\, dt \tag{46}$$

Thus, taking the Fourier transform of equation (44),

$$(-\omega^2 M + i\omega C^* + K^*)\, \bar{x}(\omega) = \bar{P}^*(\omega) \tag{47}$$

Equation (47) is solved for $\bar{x}(\omega)$ and the inverse transform used to determine the response $x(t)$,

i.e. $$x(t) = \frac{1}{2\pi} \int_{-\infty}^{\infty} \bar{x}(\omega) \exp(i\omega t)\, d\omega \tag{48}$$

In practice, the response vector $x(t)$ is obtained from $\bar{x}(\omega)$, and also $\bar{P}^*(\omega)$ from $P^*(t)$, using the fast Fourier transform (FFT) algorithm. The latter is highly efficient [6], but requires $\bar{x}(\omega)$ to be evaluated from equation (47) for a large number of discrete frequencies. This requires considerable computation for a complex structure for which the order of the matrices in equation (47) is high. For earthquakes and other practical excitations the response is usually confined to the first few modes. Methods of economizing in computation by taking advantage of the low-frequency nature of this response have been developed [7][8].

Neglecting the dynamic interaction effects would cause considerable simplification. Then the foundation is assumed to have the prescribed free-field motion, i.e. $x_F = y$. Using this in equation (42), expanding that equation above the partition and rearranging terms, an equation of the standard form of equation (21), i.e. with matrices independent of frequency, is obtained. However, for multi-storey structures and certain values of the parameters this assumption leads to gross underestimates of the foundation vibrations [7].

The vector $x_S$ in equation (41) consists of absolute displacements. However, for particular systems the equations are often formulated in terms of relative displacements (e.g. if the foundation EF in Figure 6 has displacements $x_o$ and $\theta$, the relative displacement of mass $m_j$ at height $h_j$ is $z_j = x_j - x_o - h_j\theta$). In this case the excitation vector $P^*$ in equation (44) is conveniently expressed in terms of $\ddot{y}$, the free-field acceleration vector, instead of in terms of $y$ and $\dot{y}$.

Consider a dam subjected to a prescribed horizontal acceleration of its base $\ddot{u}_g(t)$ as an example of fluid-structure interaction. If this interaction is neglected, it is assumed that the face of the dam has the acceleration $\ddot{u}_g(t)$ when determining the dynamic pressure distribution on the dam due to the water; then the vibrations of the dam are determined for the combined loading of base acceleration and dynamic pressure distribution. If, further, the water is assumed to be incompressible, the dynamic pressure distribution is determined directly in terms of $\ddot{u}_g(t)$ and analysis leads to the standard matrix equation (21). If the water is considered to be compressible, the analysis of the pressure distribution is in terms of an excitation frequency $\omega$. Thus, using equations similar to (45) to (48), $\bar{\underset{\sim}{x}}(\omega)$ can be found for an excitation $\ddot{U}_g(\omega)$ and the response of the dam $\underset{\sim}{x}(t)$ found by the inverse Fourier transform. Allowing for fluid-structure interaction results in a modified mass matrix, where the additional terms are frequency dependent (unless the water is assumed to be incompressible). For a dam-reservoir system subjected to a specific earthquake acceleration Chopra [9][10] showed that significant errors in the maximum dynamic response can be caused by neglecting dynamic interaction <u>or</u> by assuming water to be incompressible; these errors depend upon the fundamental periods of the dam and reservoir, $T_{D1}$ and $T_{R1}$ respectively, and are small for $T_{D1}/T_{R1} > 1.4$. (Other fluid-structure interaction problems are considered in Chapters 13 and 17).

For the problems of this section and for similar problems the structure and ground or fluid are treated as separate sub-structures and appropriate continuity conditions at the interface imposed to obtain the response of the complete system. This allows different methods of analysis to be used for the two parts of the system, e.g. the finite element method for the structure and an analytical solution for the ground or fluid. Even if the finite element method is used for both parts, the use of the sub-structure approach has computational advantages.

## References


1. Zienkiewicz, O.C. *The Finite Element Method in Engineering Science*, McGraw-Hill, 1971.

2. Henshell, R.D. and J.H. Ong, "Automatic masters for eigenvalue economization", *International Journal of Earthquake Engineering and Structural Dynamics*, 3, 1975, 375-383.

3. Thomson, W.T., T. Calkins and P. Caravani, "A numerical study of damping", *International Journal of Earthquake Engineering and Structural Dynamics*, 3, 1974, 97-103.

4. Bathe, K.-J. and E.L. Wilson, "Stability and accuracy analysis of direct integration methods", *International Journal of Earthquake Engineering and Structural Dynamics*, 1, 1973, 283-291.

5. Richart, F.E. Jr., J.R. Hall, Jr. and R.D. Woods, *Vibrations of Soils and Foundations*, Prentice-Hall, 1970.

6. Brigham, E.O. *The Fast Fourier Transform*, Prentice-Hall, 1974.

7. Jennings, P.C. and H. Bielak, "Dynamics of building soil interaction", *Bulletin of the Seismological Society of America,* 63, 1973, 9-48.

8. Chopra, A.K. and J.A Gutierrez, "Earthquake response analysis of multistorey buildings including foundation interaction", *International Journal of Earthquake Engineering and Structural Dynamics,* 3, 1974, 65-77.

9. Chopra, A.K. "Earthquake behaviour of reservoir-dam systems", *Journal of the Engineering Mechanics Division, ASCE,* 94, 1968, 1475-1500.

10. Chopra, A.K. "Earthquake response of concrete gravity dams", *Journal of the Engineering Mechanics Division, ASCE,* 96, 1970, 443-454.

CHAPTER 2

# FREE VIBRATION, RESONANCE AND DAMPING

by

R.R. Wilson

## 1. Introduction

In this chapter we shall examine the dynamic behaviour of a one degree of freedom system, introducing the terms and techniques which will be used in the later chapters. The equations of motion for a number of systems are derived. Then the behaviour of an undamped and a damped system are considered. Free vibration, response to sinusoidal forcing and finally the transient behaviour in response to general forcing are calculated.

First of all, what is a one degree of freedom system ? At its simplest, it is a system as shown in Fig. 1 consisting of a mass m suspended from a spring k. If the weight of the spring can be neglected then the system is completely determined by the position of the mass.

In general a one degree of freedom system can be specified by giving the value at every instant of time of only one coordinate. This coordinate often is a displacement, as in the spring-mass system, but it can equally well be the angle through which the bob of a pendulum swings or a shaft rotates as it undergoes torsional vibration.

For complicated systems, many coordinates are required to give an adequate representation. However, as will be discussed in Chapter 3, it is possible to introduce a set of generalized coordinates which are uncoupled. The behaviour of each coordinate can then be examined in turn using the methods described in the present chapter.

In the next six sections the equations of motion of different systems will be formulated. It will be shown that all the equations are of the same form, and thus it is necessary to consider the solution of only one equation when examining the vibration of any one degree of freedom system.

## 2. Spring-Mass System

Consider the system shown in Fig. 2 consisting of a mass m hanging on a spring k. When the mass is at rest, the system is in equilibrium, and the force in the spring is equal to the weight of the mass. The spring force is given by the product of

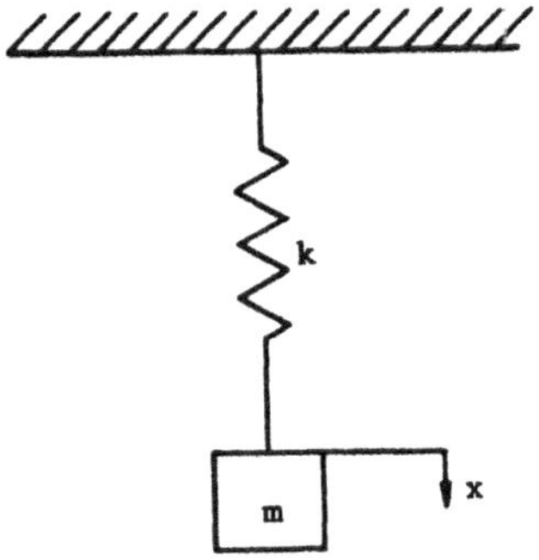


Figure 1. Spring-Mass System

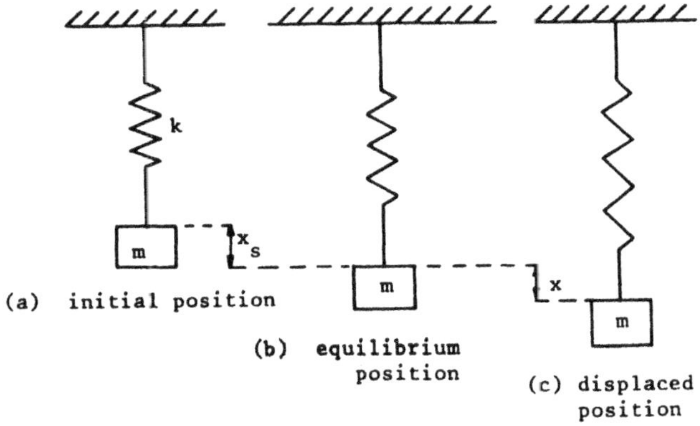


Figure 2 Displacement of a Mass on Spring

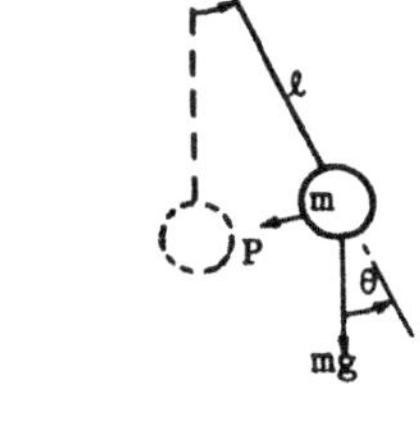


Figure 3 Simple Pendulum

the spring stiffness k and the static deflection $x_s$. Thus, taking the positive direction as downwards, we have

$$- k x_s + mg = 0 \tag{1}$$

If the mass is now displaced a further distance x then the spring force increases and there will be a net restoring force given by

$$P = - k(x_s + x) + mg \quad , \tag{2}$$

and hence from equation (1),

$$P = - kx \quad . \tag{3}$$

By Newton's 3rd Law, this will produce an acceleration given by

$$\ddot{x} = \frac{P}{m} \quad , \tag{4}$$

where a dot over the x denotes differentiation with respect to time.

Combining equations (3) and (4),

$$\ddot{x} = \frac{-kx}{m} \tag{5}$$

and the equation of motion of the system is given by

$$m \ddot{x} + kx = 0 \tag{6}$$

3. Simple Pendulum

When the pendulum of length $\ell$ as shown in Figure 3 is displaced through an angle $\theta$, a component of the weight of the mass m will act so as to restore the pendulum to its equilibrium position. This force is given by

$$P = - mg \cos(\pi/2 - \theta)$$

$$\text{i.e.} \quad P = - mg \sin \theta \tag{7}$$

For small angles, $\sin \theta \doteqdot \theta$ and so the restoring force is given approximately by

$$P = - mg\, \theta \tag{8}$$

Applying Newton's 3rd Law again, we have the equation of motion

$$\ell \ddot{\theta} = \frac{-mg\ \theta}{m}$$

$$\text{i.e.} \quad \ell \ddot{\theta} + g\theta = 0 \tag{9}$$

4. Beam with Central Load

If a force P is applied at the centre of a simply supported beam then, if the weight of the beam is neglected, the static deflection is given by

$$x_s = \frac{P \ell^3}{48EI} \tag{10}$$

where $\ell$ is the length of the beam, I is its second moment of area and E is Young's modulus. Thus the effective stiffness of the beam is given by

$$k = \frac{P}{x_s} = \frac{48EI}{\ell^3}\ , \tag{11}$$

When a mass m is placed at the centre of the beam, then the resulting deflection is given by equation (10) with P = mg. If the mass is now displaced a further distance x then a restoring force $-\frac{48EI}{\ell^3}$ x will be produced causing the mass to accelerate towards its equilibrium position. Thus the equation of motion is

$$\ddot{x} = -\frac{48EI}{\ell^3}\ x/m$$

$$\text{i.e.} \quad m\ddot{x} + \frac{48EI}{\ell^3}\ x = 0 \tag{12}$$

5. Rolling of a Ship

When a ship is floating in still water, the weight mg and the buoyancy force B are equal and both act through the centre of gravity of the ship, point G. When the ship is displaced slightly, the buoyancy force no longer acts through G but through M, the metacentre. The position of M is determined by the geometry of the ship. If M lies below G then a couple will result which will act so as to increase the displacement, capsizing the ship. Let us assume that M is a distance h above G (h is termed the metacentric height). Then a restoring couple $-mg\ h \sin\theta$ is produced, and for small angles, this couple has the approximate value $-mg\ h\ \theta$. Thus the equation of motion of the ship is given by

$$I_1 \ddot{\theta} = -mg\ h\ \theta,$$

$$\text{i.e.} \quad I_1 \ddot{\theta} + mg\ h\ \theta = 0\ , \tag{13}$$

where $I_1$ is the moment of inertia of the ship about its longitudinal axis.

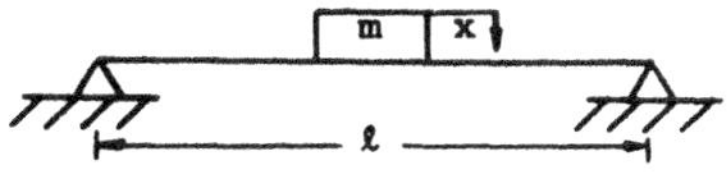


Figure 4 Beam with Central Load

(a) equilibrium position (b) displaced position

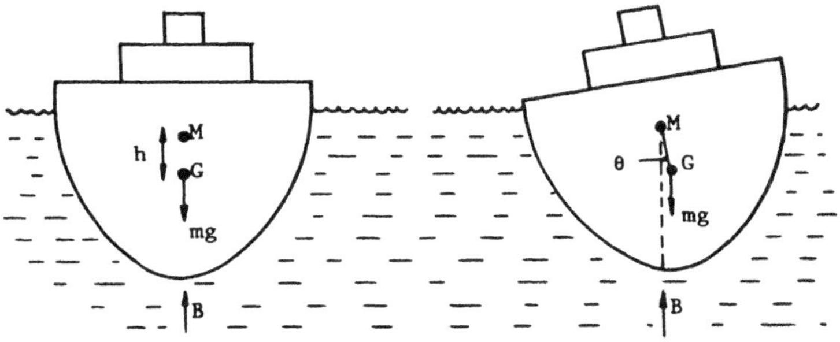


Figure 5 Rolling of a Ship

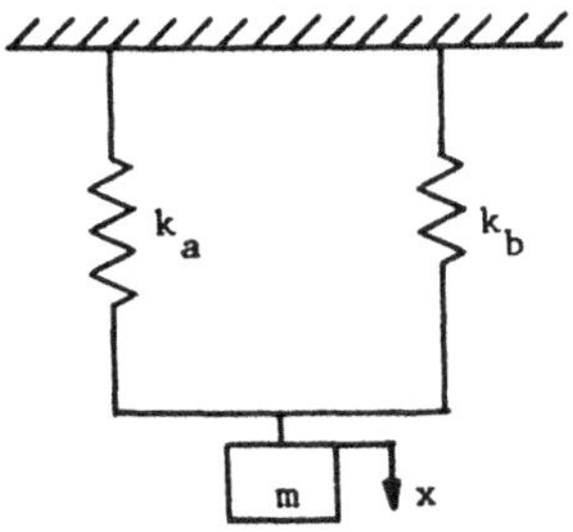


Figure 6 Springs in Parallel

## 6. Springs in Parallel

In the system shown in Figure 6, the weight of mass m is balanced by the force in two springs $k_a$ and $k_b$. Thus if the mass is displaced a distance x from its equilibrium position we have that a restoring force P is produced given by

$$P = -k_a x - k_b x$$

$$\text{i.e.} \quad P = -(k_a + k_b)x \tag{14}$$

Thus the equation of motion of the system is given by

$$\ddot{x} = -\frac{(k_a + k_b)x}{m}$$

$$\text{i.e.} \quad m\ddot{x} + (k_a + k_b)x = 0 \tag{15}$$

This equation may be written in the same form as the equation of the simple spring -mass system,

$$m\ddot{x} + k_{eq} x = 0, \tag{16}$$

where the two springs have been replaced by a single spring with equivalent stiffness $k_{eq}$ given by

$$k_{eq} = k_a + k_b \tag{17}$$

## 7. Springs in Series

In the system shown in Figure 7 the force in each of the two springs is equal, and if $x_a$ and $x_b$ are extensions of the two springs then the restoring force is given by

$$P = -k_a x_a = -k_b x_b \tag{18}$$

Thus the displacement of the mass m is given by

$$x = x_a + x_b$$

$$\text{i.e.} \quad x = -\frac{P}{k_a} - \frac{P}{k_b} \tag{19}$$

The equation of motion can be written as

$$m\ddot{x} + k_{eq} x = 0 \tag{20}$$

where the equivalent stiffness is given by

$$k_{eq} = -\frac{P}{x} \quad \text{or} \quad \frac{1}{k_{eq}} = -\frac{x}{P}$$

Thus

$$\frac{1}{k_{eq}} = \frac{1}{k_a} + \frac{1}{k_b} \quad . \tag{21}$$

## 3. Free Vibration

For all the systems considered, the form of the equation of motion is the same as for the simple spring-mass system,

$$m\ddot{x} + k x = 0 \tag{22}$$

The solution of this equation can be used to predict the behaviour of any of the systems considered. Equation (22) represents simple harmonic motion and thus its solution may be written in the form

$$x = A \cos \omega_n t + B \sin \omega_n t \tag{23}$$

Substituting in equation (22) we have that

$$m(-\omega_n^2 A \cos \omega_n t - \omega_n{}^2 B \sin \omega_n t) + k(A \cos \omega_n t + B \sin \omega_n t) = 0$$

$$\text{i.e.} -\omega_n{}^2 m + k = 0$$

$$\omega_n = \sqrt{\frac{k}{m}} \tag{24}$$

This constant $\omega_n$ (rad/sec) is the *natural frequency* of the system. The displacement is the sum of two sinusoidal functions which vary at a frequency $\omega_n$. The amplitudes of these functions are determined by the initial conditions. If at time $t = 0$, the mass has displacement $x_i$ and velocity $\dot{x}_i$ then

$$A = x_i \quad \text{and} \quad B = \frac{\dot{x}_i}{\omega_n}$$

and the displacement at any time is given by

$$x = x_i \cos \omega_n t + \frac{\dot{x}_i}{\omega_n} \sin \omega_n t \tag{25}$$

This may be written in terms of a single sinusoidal function

$$x = A' \sin (\omega t + \alpha)$$

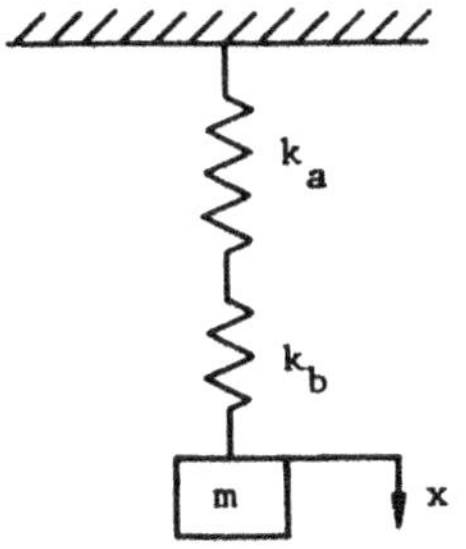


Figure 7 Springs in Series

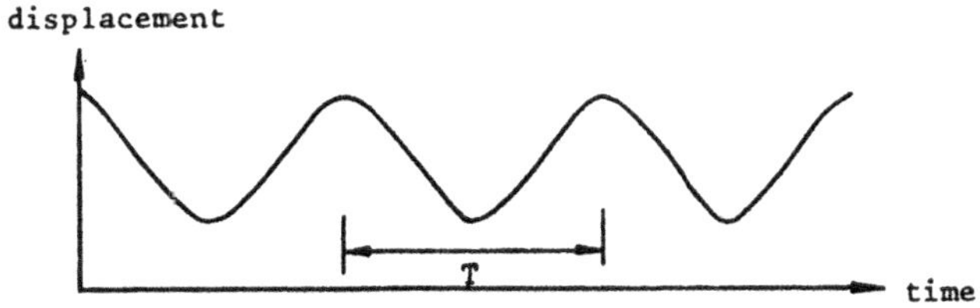


Figure 8 Sinusoidal Vibration

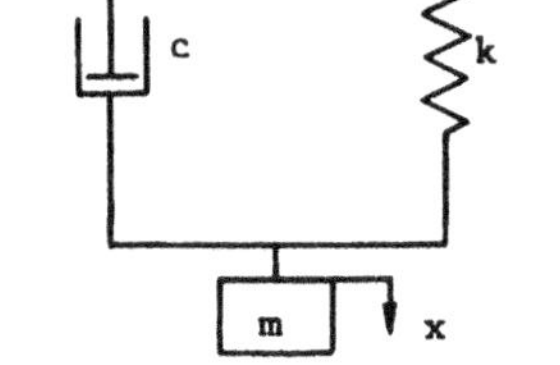


Figure 9 Spring-Mass System with Dashpot

$$\text{where} \quad A' = \sqrt{x_i^{\,2} + \left(\frac{\dot{x}_i}{\omega_n}\right)^2}$$

and

$$\tan\alpha = \frac{x_i\,\omega_n}{\dot{x}_i}$$

A' is the *amplitude* of the vibration and $\alpha$ the *phase angle* . The displacement of the mass will vary as shown in Figure 8. It can be seen from equation (25) that the time between successive peaks is given by

$$\omega_n T = 2\pi$$

$$\text{i.e.} \quad T = \frac{2\pi}{\omega_n} \qquad (26)$$

T represents the *period* of the vibrations, and the number of periods per second is called the *frequency* f in hertz or cycles per second. Thus

$$f = \frac{1}{T} = \frac{\omega_n}{2\pi} = \frac{1}{2\pi}\sqrt{\frac{k}{m}} \qquad (27)$$

9. Energy of Vibrating System

Consider the mass to be given an initial displacement $x_i$ and released so that its initial velocity is zero, then from equation (25),

$$x = x_i \cos \omega_n t \,,$$

$$\text{and} \quad \dot{x} = -\,\omega_n\, x_i \sin \omega_n t \qquad (28)$$

Thus the maximum displacement is $x_i$ and the maximum velocity $\omega_n x_i$. When the mass is at its position of maximum displacement the potential energy $V$ of the spring is given by

$$V = \tfrac{1}{2}\, k\, x_i^{\,2} \qquad (29)$$

As the mass moves from its position of maximum displacement to its equilibrium position, the potential energy stored in the spring decreases and the kinetic energy of the mass increases. At the equilibrium position the potential energy of the spring is zero and the kinetic energy of the mass is given by

$$T = \tfrac{1}{2}\, m(\dot{x}_i)^2_{max}$$

$$\text{i.e.} \quad T = \tfrac{1}{2}\, m\, \omega_n^{\,2}\, x_i^{\,2} \qquad (30)$$

Thus, by conservation of energy,

$$\tfrac{1}{2} m \omega_n^2 x_i^2 = \tfrac{1}{2} k x_i^2$$

$$\text{i.e.} \quad \omega_n = \sqrt{\frac{k}{m}} , \tag{31}$$

agreeing with equation (24).

The vibration of the system corresponds to a repeated transfer from potential energy of the spring to kinetic energy of the mass. In practice this process would not continue indefinitely - energy would be lost from the system, dissipated in the form of heat in the spring, transferred to the supporting structure or lost by friction of the air. These losses of energy are collectively termed the damping of the system.

10. Damped Free Vibration

As mentioned in the previous section, a system may lose energy by many mechanisms. In the present chapter we shall consider only the situation in which motion is resisted by a force proportional to velocity. This is *viscous damping* and may be represented by the motion of a piston in the dashpot with the motion resisted by the viscosity of the oil, as shown in Figure 9. An additional restoring force $- c \dot{x}$ is introduced. The equation of motion is given by

$$m \ddot{x} = - k x - c \dot{x}$$

$$\text{i.e.} \quad m \ddot{x} + c \dot{x} + kx = 0 \tag{32}$$

For an equation of this type a solution of the form $x = e^{\lambda t}$ can be used. Substituting in equation (32) gives

$$m \lambda^2 e^{\lambda t} + c\lambda e^{\lambda t} + k e^{\lambda t} = 0$$

$$\text{i.e.} \quad m \lambda^2 + c\lambda + k = 0$$

$$\therefore \quad \lambda = - \frac{c}{2m} \pm \sqrt{\left[\frac{c}{2m}\right]^2 - \frac{k}{m}} \tag{33}$$

The general solution is given by

$$x = A e^{\lambda_1 t} + B^{\lambda_2 t} , \tag{34}$$

where the values of $\lambda_1$ and $\lambda_2$ are found by taking the positive and negative signs in expression (33). A and B are constants whose values are determined by the initial conditions.

It is useful to consider separately the situations arising when the expression under the root sign in equation (33) is positive, zero or negative. When $(\frac{c}{2m})^2 > \frac{k}{m}$, the expression under the root sign is positive and the system is said to be *overdamped*. The displacement decays from its initial value towards zero without any vibrations; the motion is the sum of two exponential decays as given in equation (34), both values of $\lambda$ being negative.

The value of the damping constant c which corresponds to the expression under the root sign being zero is called the *critical damping* and is given by

$$\left[\frac{c_c}{2m}\right]^2 = \frac{k}{m}$$

$$\text{i.e.} \quad c_c = 2m\sqrt{\frac{k}{m}} = 2\sqrt{mk} = 2\,m\omega_n \qquad (35)$$

For this value of damping, the two roots of equation (33) are equal and the displacement decays exponentially.

When $(\frac{c}{2m})^2 < \frac{k}{m}$, the system is *underdamped*. Equation (33) can be rewritten in the form

$$\lambda = -\frac{c}{2m} \pm i\,\omega_d \qquad (36)$$

where $i = \sqrt{-1}$ and $\omega_d$ is the damped natural frequency given by

$$\omega_d = \sqrt{\frac{k}{m} - (\frac{c}{2m})} \qquad (37)$$

$$\text{i.e.} \quad \omega_d = \omega_n\sqrt{1 - \frac{c^2}{4mk}}$$

The general solution is given by

$$x = A\,e^{(-\frac{c}{2m} + i\omega_d)t} + B\,e^{(-\frac{c}{2m} - i\omega_d)t}$$

$$\text{i.e.} \quad x = e^{-\frac{c}{2m}t}\,(A\,e^{i\omega_d t} + B\,e^{-\omega_d ti}) \qquad (38)$$

This can be rewritten in the form

$$x = e^{-\frac{c}{2m}t}\,(A'\cos\omega_d t + B'\sin\omega_d t), \qquad (38a)$$

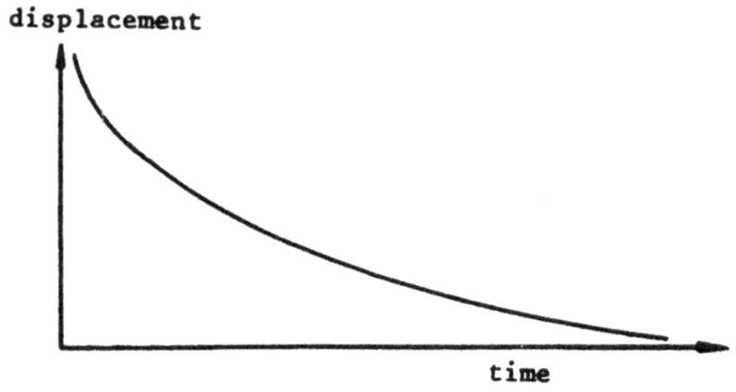


Figure 10 Displacement of an Overdamped System.

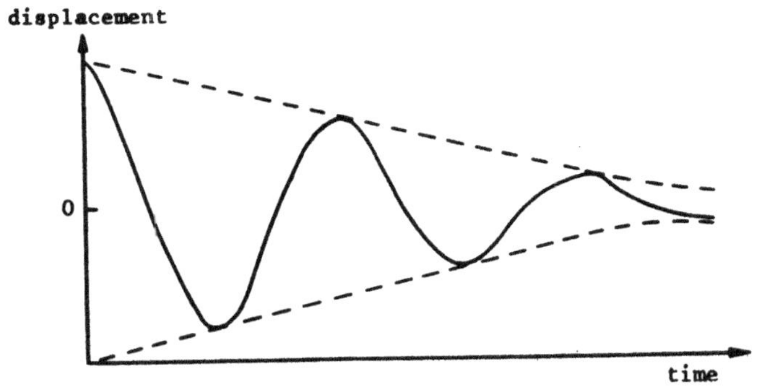


Figure 11 Displacement of Underdamped System

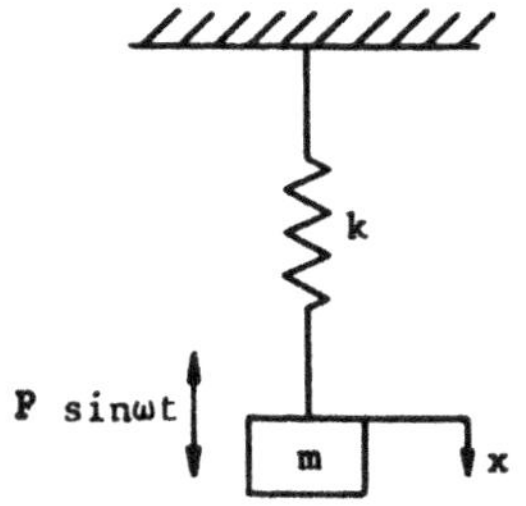


Figure 12 Undamped System with Sinusoidal Forcing

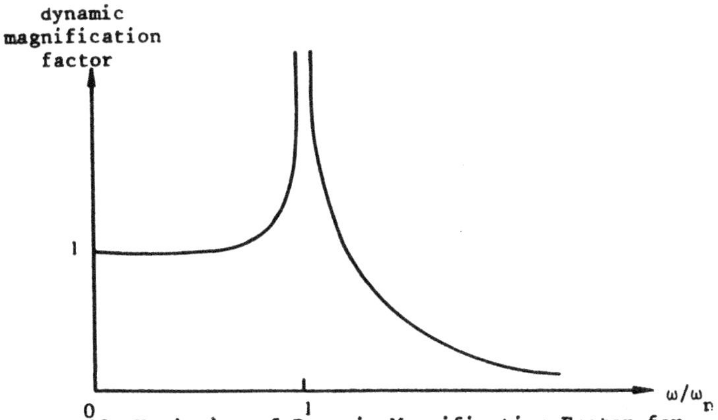


Figure 13 Variation of Dynamic Magnification Factor for Undamped system.

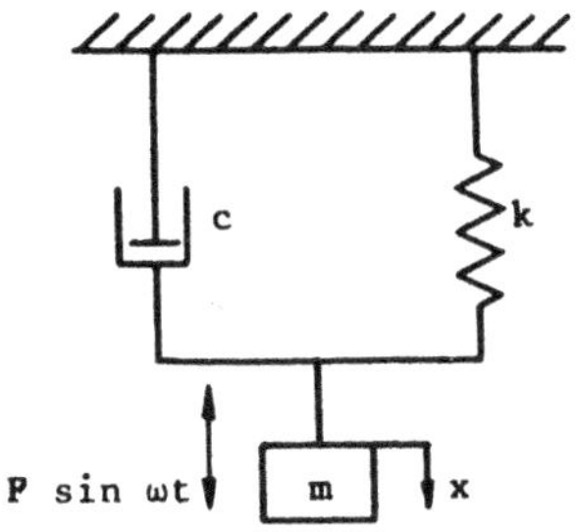


Figure 14 Damped System with Sinusoidal Forcing

or evaluating A' and B' in terms of the initial conditions

$$x = e^{-\frac{c}{2m}t}\left(x_i \cos \omega_d t + \frac{\dot{x}_i + \frac{c\,x_i}{2m}}{\omega_d} \sin \omega_d t\right) \qquad (39)$$

The displacement varies as shown in Figure 11 in the form of a damped sine wave. The time between successive maxima is determined by the period of the system $2\pi/\omega_d$. The ratio of the height of successive maxima is given by

$$e^{-\frac{c}{2m}t} \Big/ e^{-\frac{c}{2m}\left(t + \frac{2\pi}{\omega_d}\right)} = e^{\frac{c\pi}{m\omega_d}}$$

This ratio represents how quickly the vibration dies away and the logarithm of this quantity, the *logarithmic decrement* $\delta$ is often used as a measure of the damping of the system. It is given by

$$\delta = \frac{c\,\pi}{m\omega_d} \qquad (40)$$

11. Undamped Forced Response

All the systems considered so far have been undergoing free vibration. There have been no external forces applied. We shall now examine the behaviour of a mass-spring system which has a sinusoidally varying force applied to the mass as shown in Figure 12. The equation of motion is

$$m\,\ddot{x} + kx = P \sin \omega t \qquad (41)$$

where P is the amplitude of the force and $\omega$ is the frequency at which it varies.

The solution of this equation is the sum of two parts, the particular integral and the complementary function. The complementary function represents the general solution of the corresponding equation with the right hand side set to zero. In this case this is equation (22). The particular integral is any function satisfying the complete equation. It is reasonable to try a solution of the form

$$x = A \sin \omega t \qquad (42)$$

$$\therefore\ -m\,\omega^2 A \sin \omega t + kA \sin \omega t = P \sin \omega t$$

$$\text{i.e.} \qquad A = \frac{P}{k - m\omega^2}$$

Thus, a particular integral of the equation is

$$x = \frac{P}{k - m\omega^2} \sin \omega t \, , \tag{43}$$

and the complete solution is obtained by adding this to the general solution of equation (22) giving

$$x = x_i \cos \omega_n t + \frac{\dot{x}_i}{\omega_n} \sin \omega_n t + \frac{P}{k - m\omega^2} \sin \omega t \tag{44}$$

If initially the mass is at rest at its equilibrium position then $x_i = \dot{x}_i = 0$, and the solution is given by equation (43). This can be written in the form

$$x = \frac{P/k}{1 - \frac{m}{k}\omega^2} \sin \omega t$$

$$\text{i.e.} \quad x = \frac{P/k}{1 - (\frac{\omega}{\omega_n})^2} \sin \omega t \tag{45}$$

Now P/k represents the static deflection $x_s$ of the mass. Thus the system undergoes vibrations with amplitude $x_s \dfrac{1}{1 - (\frac{\omega}{\omega_n})^2}$ .

The factor $\dfrac{1}{1 - (\frac{\omega}{\omega_n})^2}$ is called the *dynamic magnification* factor and its value depends on how near the frequency of the applied force is to the natural frequency of the system. As the ratio $\omega/\omega_n$ approaches unity the vibration amplitude increases and the system is said to be at *resonance*.

## 12. Damped Force Response

In practice the amplitude of the vibration of a forced system is determined at resonance by the damping of the system. It is thus necessary to consider the system as shown in Figure 14. The equation of motion is

$$m \ddot{x} + c \dot{x} + kx = P \sin \omega t \tag{46}$$

Again the complete solution consists of a particular integral and a complementary function. The complementary function corresponds to the general solution of equation (32). For a particular integral let us try a solution of the form

$$x = A \sin \omega t + B \cos \omega t \tag{47}$$

Substituting in equation (46) gives

$$- mA\,\omega^2 \sin \omega t - mB\,\omega^2 \cos \omega t + cA\,\omega \cos \omega t$$
$$- \omega Bc \sin \omega t + Ak \sin \omega t + Bk \cos \omega t = P \sin \omega t$$

Equating coefficients of $\sin \omega t$ and $\cos \omega t$ we have

$$- mA\,\omega^2 - \omega\, cB + kA = P$$

and

$$- mB\,\omega^2 + cA\,\omega + kB = 0$$

These equations can be solved to give

$$A = \frac{P(k - m\omega^2)}{(k - m\omega^2)^2 + \omega^2 c^2}$$

and

$$B = \frac{- P\omega c}{(k - m\omega^2)^2 + \omega^2 c^2} \qquad (48)$$

The particular integral given by substituting equations (48) into (47) can be combined with the complementary function as given in equation (38a) to give the general solution

$$x = e^{-\frac{c}{2m}t}(A' \cos \omega_d t + B' \sin \omega_d t)$$
$$+ \frac{P(k - m\omega^2)}{(k - m\omega^2)^2 + \omega^2 c^2} \sin \omega t - \frac{P\omega c}{(k - m\omega^2)^2 + \omega^2 c^2} \cos \omega t \qquad (49)$$

where $A'$ and $B'$ are evaluated from the initial conditions. Because of the factor $e^{-c/2m\ t}$, the first term, called the *transient component* , will die away, and the displacement will be determined after a sufficiently long time by the second and third terms, the *steady state components* . The steady state displacement is given by

$$x = \frac{P(k - m\omega^2)}{(k - m\omega^2)^2 + \omega^2 c^2} \sin \omega t - \frac{P\omega c}{(k - m\omega^2)^2 + \omega^2 c^2} \cos \omega t$$

i.e. $$x = \frac{P}{\sqrt{(k - m\omega^2)^2 + \omega^2 c^2}} \sin(\omega t - \alpha) \qquad (50)$$

where $$\tan \alpha = \frac{\omega c}{k - m\omega^2} \qquad (51)$$

The amplitude of the steady state response can be written in the form

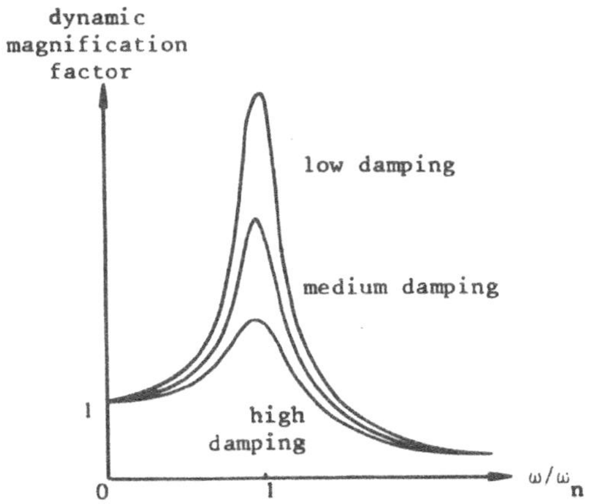


**Figure 15** **Variation of Dynamic Magnification Factor for Damped System**

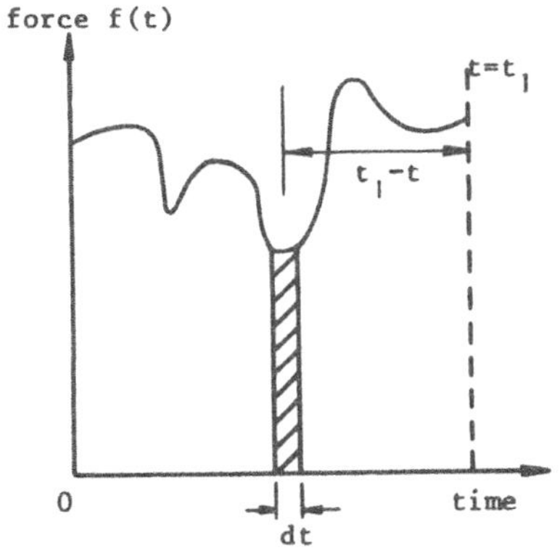


**Figure 16** **Time-varying Force**

$$P/k \cdot \frac{1}{\sqrt{\left(1 - \left(\frac{\omega}{\omega_n}\right)^2\right)^2 + \frac{\omega^2 c^2}{k^2}}}$$

The static deflection P/k is multiplied by the dynamic magnification factor

$$\frac{1}{\sqrt{\left(1 - \left(\frac{\omega}{\omega_n}\right)^2\right)^2 + \frac{\omega^2 c^2}{k^2}}} \quad .$$

This factor varies with the ratio of $\omega/\omega_n$ and the value of damping, as shown in Figure 15. The frequency for which the value of the dynamic magnification factor is largest, corresponding to the maximum steady state vibration amplitude, is given by

$$\omega_{max} = \omega_n \sqrt{1 - \frac{c}{2km}^2} \qquad (52)$$

13. Undamped Transient Vibration

Consider the situation when the applied force varies arbitrarily with time as shown in Figure 16. The force is applied at time t = 0 and we wish to calculate the displacement at time $t = t_1$. The equation of motion is given by

$$m\ddot{x} + kx = f(t) \qquad (53)$$

In a small time interval dt, the mass receives an impulse fdt. Because of this impulse the mass gains momentum given by

$$m\,d\dot{x} = fdt \qquad (54)$$

From equation (25), the displacement at a time t after an initial velocity of $\dot{x}_1$ has been given to a mass is

$$x = \frac{\dot{x}_i}{\omega_n} \sin \omega_n t$$

Thus, the contribution to the displacement of the mass at a time $\tau$ after it has been given a velocity $d\dot{x}$ is

$$dx = \frac{d\dot{x}}{\omega_n} \sin \omega_n \tau$$

i.e.

$$dx = \frac{fdt}{m\omega_n} \sin \omega_n \tau \qquad (55)$$

The total displacement of the mass at time $t_1$ due to application of the force is the sum of all the contributions during time intervals dt at times $\tau (= t_1 - t)$

for all values of $t < t_1$. Hence

$$x(t_1) = \int_0^{t_1} \frac{f(t)}{m\omega_n} \sin \omega_n (t_1 - t)dt \tag{56}$$

If at $t = 0$, the mass has displacement $x_i$ and velocity $\dot{x}_i$ then the displacement is

$$x(t_1) = x_i \cos \omega_n t_1 + \frac{\dot{x}_i}{\omega_n} \sin \omega_n t_1 + \int_0^{t_1} \frac{f(t)}{m\omega_n} \sin \omega_n (t_1 - t)dt \tag{57}$$

As an example of an application of this equation, consider the situation when a constant force P is suddenly applied to the mass which is initially at rest in its equilibrium position then

$$x(t_1) = \int_0^{t_1} \frac{P}{m\omega_n} \sin \omega_n (t_1 - t)dt$$

i.e. $$x(t_1) = \frac{P}{m\omega_n^2} (1 - \cos \omega_n t_1) \tag{58}$$

The maximum displacement is

$$x_{max} = \frac{2P}{m\omega_n^2} = \frac{2P}{k} \tag{59}$$

Hence the maximum displacement produced when the force is applied suddenly is twice the static displacement caused by the same force.

## 14. Damped Transient Vibration

The case of time varying force applied to a damped system can be treated as in the previous section. The equation of motion is

$$m\ddot{x} + c\dot{x} + kx = f(t) \tag{60}$$

From equation (39), an initial velocity of $\dot{x}_i$ results in a displacement given by

$$x = e^{-\frac{c}{2m}t} \frac{\dot{x}_i}{\omega_d} \sin \omega_d t$$

Hence the contribution to the displacement after a time $\tau$, given by the application of the force during the interval dt, is

$$dx = e^{-\frac{c}{2m}\tau} \frac{fdt}{m\omega_d} \sin \omega_d \tau, \tag{61}$$

and the total contribution to the displacement at time $t_1$ from the applied force is

$$x(t_1) = \int_0^{t_1} \frac{f(t)}{m\omega_d} e^{-\frac{c}{2m}(t_1-t)} \sin \omega_d(t_1-t)dt \qquad (62)$$

The complete displacement is

$$x(t_1) = e^{-\frac{c}{2m}t}\left(x_1 \cos \omega_d t_1 + \frac{\dot{x}_i + \frac{cx_i}{2m}}{\omega_d} \sin \omega_d t_1\right)$$

$$+ \int_0^{t_1} \frac{f(t)}{m\omega_d} e^{-\frac{c}{2m}(t_1-t)} \sin \omega_d(t_1-t)dt \qquad (63)$$

## 15. Summary of Results

Undamped natural frequency $\omega_n = \sqrt{\frac{k}{m}}$

Damped natural frequency $\omega_n = \omega_n \sqrt{1 - \frac{c^2}{4mk}}$

Frequency of maximum response for forced vibration $\omega_{max} = \omega_n \sqrt{1 - \frac{c^2}{2mk}}$

Critical damping $c_c = 2\sqrt{mk}$

Logarithmic decrement $\delta = \frac{c\,\pi}{m\omega_d}$

## Bibliography


1. Timoshenko, S. and D.H. Young *Vibration Problems in Engineering*, Van Nostrand, New York, 3rd edition, 1955.

2. Den Hartog, J.P. *Mechanical Vibrations*, McGraw Hill, New York, 1956.

3. Warburton, G.B. *The Dynamical Behaviour of Structures*, Pergamon, Oxford, 1964.

4. Bishop, R.E.D. and D.C. Johnson, *The Mechanics of Vibrations*, Cambridge University Press, London, 1960.

5. Hurty, W.C. and M.F. Rubinstein, *Dynamics of Structures*, Prentice-Hall, 1964.

# CHAPTER 3

# VIBRATIONS OF MULTI-DEGREE OF FREEDOM SYSTEMS

by

J. Wilson

## 1. Introduction

In Chapter 2, the vibration of a single degree of freedom system was studied. This is the simplest structural system. Actual structures are, of course, not as simple as this. They behave as lightly damped multi-degree of freedom systems. However, we can apply much from our study of single-degree of freedom systems to multi-degree of freedom systems (or real structures).

For instance a single degree of freedom system has a single natural frequency. (Real structures have many, only a few of which are important in general). A single degree of freedom system resonates when excited by a sinusoidal, time varying force having a frequency near the natural frequency of the system. The effect of light damping is to reduce the amplitude of the resonance at the natural frequency from infinity to a large but finite value. (Real structures resonate near each of their natural frequencies, the magnitude of the resonant oscillation depending on the amount of damping present in the structure). Thus in order to study the behaviour of real structures we must first look at multi-degree of freedom systems.

The simplest multi-degree of freedom system is the 2-degree of freedom system. We shall find that this system has two natural frequencies and associated with each natural frequency a mode shape defining the relative amplitudes of vibration of the two masses which constitute the system. Then we shall go on to a multi-degree of freedom system with n degrees of freedom. We shall find that there exist n natural frequencies each associated with its own mode shape. So in order to define the vibrations we must specify not only the natural frequency but also the mode shape.

We also find that the mode shapes exhibit a property known as *orthogonality*. This property enables us to simplify the analysis of the behaviour of multi-degree freedom systems which are representative of our real structures. If we express the displacements of a multi-degree structural system in terms of the mode shapes as coordinates, then we can transform a set of n coupled equations into a set of n uncoupled equations each of which can be solved independently of the other.

## 2. Free Vibrations of 2-degree of Freedom System

A two-degree-of-freedom system is represented as shown in Figure 1 by two masses and two springs. It is assumed that the masses can only move in the x direction. We denote the displacement of the first mass, $m_1$, from its static equilibrium position by $u_1$ and similarly for $m_2$ and $u_2$. These two displacements completely define the state (or position) of the system.

We write down the equations of motion for $m_1$ for any displaced position $u_1$ and $u_2$;

$$m_1 \ddot{u}_1 = -k_1 u_1 + k_2(u_2 - u_1) \tag{1a}$$

Similarly for $m_2$;

$$m_2 \ddot{u}_2 = -k_2(u_2 - u_1) \tag{1b}$$

Rearranging equations (1a) and (1b) we obtain

$$m_1 \ddot{u}_1 + (k_1 + k_2)u_1 - k_2 u_2 = 0 \tag{2a}$$

$$m_2 \ddot{u}_2 + k_2 u_2 - k_2 u_1 = 0 \tag{2b}$$

We can write equations (2a) and (2b) in matrix form as

$$\begin{bmatrix} m_1 & 0 \\ 0 & m_1 \end{bmatrix} \begin{bmatrix} \ddot{u}_1 \\ \ddot{u}_2 \end{bmatrix} + \begin{bmatrix} k_1+k_2 & -k_2 \\ -k_2 & k_2 \end{bmatrix} \begin{bmatrix} u_1 \\ u_2 \end{bmatrix} = \begin{bmatrix} 0 \\ 0 \end{bmatrix} \tag{2c}$$

or

$$\underset{(2x2)}{\underset{\sim}{M}} \; \underset{(2x1)}{\underset{\sim}{\ddot{U}}} + \underset{(2x2)}{\underset{\sim}{K}} \; \underset{(2x1)}{\underset{\sim}{U}} = \underset{(2x1)}{\underset{\sim}{0}} \tag{2d}$$

where $\underset{(2x2)}{\underset{\sim}{K}}$ is the stiffness matrix

$\underset{(2x2)}{\underset{\sim}{M}}$ is the mass matrix

$\underset{(2x1)}{\underset{\sim}{U}}$ is the vector of displacements

Note that this is similar to the equation for a single degree of freedom system

$$m \ddot{u} + k u = 0 \tag{2e}$$

Let us assume that a solution to equations (2a) and (2b) is of the form

$$u_1 = a_1 e^{i\omega t} \tag{3a}$$

$$u_2 = a_2 e^{i\omega t} \tag{3b}$$

where $a_1$ and $a_2$ are constants and $\omega$ is the natural frequency. (By using the expression

$$e^{i\omega t} = \cos \omega t + i \sin \omega t$$

we can represent any sinusoidal wave composed of cosines or sines by separating real and imaginary parts). On substitution of equations (3) in (2a) and (2b) we obtain

$$\{- m_1\omega^2 a_1 + (k_1+k_2)a_1 - k_2a_2\} e^{i\omega t} = 0$$

$$\{- m_2\omega^2 a_2 + k_2a_2 - k_2a_1\} e^{i\omega t} = 0$$

On simplifying these become

$$\{ (k_1+k_2) - \omega^2 m_1\} a_1 - k_2a_2 = 0$$

$$- k_2a_1 + (k_2-\omega^2 m_2)a_2 = 0 \tag{4}$$

the solution to which must be true for arbitrary values of $a_1$ and $a_2$. (Obviously one possible solution is that $a_1 = a_2 = 0$ but this would give no motion at all). If we divide through by $a_1$ we obtain equations in terms of the ratio $a_2/a_1$ and $\omega^2$.

$$\{ (k_1+k_2) - \omega^2 m_1\} - k_2 \frac{a_2}{a_1} = 0$$

$$- k_2 + (k_2-\omega^2 m_2) \frac{a_2}{a_1} = 0 \tag{5}$$

We can eliminate $a_2/a_1$ between these equations and obtain,

$$\{ (k_1+k_2) - \omega^2 m_1\} (k_2-\omega^2 m_2) - k_2^2 = 0$$

or multiplying out

$$m_1 m_2\omega^4 - \{m_1k_2 + m_2(k_1+k_2)\} \omega^2 + k_1k_2 = 0 \tag{6}$$

This type of equation always has two solutions $\omega_1{}^2$ and $\omega_2{}^2$ which are positive and such that $\omega_1$ and $\omega_2$ are the two natural frequencies of the system. If we now go back to equations (5) and substitute for $\omega^2$ by $\omega_1{}^2$ or $\omega_2{}^2$ we will find two

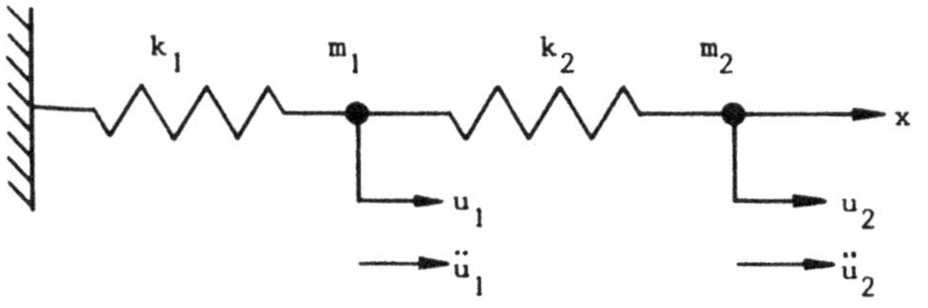


Figure 1 Two Degree of Freedom System

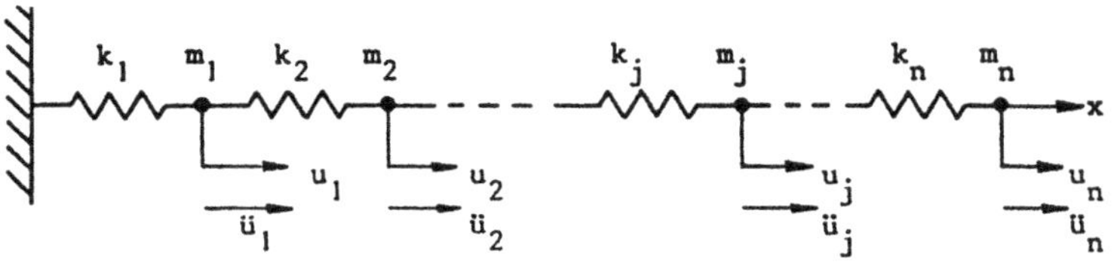


Figure 2 Multi-degree of Freedom System

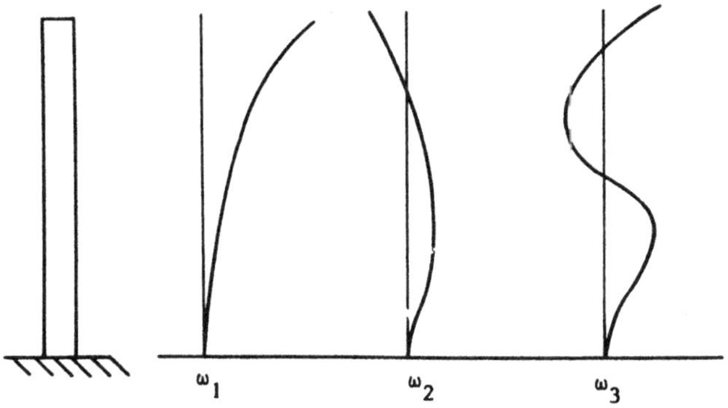


**Figure 3 Mode Shapes of a Cantilever Beam**

different ratios for $a_2/a_1$ corresponding to each natural frequency. These ratios are the mode shapes.

Example 1

Suppose $m_1 = m_2 = 1$ and $k_1 = k_2 = 1$ in consistent units. Equation (6) gives

$$\omega^4 - 3\omega^2 + 1 = 0$$

$$\omega^2 = \frac{3 \pm \sqrt{9-4}}{2} = \frac{3 \pm \sqrt{5}}{2}$$

$$\omega_1 = 0.618 \text{ rad/sec}; \qquad \omega_2 = 1.62 \text{ rad/sec.}$$

$\omega_1$ is known as the fundamental (lowest) natural frequency.

Substituting for $\omega$ by $\omega = \omega_1 = 0.618$ and $\omega = \omega_2 = 1.62$ in either of the expressions (5) gives

$$\left(\frac{a_2}{a_1}\right)_1 = 2 - \frac{3 - \sqrt{5}}{2} = 1.62$$

$$\left(\frac{a_2}{a_1}\right)_2 = 2 - \frac{3 + \sqrt{5}}{2} = -0.62$$

What does all this algebra mean? It shows that for a two-degree of freedom system there exist two natural modes of vibration (not necessarily distinct) such that for each frequency the displacements $u_1$ and $u_2$ are in phase and there is a constant ratio between the displacements. The ratio of these displacements is the mode shape.

We can write the equations (4) in matrix form as

$$\begin{bmatrix} (k_1+k_2)-\omega^2 m_1 & -k_2 \\ -k_2 & k_2-\omega^2 m_2 \end{bmatrix} \begin{bmatrix} a_1 \\ a_2 \end{bmatrix} = \begin{bmatrix} 0 \\ 0 \end{bmatrix}$$

or

$$\begin{bmatrix} k_1+k_2 & -k_2 \\ -k_2 & k_2 \end{bmatrix} \begin{bmatrix} a_1 \\ a_2 \end{bmatrix} - \omega^2 \begin{bmatrix} m_1 & 0 \\ 0 & m_1 \end{bmatrix} \begin{bmatrix} a_1 \\ a_2 \end{bmatrix} = \begin{bmatrix} 0 \\ 0 \end{bmatrix}$$

or

$$\left(\underset{(2x2)}{\underset{\sim}{K}} - \omega^2 \underset{(2x2)}{\underset{\sim}{M}}\right) \underset{(2x1)}{\underset{\sim}{A}} = \underset{(2x1)}{\underset{\sim}{0}} \qquad (7)$$

where $\underset{(2x2)}{\underset{\sim}{K}}$ is the stiffness matrix

$\underset{(2x2)}{\underset{\sim}{M}}$ is the mass matrix

$\underset{(2x1)}{\underset{\sim}{A}}$ is a vector of amplitudes

Notice the similarity between (7), a set of equations in matrix form, and the frequency equation for a single degree of freedom system,

$$(k - \omega_n^2 m)a = 0$$

giving
$$\omega_n^2 = \frac{k}{m}$$

or
$$\omega_n = \sqrt{\frac{k}{m}}$$

3. Free Vibrations of a Multi-Degree of Freedom System

Consider now a multi-degree of freedom system consisting of n discrete masses constrained to move only in the x direction and connected by elastic springs (Fig. 2).

If we write down the equations of motion for the n masses, we end up with n-2 equations of the type,

$$m_j\ddot{u}_j = k_{j+1}(u_{j+1} - u_j) - k_j(u_j - u_{j-1}), \qquad j = 2, n-1 \tag{8a}$$

$$m_1\ddot{u}_1 = k_2(u_2 - u_1) - k_1u_1 \tag{8b}$$

and
$$m_n\ddot{u}_n = -k_n(u_n - u_{n-1}) \tag{8c}$$

We can rearrange these equations in the form

$$m_1\ddot{u}_1 + (k_1 + k_2)u_1 - k_2u_2 = 0$$

$$m_j\ddot{u}_j + (k_j + k_{j+1})u_j - k_ju_{j-1} - k_{j+1}u_{j+1} = 0 \tag{9}$$

$$m_n\ddot{u}_n + k_nu_n - k_nu_{n-1} = 0$$

As before we assume a solution of the type $u_j = a_je^{i\omega t}$ where $a_j$ are constants. Note that $\omega$ and the phase is the same for all displacements. The equations (9) can then be written in matrix form as

$$\underset{(n\times n)}{\underset{\sim}{K}} - \omega^2 \underset{(n\times n)}{\underset{\sim}{M}} \;\; \underset{(n\times 1)}{\underset{\sim}{A}} = \underset{(n\times 1)}{\underset{\sim}{0}} \qquad (10)$$

where $\underset{(n\times n)}{\underset{\sim}{K}}$ is the stiffness matrix, $\underset{(n\times n)}{\underset{\sim}{M}}$ is the mass matrix and $\underset{(n\times 1)}{\underset{\sim}{A}}$ is the vector of displacement amplitudes. For example, when n = 5, the stiffness matrix, $\underset{(5\times 5)}{\underset{\sim}{K}}$, becomes

$$\begin{bmatrix} k_1+k_2 & -k_2 & 0 & 0 & 0 \\ -k_2 & k_2+k_3 & -k_3 & 0 & 0 \\ 0 & -k_3 & k_3+k_4 & -k_4 & 0 \\ 0 & 0 & -k_4 & k_4+k_5 & -k_5 \\ 0 & 0 & 0 & -k_5 & k_5 \end{bmatrix}$$

and the mass matrix, $\underset{(5\times 5)}{\underset{\sim}{M}}$

$$\begin{bmatrix} m_1 & 0 & 0 & 0 & 0 \\ 0 & m_2 & 0 & 0 & 0 \\ 0 & 0 & m_3 & 0 & 0 \\ 0 & 0 & 0 & m_4 & 0 \\ 0 & 0 & 0 & 0 & m_5 \end{bmatrix}$$

Equation (10) is of the same general form as equation (7) except that the matrices are of order n instead of 2. It is known as the eigenvalue equation. It turns out that n values of $\omega^2(\omega_1^2, \omega_2^2, \ldots \omega_r^2 \ldots \omega_n^2)$ will satisfy this equation corresponding to n natural frequencies. The roots $\omega_r^2$ always turn out to be positive and real.

Note that the matrix $\underset{(n\times n)}{\underset{\sim}{K}}$ is always square (order n) and is symmetric, i.e. if we swap rows and columns the matrix is unchanged. This is true for all linear elastic structural systems. The mass matrix is also square (order n) and symmetric, but need not necessarily be diagonal as shown in this case.

Corresponding to each of the n values of $\omega_r$ we find on substituting in each of equations (10) that we can obtain a set of simultaneous equations from which it is possible to find the ratios $a_{1r}:a_{2r}: \ldots :a_{n-1r}:a_{nr}$. This set of ratios represents

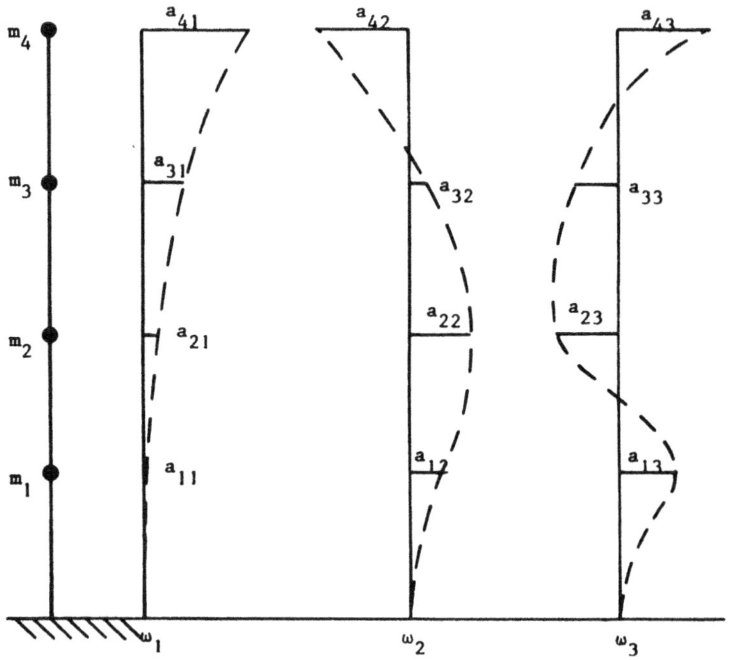


Figure 4 Cantilever Beam Represented by Four Masses

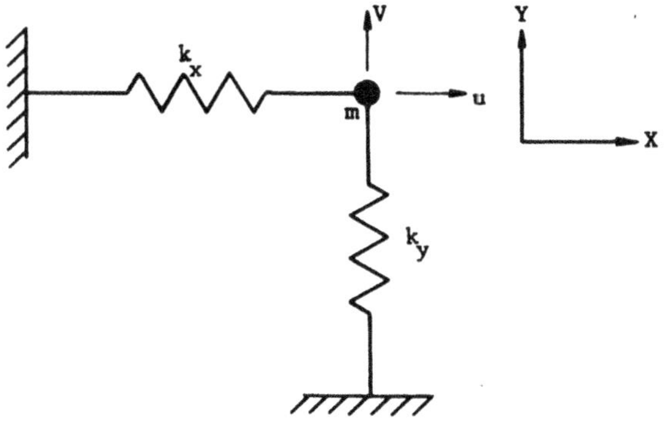


Figure 5 Two Degrees of Freedom System

the mode shape associated with the $r^{th}$ frequency.

Thus from equation (10) we can find the natural frequencies and mode shapes of a discrete multi-degree of freedom elastic structural system. You might well ask what is the use of this if our structure is continuous, as say a uniform cantilever beam. For such a simple structure, analytical methods of determining the natural frequencies may be used. However, in general, it will not be possible to find an analytical solution for a continuous structure of rather more complex nature. Thus what we do is idealize the structure in some way by a discretization process, reducing the number of degrees of freedom from infinity to n where n is as large a number as we like.

Let us consider the case of the mode shapes of a uniform cantilever beam (Fig. 3). If we replaced the cantilever by 4 masses, connected by four light beams, we might find the mode shapes as in Figure 4. These are fairly close to the actual mode shapes and we can make good approximations by drawing a curve between the discrete values. Obviously for higher frequencies it becomes much more difficult to define mode shapes unless we increase the number of points which we take to define the mode shapes (i.e. increase the number of degrees of freedom).

4. Orthogonality of Mode Shapes

Let us take a discretized structure vibrating in its $r^{th}$ mode shape

$$\underset{(n\times 1)}{\underset{\sim}{\phi}_r} = \{a_{1r}\ a_{2r}\ \ldots\ldots\ a_{jr}\ \ldots\ a_{n-1r}\ a_{nr}\}$$

and put this in place of $\underset{\sim}{A}$ in equation (10). (Since $\underset{\sim}{A}$ is arbitrary we can do this). The frequency of vibration will of course be $\omega_r$, so that equation (10) becomes

$$(\underset{(n\times n)}{\underset{\sim}{K}} - \omega_r^2 \underset{(n\times n)}{\underset{\sim}{M}})\ \underset{(n\times 1)}{\underset{\sim}{\phi}_r} = \underset{(n\times 1)}{\underset{\sim}{0}}$$

which we can write as

$$\underset{(n\times n)}{\underset{\sim}{K}}\ \underset{(n\times 1)}{\underset{\sim}{\phi}_r} = \omega_r^2\ \underset{(m\times n)}{\underset{\sim}{M}}\ \underset{(n\times 1)}{\underset{\sim}{\phi}_r} \qquad (11a)$$

For the same system vibrating in any other mode shape, $\underset{(n\times 1)}{\underset{\sim}{\phi}_s}$ , at frequency $\omega_s$

(where $\omega_s \neq \omega_r$) we obtain a similar expression

$$\underset{(n\times n)}{\tilde{K}} \; \underset{(n\times 1)}{\tilde{\phi}_s} = \omega_s^2 \; \underset{(n\times n)}{\tilde{M}} \; \underset{(n\times 1)}{\tilde{\phi}_s} \tag{11b}$$

Both sides of equation (11a) can be transposed without altering the equation to give

$$(\underset{(n\times n)}{\tilde{K}} \; \underset{(n\times 1)}{\tilde{\phi}_r})^T = \omega_r^2 (\underset{(n\times n)}{\tilde{M}} \; \underset{(n\times 1)}{\tilde{\phi}_r})^T \tag{12a}$$

or

$$\underset{(1\times n)}{\tilde{\phi}_r^T} \; \underset{(n\times n)}{\tilde{K}^T} = \omega_r^2 \; \underset{(1\times n)}{\tilde{\phi}_r^T} \; \underset{(n\times n)}{\tilde{M}^T}$$

where T denotes the transpose of a matrix or vector and the rule for reversing the order of multiplication of transposed matrices has been used.

But matrices $\underset{(n\times n)}{\tilde{K}}$ and $\underset{(n\times n)}{\tilde{M}}$ are symmetric so that

$$\underset{(n\times n)}{\tilde{K}^T} = \underset{(n\times n)}{\tilde{K}} \quad \text{and} \quad \underset{(n\times n)}{\tilde{M}^T} = \underset{(n\times n)}{\tilde{M}}$$

Hence equation (12a) becomes

$$\underset{(1\times n)}{\tilde{\phi}_r^T} \; \underset{(n\times n)}{\tilde{K}} = \omega_r^2 \; \underset{(1\times n)}{\tilde{\phi}_r^T} \; \underset{(n\times n)}{\tilde{M}} \tag{12b}$$

Now let us post-multiply equation (12b) by $\tilde{\phi}_s$ to give

$$\underset{(1\times n)}{\tilde{\phi}_r^T} \; \underset{(n\times n)}{\tilde{K}} \; \underset{(n\times 1)}{\tilde{\phi}_s} = \omega_r^2 \; \underset{(1\times n)}{\tilde{\phi}_r^T} \; \underset{(n\times n)}{\tilde{M}} \; \underset{(n\times 1)}{\tilde{\phi}_s} = c_{rs} \tag{13}$$

where $c_{rs}$ is a scalar quantity as yet unknown.

Now let us take equation (11b) and pre-multiply by $\underset{(1\times n)}{\tilde{\phi}_r^T}$ to give

$$\underset{(1\times n)}{\tilde{\phi}_r^T} \; \underset{(n\times n)}{\tilde{K}} \; \underset{(n\times 1)}{\tilde{\phi}_s} = \omega_s^2 \; \underset{(1\times n)}{\tilde{\phi}_r^T} \; \underset{(n\times n)}{\tilde{M}} \; \underset{(n\times 1)}{\tilde{\phi}_s} = c_{rs} \tag{14}$$

Clearly by subtracting (13) from (14)

$$0 \quad = \quad (\omega_s^2 - \omega_r^2) \underset{(1xn)}{\underset{\sim}{\phi}_r^T} \underset{(nxn)}{\underset{\sim}{M}} \underset{(nx1)}{\underset{\sim}{\phi}_s} \qquad (15)$$

Now we said earlier that $\omega_s \neq \omega_r$. The only possible explanation for equation (15) is that

$$\underset{(1xn)}{\underset{\sim}{\phi}_r^T} \underset{(nxn)}{\underset{\sim}{M}} \underset{(nx1)}{\underset{\sim}{\phi}_s} = 0 \qquad \text{for } r \neq s. \qquad (16a)$$

Thus $c_{rs} = 0$ and also from (13) or (14)

$$\underset{(1xn)}{\underset{\sim}{\phi}_r^T} \underset{(nxn)}{\underset{\sim}{K}} \underset{(nx1)}{\underset{\sim}{\phi}_s} = 0 \qquad (16b)$$

Equations (16) are true for any pair of mode shapes we care to choose. This property exhibited by the mode shapes is known as orthogonality and it has considerable importance. The reason for this terminology may not seem obvious. However it can be demonstrated by a simple example. Suppose we have the system shown in Figure 5 which has two degrees of freedom. Then the vibration in the x direction is completely independent of the motion in the y direction because the system is physically orthogonal. The mode shape vectors for the vibration in the x and y directions will be

$$\begin{bmatrix} a \\ 0 \end{bmatrix} \quad \text{and} \quad \begin{bmatrix} 0 \\ b \end{bmatrix}$$

where a and b are respectively amplitudes in the x and y directions. Thus if we take the product

$$\begin{bmatrix} a & 0 \end{bmatrix} \begin{bmatrix} m & 0 \\ 0 & m \end{bmatrix} \begin{bmatrix} 0 \\ b \end{bmatrix}$$

corresponding to $\underset{\sim}{\phi}_r^T \underset{\sim}{M} \underset{\sim}{\phi}_s$ it will give zero. This property of orthogonality of the modes shapes is important because it means that we can always choose our degrees of freedom, which define the motion of the structure, such that the equations of motion are in terms of independent variables.

## 5. Modal Decomposition

Let us choose a normalizing vector, $\underset{\sim}{z}_r$, for the mode shape, $\underset{\sim}{\phi}_r$, which is defined by ,

$$\underset{(n\times1)}{\tilde{z}_r} = \frac{1}{c_r} \underset{(n\times1)}{\tilde{\phi}_r} \tag{17}$$

where $c_r$ is the scalar constant such that for each value of r

$$\underset{(1\times n)}{\tilde{z}_r^T} \underset{(n\times n)}{\tilde{M}} \underset{(n\times1)}{\tilde{z}_r} = 1 \tag{18}$$

Now we make up a *modal matrix* $\underset{(n\times n)}{\tilde{Z}}$ consisting of n normalizing vectors as columns

$$\underset{(n\times n)}{\tilde{Z}} = [ \underset{(n\times1)}{\tilde{z}_1} \; \underset{(n\times1)}{\tilde{z}_2} \; \cdots\cdots \; \underset{(n\times1)}{\tilde{z}_r} \; \cdots\cdots \; \underset{(n\times1)}{\tilde{z}_{n-1}} \; \underset{(n\times1)}{\tilde{z}_n} ]$$

From the orthogonality condition (16a) and equation (18)

$$\tilde{z}_r^T \tilde{M} \tilde{z}_s = 0 \quad \text{if } r \neq s \text{ for } r = 1,n; \; s = 1,n. \tag{19a}$$

$$\tilde{z}_r^T \tilde{M} \tilde{z}_r = 1 \quad \text{if } r = 1,n \tag{19b}$$

Hence

$$\tilde{Z}^T \tilde{M} \tilde{Z} = \tilde{I} \tag{19c}$$

where $\tilde{I}$ is the unit matrix. Furthermore from expressions like equations (11a) and (11b)

$$\tilde{z}_r^T \tilde{K} \tilde{z}_s = \omega_r^2 \tilde{z}_r^T \tilde{M} \tilde{z}_s = 0 \qquad r \neq s$$

and

$$\tilde{z}_r^T \tilde{K} \tilde{z}_r = \omega_r^2 \tilde{z}_r^T \tilde{M} \tilde{z}_r = \omega_r^2 \tag{20}$$

Hence if we take the matrix $\underset{(n\times n)}{\tilde{K}}$, pre-multiply it by $\underset{(n\times n)}{\tilde{Z}^T}$ and post-multiply it by $\underset{(n\times n)}{\tilde{Z}}$ we shall obtain a diagonal matrix $\underset{(n\times n)}{\tilde{\Omega}}$ containing n elements which are the squares of the n natural frequencies.

$$\begin{bmatrix} \underset{(1\times n)}{\tilde{z}_1^T} \\ \underset{(1\times n)}{\tilde{z}_2^T} \\ \vdots \\ \underset{(1\times n)}{\tilde{z}_r^T} \\ \vdots \\ \underset{(1\times n)}{\tilde{z}_{n-1}^T} \\ \underset{(1\times n)}{\tilde{z}_n^T} \end{bmatrix} \underset{(n\times n)}{\tilde{K}} \left[ \underset{(n\times 1)}{\tilde{z}_1} \; \underset{(n\times 1)}{\tilde{z}_2} \; \cdots \; \underset{(n\times 1)}{\tilde{z}_r} \; \cdots \; \underset{(n\times 1)}{\tilde{z}_{n-1}} \; \underset{(n\times 1)}{\tilde{z}_n} \right]$$

$$= \begin{bmatrix} \omega_1^2 & & & & & \tilde{0} \\ & \omega_2^2 & & & & \\ & & \ddots & & & \\ & & & \omega_r^2 & & \\ & & & & \ddots & \\ & \tilde{0} & & & \omega_{n-1}^2 & \\ & & & & & \omega_n^2 \end{bmatrix}$$

or $\underset{(n\times n)}{\tilde{Z}^T} \; \underset{(n\times n)}{\tilde{K}} \; \underset{(n\times n)}{\tilde{Z}} = \underset{(n\times n)}{\tilde{\Omega}}$

since the $r^{th}$ diagonal element is $\omega_r^2$ and all other elements are zero.

Thus

$$\underset{(n\times n)}{\tilde{Z}^T} \; \underset{(n\times n)}{\tilde{K}} \; \underset{(n\times n)}{\tilde{Z}} = \underset{(n\times n)}{\tilde{\Omega}} = \underset{(n\times n)}{\tilde{Z}^T} \; \underset{(n\times n)}{\tilde{M}} \; \underset{(n\times n)}{\tilde{Z}} \; \underset{(n\times n)}{\tilde{\Omega}} \tag{21}$$

where $\underset{(n\times n)}{\underset{\sim}{\Omega}}$ is the matrix of the squares of the natural frequencies.

If we return to the original equations of motion (equation (10)) and pre-multiply these by $\underset{(n\times n)}{\underset{\sim}{Z}^T}$, we obtain

$$\underset{(n\times n)}{\underset{\sim}{Z}^T} \left( \underset{(n\times n)}{\underset{\sim}{K}} - \omega^2 \underset{(n\times n)}{\underset{\sim}{M}} \right) \underset{(n\times 1)}{\underset{\sim}{A}} = \underset{(n\times 1)}{0} \tag{22}$$

We can now carry out a coordinate transformation and put

$$\underset{(n\times 1)}{\underset{\sim}{A}} = \underset{(n\times n)}{\underset{\sim}{Z}} \underset{(n\times 1)}{\underset{\sim}{Q}} \tag{23}$$

where $\underset{(n\times 1)}{\underset{\sim}{Q}}$ is a vector of normal coordinates. This is a change of coordinates from the original set to a new set which represents each degree of freedom of the original set by a combination of proportions of each of the normalized mode shapes. Substituting equation (23) in (22) we have

$$\underset{(n\times n)}{\underset{\sim}{Z}^T} \left( \underset{(n\times n)}{\underset{\sim}{K}} - \omega^2 \underset{(n\times n)}{\underset{\sim}{M}} \right) \underset{(n\times n)}{\underset{\sim}{Z}} \underset{(n\times 1)}{\underset{\sim}{Q}} = \underset{(n\times 1)}{\underset{\sim}{0}}$$

which on expanding gives

$$\underset{(n\times n)}{\underset{\sim}{Z}^T} \underset{(n\times n)}{\underset{\sim}{K}} \underset{(n\times n)}{\underset{\sim}{Z}} \underset{(n\times 1)}{\underset{\sim}{Q}} - \omega^2 \underset{(n\times n)}{\underset{\sim}{Z}^T} \underset{(n\times n)}{\underset{\sim}{M}} \underset{(n\times n)}{\underset{\sim}{Z}} \underset{(n\times 1)}{\underset{\sim}{Q}} = \underset{(n\times 1)}{\underset{\sim}{0}} \tag{24}$$

Making use of the relationships expressed in equations (19) and (21), equation (24) reduces to

$$\underset{(n\times n)}{\underset{\sim}{\Omega}} \underset{(n\times 1)}{\underset{\sim}{Q}} - \omega^2 \underset{(n\times 1)}{\underset{\sim}{Q}} = \underset{(n\times 1)}{\underset{\sim}{0}} \tag{25}$$

or

$$\left[ \underset{(n\times n)}{\underset{\sim}{\Omega}} - \omega^2 \underset{(n\times n)}{\underset{\sim}{I}} \right] \underset{(n\times 1)}{\underset{\sim}{Q}} = \underset{(n\times 1)}{\underset{\sim}{0}}$$

This corresponds to an original set of equations

$$\underset{(n\times 1)}{\underset{\sim}{\ddot{q}}} + \underset{(n\times n)}{\underset{\sim}{\Omega}} \underset{(n\times 1)}{\underset{\sim}{q}} = \underset{(n\times 1)}{\underset{\sim}{0}} \tag{26}$$

where $$\underset{(n\times1)}{\tilde{q}} = \underset{(n\times1)}{\tilde{Q}} e^{i\omega t} \qquad (27)$$

Equations (26) represent a set of n independent differential equations, each of which can be written in the form

$$\ddot{q}_r + \omega_r^2 q_r = 0 \qquad (28)$$

Thus by use of normal coordinates we can reduce the coupled equations of free vibration of a multi-degree of freedom system to a set of uncoupled equations, each involving just a single degree of freedom. For free vibration a multi-degree of freedom system behaves in exactly the same way as a set of independent single degree of freedom systems if we express the physical displacements in terms of a normal coordinate system.

Example 2

Let us go back to the 2-degree of freedom example. The mass matrix is

$$\begin{bmatrix} 1 & 0 \\ 0 & 1 \end{bmatrix}$$

and the mode shapes are $\begin{bmatrix} 1 \\ 1.62 \end{bmatrix}$ and $\begin{bmatrix} 1 \\ -0.62 \end{bmatrix}$

They can be normalized as shown in equations (17) and (18). For the first mode shape, the product

$$\underset{(1\times2)}{\tilde{\phi}_1^T} \; \underset{(2\times2)}{\tilde{M}} \; \underset{(2\times1)}{\tilde{\phi}_1} \text{ gives } \begin{bmatrix} 1 & 1.62 \end{bmatrix} \begin{bmatrix} 1 & 0 \\ 0 & 1 \end{bmatrix} \begin{bmatrix} 1 \\ 1.62 \end{bmatrix} = 1^2 + 1.62^2 = 3.62$$

Hence the constant $c_1$ in equation (17) is given by

$$c_1 = \sqrt{3.62} = 1.90$$

Thus the normalized mode shape, $\underset{(2\times1)}{\tilde{z}}$, is given by

$$\underset{(2\times1)}{\tilde{z}_1} = \frac{1}{1.90} \underset{(2\times1)}{\tilde{\phi}_1} = \begin{bmatrix} 0.526 \\ 0.851 \end{bmatrix}$$

By a similar procedure,

$$c_2 = \sqrt{1^2 + 0.62^2} = \sqrt{1.384} = 1.18$$

and $$\underset{(2\times 1)}{\underset{\sim}{z}_2} = \begin{bmatrix} 0.851 \\ -0.526 \end{bmatrix}$$

Thus the matrix $\underset{(2\times 2)}{\underset{\sim}{Z}}$ becomes $\begin{bmatrix} 0.526 & 0.851 \\ 0.851 & -0.526 \end{bmatrix}$

Now we can carry out a coordinate transformation to represent the vector $\begin{bmatrix} 1 \\ 0 \end{bmatrix}$ corresponding to unit displacement $u_1$ and zero displacement $u_2$ in terms of the generalized coordinates and the matrix $\underset{(2\times 2)}{\underset{\sim}{Z}}$. It will be seen that

$$\begin{bmatrix} 1 \\ 0 \end{bmatrix} = \begin{bmatrix} 0.526 & 0.851 \\ 0.851 & -0.526 \end{bmatrix} \begin{bmatrix} 0.526 \\ 0.851 \end{bmatrix}$$

Thus the normal coordinates corresponding to a displacements

$$\left.\begin{array}{l} u_1 = 1 \\ u_2 = 0 \end{array}\right\} \begin{array}{l} \text{are } 0.526 \times \text{first normalized mode shape} \\ \quad 0.851 \times \text{second} \quad '' \quad '' \quad '' \end{array}$$

Similarly the vector corresponding to $u_2 = 1$ and $u_1 = 0$ can be transformed as

$$\begin{bmatrix} 0 \\ 1 \end{bmatrix} = \begin{bmatrix} 0.526 & 0.851 \\ 0.851 & -0.526 \end{bmatrix} \begin{bmatrix} 0.851 \\ -0.526 \end{bmatrix}$$

The normal coordinates corresponding to these displacements

$$\left.\begin{array}{l} u_1 = 0 \\ u_2 = 1 \end{array}\right\} \begin{array}{l} \text{are } 0.851 \times \text{first normalized mode shape} \\ \quad -0.526 \times \text{second} \quad '' \quad '' \quad '' \end{array}$$

## 6. Damped Free Vibrations of Multi-Degree of Freedom Systems

Real structures behave as lightly damped multi-degree of freedom systems. (The nature of damping and the means of representing damping in structures are discussed more fully in Chapter 9). Damping is observed as a result of energy being dissipated in a structural system. We can in general *represent* this dissipation of energy by an extra set of forces acting on the structural system. Thus the equations of motion in matrix form (2d) become

$$\underset{(n \times n)}{\underset{\sim}{M}} \; \underset{(n \times 1)}{\underset{\sim}{\ddot{U}}} + \underset{(n \times n)}{\underset{\sim}{K}} \; \underset{(n \times 1)}{\underset{\sim}{U}} = - \underset{(n \times 1)}{\underset{\sim}{P}_d} \qquad (29)$$

where $\underset{(n \times 1)}{\underset{\sim}{P}_d}$ is a vector of damping forces which is time dependent. The minus sign indicates that damping tends to reduce the magnitude of the motion.

Conventionally, the vector $\underset{(n \times 1)}{\underset{\sim}{P}_d}$ is represented as

$$\underset{(n \times 1)}{\underset{\sim}{P}_d} = \underset{(n \times n)}{\underset{\sim}{C}} \; \underset{(n \times 1)}{\underset{\sim}{\dot{U}}} \qquad (30)$$

where $\underset{(n \times n)}{\underset{\sim}{C}}$ is the damping matrix which is square and symmetric and $\underset{(n \times 1)}{\underset{\sim}{\dot{U}}}$ is the velocity vector. Equation (29) can then be written as

$$\underset{(n \times n)}{\underset{\sim}{M}} \; \underset{(n \times 1)}{\underset{\sim}{\ddot{U}}} + \underset{(n \times n)}{\underset{\sim}{C}} \; \underset{(n \times 1)}{\underset{\sim}{\dot{U}}} + \underset{(n \times n)}{\underset{\sim}{K}} \; \underset{(n \times 1)}{\underset{\sim}{U}} = \underset{(n \times 1)}{\underset{\sim}{O}} \qquad (31)$$

Equation (31) is the equation of motion of a free damped multi-degree of freedom system. Under certain circumstances, it is possible to make use of the orthogonal properties of the mode shapes to simplify the solution of equation (31). (See Chapter 9). In general a solution to equation (31) can be found by assuming the form

$$\underset{(n \times 1)}{\underset{\sim}{U}} = \underset{(n \times 1)}{\underset{\sim}{A}} \; e^{i\Omega t} \qquad (32)$$

where $\underset{(n \times 1)}{\underset{\sim}{A}}$ is a vector of complex elements, a + ib, and $\Omega$ is a complex frequency. The real and imaginary parts of $\Omega$ are the damped natural frequency, $\omega_d$, and a quantity, $\Delta$, which is the decay constant of the oscillation. Thus

$$\Omega = \omega_d + i\Delta \qquad (33)$$

The real and imaginary parts of the amplitude vector represent the phase relationships between the amplitudes of vibration of different parts of the structure (See reference [1]). Substituting equations (32) in (31) and dividing by $e^{i\Omega t}$ we obtain

$$\left[ -\Omega^2 \underset{(nxn)}{\underset{\sim}{M}} + i\Omega \underset{(nxn)}{\underset{\sim}{C}} + \underset{(nxn)}{\underset{\sim}{K}} \right] \underset{(nx1)}{\underset{\sim}{A}} = \underset{(nx1)}{\underset{\sim}{O}} \qquad (34)$$

The solution to equation (34) is obtained by finding the complex roots, $\Omega_r$, which make the determinant of $\left[ -\Omega^2 \underset{(nxn)}{\underset{\sim}{M}} + i\Omega \underset{(nxn)}{\underset{\sim}{C}} + \underset{(nxn)}{\underset{\sim}{K}} \right]$ zero. This may be done using standard algorithms such as the method developed by Nelder and Mead [2] or modifications of it.

Equation (33) may then be used to obtain the damped natural frequencies $\omega_{dr}$ and associated logarithmic decrements, $\delta_r$, for the $r^{th}$ mode of vibration, using $\Omega_r = \omega_{dr} + i\Delta_r$. By definition

$$\delta_r = \frac{2\pi\Delta_r}{\omega_{dr}} \simeq \frac{2\pi\Delta_r}{\omega_{nr}} \qquad (35)$$

for light damping, where $\omega_{nr}$ is the $r^{th}$ natural frequency.

## 7. Forced Vibration of Multi-Degree of Freedom Systems

Structures vibrate when they are subject to time varying forces. These forces may be of a periodic nature or continuously variable and random depending on the source of loading. We will confine the analysis in this section to periodic forces only and leave the analysis of transient responses till later (Chapter 9).

Any periodic non-harmonic function can be expressed in terms of a series of sine and cosine terms. Thus a periodic non-harmonic force can be expressed as a set of harmonic forces of constant amplitude (See reference [1]). If the amplitude is made complex, then sine and cosine terms can be collected together. The equations of motion for a lightly damped structure vibrating under a variable set of forces, expressed by the vector $\underset{(nx1)}{\underset{\sim}{P}(t)}$ become

$$\underset{(nxn)}{\underset{\sim}{M}} \underset{(nx1)}{\underset{\sim}{\ddot{U}}} + \underset{(nxn)}{\underset{\sim}{C}} \underset{(nx1)}{\underset{\sim}{\dot{U}}} + \underset{(nxn)}{\underset{\sim}{K}} \underset{(nx1)}{\underset{\sim}{U}} = \underset{(nx1)}{\underset{\sim}{P}(t)} \qquad (36)$$

If each of the forces making up the vector $\underset{(nx1)}{\underset{\sim}{P}(t)}$ is periodic with period, T, then $\underset{(nx1)}{\underset{\sim}{P}(t)}$ may be written as

$$\underset{\sim}{P}(t) = \underset{\sim o}{P} + \sum_{j=1}^{M} \underset{\sim j}{P} \, e^{i \frac{2\pi j}{T} t} \qquad (37)$$

where $\underset{(n\times 1)}{\underset{\sim}{P}_j}$ is a vector of complex force amplitudes. The total response of the system can be found by summing the responses due to each of the sets of harmonic forces acting separately. Let us consider the $j^{th}$ set and put $\omega_j = \frac{2\pi j}{T}$. The equations of motion (36) become

$$\underset{(n\times n)}{\underset{\sim}{M}} \; \underset{(n\times 1)}{\ddot{\underset{\sim}{U}}_j} + \underset{(n\times n)}{\underset{\sim}{C}} \; \underset{(n\times 1)}{\dot{\underset{\sim}{U}}_j} + \underset{(n\times n)}{\underset{\sim}{K}} \; \underset{(n\times 1)}{\underset{\sim}{U}_j} = \underset{(n\times 1)}{\underset{\sim}{P}_j} \; e^{i\omega_j t} \tag{38}$$

Assume a solution of the form

$$\underset{(n\times 1)}{\underset{\sim}{U}_j} = \underset{(n\times 1)}{\underset{\sim}{A}_j} \; e^{i\omega_j t} \tag{39}$$

By substituting (39) in (38), cancelling $e^{i\omega_j t}$ and collecting real and imaginary parts, we obtain

$$\left[ \left( \underset{(n\times n)}{\underset{\sim}{K}} - \omega_j^2 \underset{(n\times n)}{\underset{\sim}{M}} \right) + i\omega_j \underset{(n\times n)}{\underset{\sim}{C}} \right] \underset{(n\times 1)}{\underset{\sim}{A}_j} = \underset{(n\times 1)}{\underset{\sim}{P}_j} \tag{40}$$

Equation (40) can be solved for $\underset{(n\times 1)}{\underset{\sim}{A}_j}$ by Gaussian elimination using a complex version of any standard method. The vector $\underset{(n\times 1)}{\underset{\sim}{A}_j}$ will be complex. The modulus of each vector component will correspond to the amplitude of vibration and the argument will give the phase angle. In general the phase angle of the components of the vector $\underset{(n\times 1)}{\underset{\sim}{A}_j}$ will be different from those of force vector $\underset{(n\times 1)}{\underset{\sim}{P}_j}$, depending on the amount of damping.

Under certain circumstances, (see reference [3] and Chapter 9), equations (38) can be uncoupled by using the orthogonal property of the mode shapes to give a set of n independent single degree of freedom systems which can be solved by the methods given in Chapter 2.

References

1. Hurty, W.C. and M.F. Rubinstein, *Dynamics of Structures*, Prentice-Hall, Englewood Cliffs, New Jersey, 1964.

2. Nelder, J.A. and R. Mead, *Computer J.* 7, (4), 308-311, 1965.

3. Meirovitch, L. *Computational Methods in Structural Dynamics*, Sijthoff & Noordhoff, Netherlands, 1980.

CHAPTER 4

# EIGENVALUE-EIGENVECTOR SOLUTION

by

R.R. Wilson

## 1. Introduction

In this chapter, methods are presented for solving the free vibration equation for a multi-degree of freedom system. This equation is called the eigenvalue equation. Its solution not only gives the natural frequencies and mode shapes but also allows the set of coupled equations representing the dynamic response of a system to be uncoupled and hence solved. Thus the technique is at the heart of any general solution procedure for vibration problems.

Firstly, why is it necessary to consider sophisticated solution techniques for vibration problems ? Many standard subroutine packages contain eigenvalue solutions but unfortunately these are quite unsuitable for large structural vibration problems. Unlike static problems it is not possible to use a method, such as Gaussian elimination, in which only a small part of the system of equations needs to be in the computer at the one time. Care has to be taken to ensure that as large a problem as possible can be solved in the computer store available without at the same time requiring excessive computer time.

A three degree of freedom system is considered to introduce the terms used. This example is then used to illustrate solution by the evaluation of the zeros of a determinant. Three other methods for large systems are presented. The first, reduction of the equation to standard form and solution using a Stürm sequence technique, is considered in some detail and the other two, simultaneous iteration and the application of the Stürm sequence technique to the unreduced equations, are introduced briefly.

Two methods for reducing the total number of unknowns in a problem are considered, node condensation and substructure analysis. Finally, it is shown how, once the solution for a particular system has been found, it is possible to estimate the effect of a small change without having to solve an eigenvalue equation for the modified system.

## 2. Three Degree of Freedom System

The system shown in Figure 1 can be fully described by the three coordinates $u_1, u_2,$

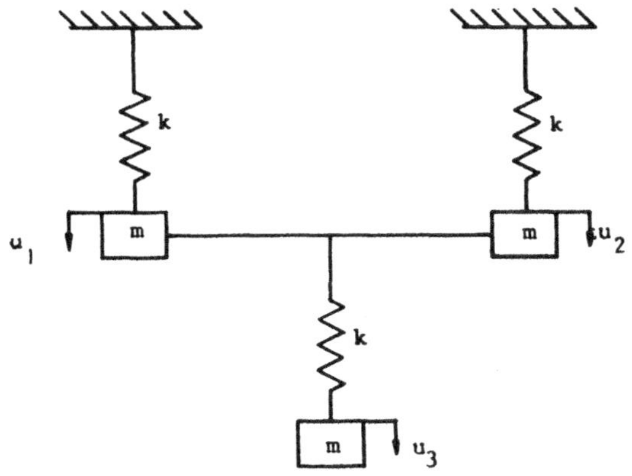


Figure 1. Three Degree of Freedom System.

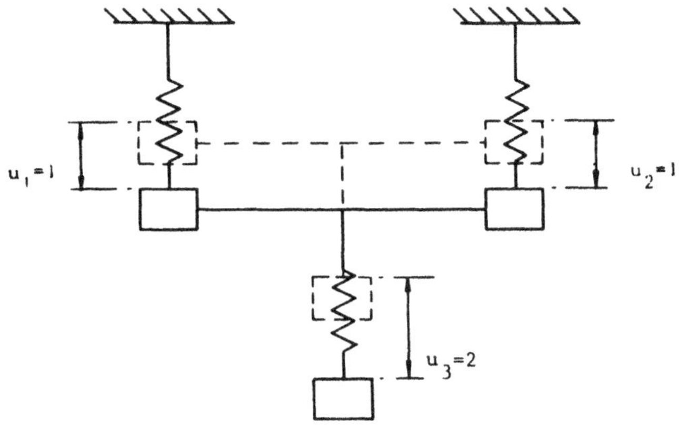


Figure 2 Mode Shape for 1st Natural Frequency

$u_3$, if it is assumed that the motion is only in the vertical direction. Applying Newton's Law, we have

$$m\ddot{u}_1 = -ku_1 + \tfrac{1}{2}k\,[u_3 - \tfrac{1}{2}(u_1+u_2)]$$

$$m\ddot{u}_2 = -ku_2 + \tfrac{1}{2}k\,[u_3 - \tfrac{1}{2}(u_1+u_2)] \qquad (1)$$

$$m\ddot{u}_3 = k\,[-u_3 + \tfrac{1}{2}(u_1+u_2)]$$

i.e.
$$\underset{\sim}{M}\,\underset{\sim}{U} + \underset{\sim}{K}\,\underset{\sim}{U} = \underset{\sim}{0} \qquad (2)$$

where
$$\underset{\sim}{M} = \begin{bmatrix} m & 0 & 0 \\ 0 & m & 0 \\ 0 & 0 & m \end{bmatrix}, \qquad \underset{\sim}{K} = \begin{bmatrix} \frac{5}{4}k & \frac{1}{4}k & -\frac{1}{2}k \\ \frac{1}{4}k & \frac{5}{4}k & -\frac{1}{2}k \\ -\frac{1}{2}k & -\frac{1}{2}k & k \end{bmatrix},$$

and
$$\underset{\sim}{U} = \begin{bmatrix} u_1 \\ u_2 \\ u_3 \end{bmatrix}.$$

If we assume a solution of the form $\underset{\sim}{U} = \underset{\sim}{\phi}\sin\omega t$, where $\underset{\sim}{\phi}$ gives the amplitude of vibration of each coordinate, then equation (2) becomes

$$(\underset{\sim}{K} - \omega^2\,\underset{\sim}{M})\underset{\sim}{\phi} = \underset{\sim}{0} \qquad (3)$$

Substituting $\lambda = \omega^2$, we have the eigenvalue equation,

$$(\underset{\sim}{K} - \lambda\,\underset{\sim}{M})\underset{\sim}{\phi} = \underset{\sim}{0} \qquad (4)$$

In order that this set of equations has a non-trivial solution (i.e. a solution for which $\underset{\sim}{\phi} \neq \underset{\sim}{0}$), we must have the determinant,

$$|\,\underset{\sim}{K} - \lambda\,\underset{\sim}{M}\,|\underset{\sim}{\phi} = \underset{\sim}{0} \qquad (5)$$

i.e.
$$\begin{vmatrix} \frac{5}{4}k - \lambda m & \frac{1}{4}k & -\frac{1}{2}k \\ \frac{1}{4}k & \frac{5}{4}k - \lambda m & -\frac{1}{2}k \\ -\frac{1}{2}k & -\frac{1}{2}k & k - \lambda m \end{vmatrix} = 0 \qquad (6)$$

Expanding the determinant, we have

$$(m\lambda)^3 - \frac{7}{2}(m\lambda)^2 + \frac{7}{2}m\lambda - k = 0$$

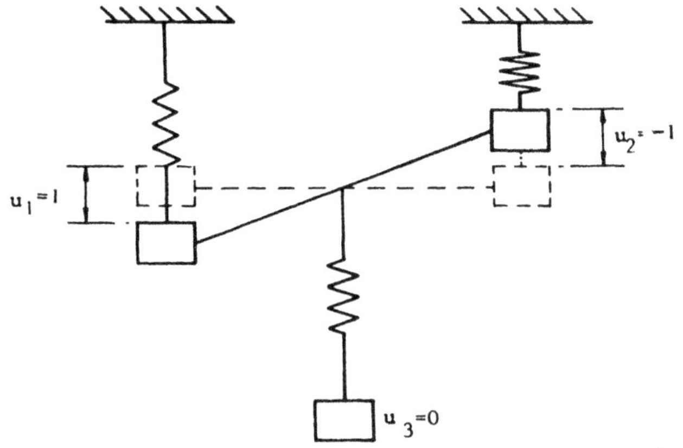


Figure 3  Mode Shape for 2nd Natural Frequency

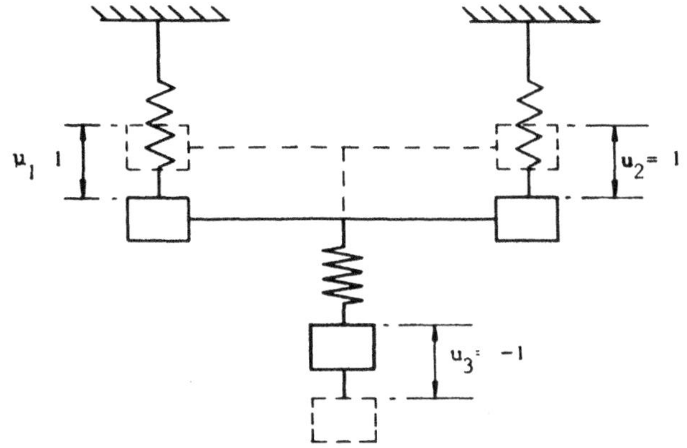


Figure 4  Mode Shape for 3rd Natural Frequency

This is a cubic equation with three distinct roots. Thus equation (4) has three non-trivial solutions given by

$$\lambda_1 = \frac{k}{2m}, \quad \lambda_2 = \frac{k}{m} \quad \text{and} \quad \lambda_3 = \frac{2k}{m} \tag{7}$$

These are the *eigenvalues* (or *latent roots*) of the equation and correspond to natural frequencies of the system

$$\omega_1 = \sqrt{\frac{k}{2m}}, \quad \omega_2 = \sqrt{\frac{k}{m}} \quad \text{and} \quad \omega_3 = \sqrt{\frac{2k}{m}} \tag{8}$$

Substituting each of the eigenvalues in turn into equation (4), the corresponding $\underset{\sim}{\phi}$ may be found.

$$\text{For} \quad \lambda = \lambda_1 = \frac{k}{2m}, \quad \underset{\sim}{\phi} = \underset{\sim}{\phi}_1 = \begin{bmatrix} 1 \\ 1 \\ 2 \end{bmatrix},$$

$$\text{for} \quad \lambda = \lambda_2 = \frac{k}{m}, \quad \underset{\sim}{\phi} = \underset{\sim}{\phi}_2 = \begin{bmatrix} 1 \\ -1 \\ 0 \end{bmatrix}, \tag{9}$$

$$\text{and for} \quad \lambda = \lambda_3 = \frac{2k}{m}, \quad \underset{\sim}{\phi} = \underset{\sim}{\phi}_3 = \begin{bmatrix} 1 \\ 1 \\ -1 \end{bmatrix}.$$

The vectors $\underset{\sim}{\phi}_i$ are called the *eigenvectors* (or *latent vectors* ) of the equation. They describe the relative magnitudes of the amplitude of vibration of each degree of freedom giving the (normal) mode shapes of the system when vibrating at each of its natural frequencies as shown in Figures 2, 3 and 4.

The eigenvectors give the ratios between the different components; the absolute value of the displacements is determined by the magnitude of the applied forces. It is often useful to express the eigenvectors as *normalized vectors* by scaling them so that, for each eigenvector, the sum of the squares of its components is 1. The normalized eigenvectors for the present system are

$$\underset{\sim}{\phi}_{1n} = \begin{bmatrix} 1/\sqrt{6} \\ 1/\sqrt{6} \\ 2/\sqrt{6} \end{bmatrix}, \quad \underset{\sim}{\phi}_{2n} = \begin{bmatrix} 1/\sqrt{2} \\ -1/\sqrt{2} \\ 0 \end{bmatrix}, \quad \underset{\sim}{\phi}_{3n} = \begin{bmatrix} 1/\sqrt{3} \\ 1/\sqrt{3} \\ -1/\sqrt{3} \end{bmatrix} \tag{10}$$

It should be noted that the eigenvectors are mutually *orthogonal*

$$\text{i.e.} \quad \underset{\sim}{\phi}_i^T \underset{\sim}{\phi}_j = 0, \quad i \neq j \tag{11}$$

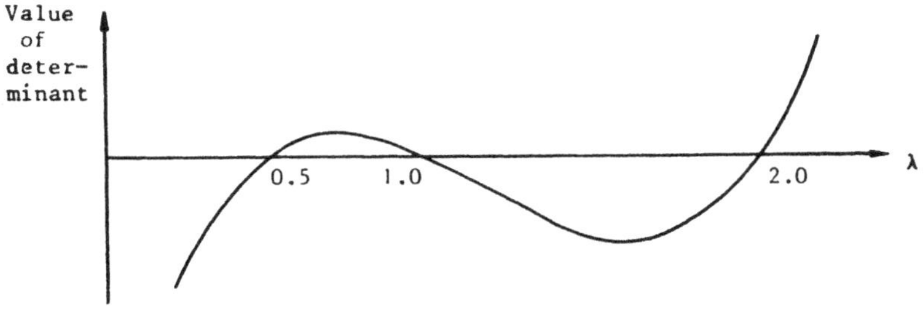


Figure 5. Zeros of the Determinant for 3 Degree of Freedom System.

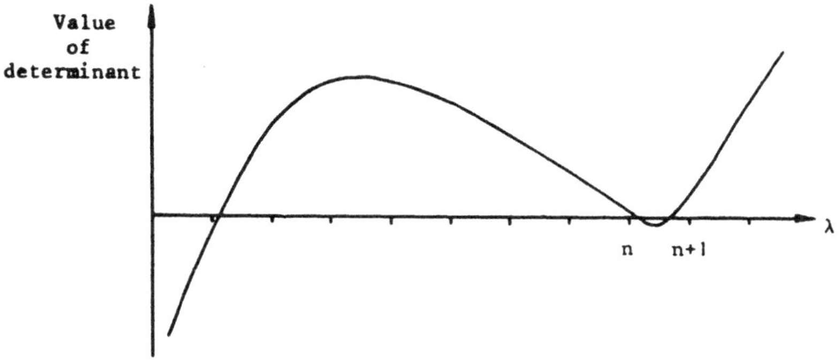


**Figure 6 Determinant for System with Two Close Eigenvalues.**

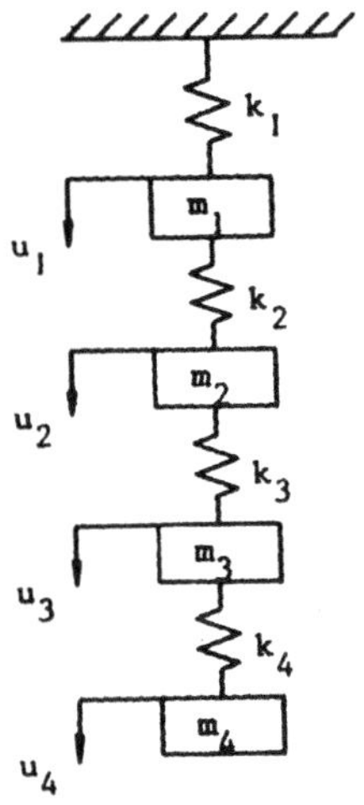


Figure 7 Four Degree of Freedom System.

$$\begin{array}{c} (1) \\ (2) \\ (3) \\ (4) \\ (5) \end{array} \left[ \begin{array}{c:c:c:c:c} a_{11} & a_{12} & a_{13} & a_{15} & a_{16} \\ \hdashline a_{21} & a_{22} & a_{23} & a_{24} & a_{25} \\ \hdashline a_{31} & a_{32} & a_{33} & a_{34} & a_{35} \\ \hdashline a_{41} & a_{42} & a_{43} & a_{44} & a_{45} \\ \hdashline a_{51} & a_{52} & a_{53} & a_{54} & a_{55} \end{array} \right]$$

The successive minors are the determinants of the series of submatrices shown.

Figure 8 Successive Minors of a Matrix.

and thus for normalized eigenvectors we have

$$\underset{\sim}{\phi}_{in}^{T}\ \underset{\sim}{\phi}_{jn} = \delta_{ij} \tag{12}$$

where
$$\delta_{ij} = \begin{cases} 1 & i = j \\ 0 & i \neq j \end{cases}$$

## 3. Zeros of Determinants

For the three degree of freedom system discussed in the previous section it was possible to determine the zeros of the determinant directly, by solving the cubic equation (6). For larger systems other techniques have to be used. By plotting the value of the determinant against $\lambda$, the eigenvalues may be found graphically. This is illustrated in Figure 3. Where it is not practical to evaluate the determinant by hand, root search techniques can be used on the computer [1].

One such procedure consists of evaluating the determinant for a range of values of $\lambda$. When there is a change of sign between successive evaluations, there must be a root lying in this interval. This root can be found to any required degree of accuracy by successively bisecting the interval containing the change of sign.

This process can be made more efficient by using the magnitude of the determinant to predict a root by interpolation. However this method has two major drawbacks. Firstly, when there are two identical or nearly identical roots, it is possible that the technique will fail to find either of them. Consider a system in which the plot of determinant against $\lambda$ has the form shown in Figure 6.

If for example, the determinant is evaluated at the points shown, then there will be no change of sign between the values at n and n+1. Hence, the two roots between these points will not be found. To ensure that this situation does not arise it is necessary to use very small intervals, resulting in excessive computing times. A second difficulty is caused by the fact that, for large systems, the absolute value of the determinant can become either very small or very large. Sophisticated scaling procedures are required to prevent overflow or underflow occurring during the computation.

## 4. Banded and Symmetric Matrices

Consider the system shown in Figure 7. The equation of motion for free vibration is given by

$$[\ K - \lambda\ M\ ]\phi = 0 \tag{13}$$

where $\underset{\sim}{M} = \begin{bmatrix} m_1 & 0 & 0 & 0 \\ 0 & m_2 & 0 & 0 \\ 0 & 0 & m_3 & 0 \\ 0 & 0 & 0 & m_4 \end{bmatrix}$ and $\underset{\sim}{K} = \begin{bmatrix} k_1+k_2 & -k_2 & 0 & 0 \\ -k_2 & k_2+k_3 & -k_3 & 0 \\ 0 & -k_3 & k_3+k_4 & -k_4 \\ 0 & 0 & -k_4 & k_4 \end{bmatrix}$

It can be seen that the mass matrix is a *diagonal matrix*; only those terms on the leading diagonal are non-zero. This reflects the fact that the system has lumped masses. In the finite element method discussed later, the mass matrices have non-zero off-diagonal entries. However, this reduces the size of the problem that can be considered and, for many applications, it is doubtful if the additional accuracy of representation gained by introducing these off-diagonal terms always justifies the increased complexity of solution.

In all the examples considered and in general for all structural problems, the mass and stiffness matrices are *symmetric*; the (i,j)th entry is equal to the (j,i)th entry. It is important that solution procedures take advantage of this. For an n x n symmetric matrix, it is only necessary to store n(n+1)/2 numbers.

The stiffness matrix K is *banded* with a semi-bandwidth of b = 1. It has non-zero terms only on the main diagonal and on b diagonal rows on either side of it. This reflects the fact that there is no interaction between for example, degrees of freedom 1 and 4. This situation often arises in practice - in a structure a degree of freedom interacts only with those freedoms in the adjacent parts of the structure. It is frequently useful to remember the degrees of freedom so that all the non-zero terms are grouped near the main diagonal and automatic schemes have been formulated for carrying this out [2,3]. For a symmetric banded matrix it is necessary to store only (b+1) n numbers.

## 5. Reduction of Eigenvalue Equation to Standard Form

Instead of directly solving the eigenvalue equation,

$$(\underset{\sim}{K} - \lambda \underset{\sim}{M})\underset{\sim}{\phi} = \underset{\sim}{0} \quad , \qquad (14)$$

it is often preferable to reduce it first to a simpler form.

Let
$$\underset{\sim}{y} = \underset{\sim}{M}\,\underset{\sim}{\phi} \qquad (15)$$

i.e.
$$\underset{\sim}{\phi} = \underset{\sim}{M}^{-1}\,\underset{\sim}{y} \qquad (16)$$

Then equation (14) can be written in the form

$$(\underset{\sim}{A} - \lambda \underset{\sim}{I}) \underset{\sim}{y} = \underset{\sim}{0} \tag{17}$$

where $\underset{\sim}{I}$ is the identity matrix

$$\begin{bmatrix} 1 & 0 & 0 & \cdots & 0 & 0 \\ 0 & 1 & 0 & \cdots & 0 & 0 \\ 0 & 0 & 1 & \cdots & 0 & 0 \\ \cdots & \cdots & \cdots & \cdots & \cdots & \cdots \\ 0 & 0 & 0 & \cdots & 1 & 0 \\ 0 & 0 & 0 & & 0 & 1 \end{bmatrix}$$

and $\underset{\sim}{A} = \underset{\sim}{K}\,\underset{\sim}{M}^{-1}$.

The solution of equation (17) may be regarded as finding the eigenvalues of the matrix $\underset{\sim}{A}$.

Because of the need to invert a matrix, this method is only suitable for small systems. In addition matrix $\underset{\sim}{A}$ is not symmetric and so the full matrix would require to be stored.

If matrix $\underset{\sim}{M}$ is positive definite (this corresponds to the requirement that the kinetic energy of the system is always positive) then, by using a technique called Choleski decomposition [4,5], a matrix $\underset{\sim}{L}$ can be found such that

$$\underset{\sim}{M} = \underset{\sim}{L}\,\underset{\sim}{L}^T , \tag{18}$$

where $\underset{\sim}{L}$ is lower triangular i.e. it has non-zero entries only on and below the main diagonal.

Thus equation (14) can be written

$$\underset{\sim}{K}\,\underset{\sim}{\phi} = \lambda\, \underset{\sim}{L}\,\underset{\sim}{L}^T \underset{\sim}{\phi} \tag{19}$$

Now $\underset{\sim}{L}^{-T}\,\underset{\sim}{L}^T = \underset{\sim}{I}$, and so

$$\underset{\sim}{K}\,(\underset{\sim}{L}^{-T}\,\underset{\sim}{L}^T)\underset{\sim}{\phi} = \underset{\sim}{L}\;\lambda\;(\underset{\sim}{L}^T\,\underset{\sim}{\phi}) \tag{20}$$

Premultiplying by $\underset{\sim}{L}^{-1}$ we have

$$(\underset{\sim}{L}^{-1}\,\underset{\sim}{K}\,\underset{\sim}{L}^{-T})(\underset{\sim}{L}^T\,\underset{\sim}{\phi}) = \lambda\;(\underset{\sim}{L}^T\,\underset{\sim}{\phi}) \tag{21}$$

This can be written as

$$(\underset{\sim}{B} - \lambda\,\underset{\sim}{I})\,\underset{\sim}{Z} = \underset{\sim}{0} \tag{22}$$

where $$\underset{\sim}{B} = \underset{\sim}{L}^{-1} \underset{\sim}{K} \underset{\sim}{L}^{-T} \quad \text{and} \quad \underset{\sim}{Z} = \underset{\sim}{L}^{T} \underset{\sim}{\Phi} \tag{22a}$$

Thus we now have a new eigenvalue equation in standard form, which has the same eigenvalues $\lambda$ as equation (14). The matrix $\underset{\sim}{B}$ is symmetric and so it is possible to use solution techniques which take advantage of this.

## 6. Solution of Standard Eigenvalue Equation by Stürm Sequence Technique

It would be possible to calculate the eigenvalues of the matrix $\underset{\sim}{B}$ directly but since unlike $\underset{\sim}{M}$ and $\underset{\sim}{K}$ it is not banded, considerable calculation would be required. It is preferable to find a matrix which has the same eigenvalues as $\underset{\sim}{B}$ but which is of simpler form. Householder's method [6] provides a means for doing this. Matrix $\underset{\sim}{B}$ can be reduced to a tridiagonal matrix which has the same eigenvalues as $\underset{\sim}{B}$ (a tridiagonal matrix is n one which has non-zero entries only on the main diagonal and the two adjacent rows).

A series of mutually orthogonal matrices $\underset{\sim}{P}_i$ can be found, defining a series $\underset{\sim}{B}_i$ with

$$\underset{\sim}{B}_{i+1} = \underset{\sim}{P}_i^{T} \underset{\sim}{B}_i \underset{\sim}{P}_i \quad , \tag{23}$$

where $$\underset{\sim}{B}_1 = \underset{\sim}{B}$$

The $\underset{\sim}{P}_i$ are chosen to make zero certain entries of $\underset{\sim}{B}_{i+1}$. Equation (22) can be written as

$$\underset{\sim}{B}_1 \underset{\sim}{Z} = \lambda \underset{\sim}{Z} \tag{24}$$

Now $\underset{\sim}{P}_1 \underset{\sim}{P}_1^{T} = \underset{\sim}{I}$, and so

$$\underset{\sim}{B}_1 \underset{\sim}{P}_1 \underset{\sim}{P}_1^{T} \underset{\sim}{Z} = \lambda \underset{\sim}{P}_1 \underset{\sim}{P}_1^{T} \underset{\sim}{Z} \tag{25}$$

$$\underset{\sim}{P}_1^{T} \underset{\sim}{B}_1 \underset{\sim}{P}_1 (\underset{\sim}{P}_1^{T} \underset{\sim}{Z}) = \lambda \underset{\sim}{P}_1^{T} \underset{\sim}{Z} \tag{26}$$

$$\text{i.e.} \quad (\underset{\sim}{B}_2 - \lambda \underset{\sim}{I})(\underset{\sim}{P}_1^{T} \underset{\sim}{Z}) = \underset{\sim}{0} \tag{27}$$

Hence the eigenvalues are unchanged but we have new eigenvectors $\underset{\sim}{P}_1^{T} \underset{\sim}{Z}$. This process can be repeated until a tri-diagonal matrix is obtained. This will still have the same eigenvalues as our original equation but eigenvectors $\underset{\sim}{P}_{n-2}^{T} \; \underset{\sim}{P}_{n-3}^{T} \ldots \underset{\sim}{P}_2^{T} \underset{\sim}{P}_1^{T} \underset{\sim}{Z}$.

The eigenvalues of this tridiagonal matrix can be found by bisection [7] using a result from the theory of polynomials that the successive minors of the matrix (see Figure 8) form what is called a Stürm sequence. The number of roots (i.e.

eigenvalues) greater than $\lambda_a$ is given by the number of changes in sign in the sequence when the minors are evaluated with $\lambda = \lambda_a$. By calculating the number of eigenvalues greater than $\lambda_a$ for a range of values of $\lambda_a$ and bisecting any interval containing more than one eigenvalue, it is possible to isolate each of the required roots. These may then be determined to any required accuracy by successively bisecting each interval containing a root. This method allows all the eigenvalues in a given range to be found or, for example, the ten lowest eigenvalues.

$$\begin{array}{c} (1) \\ (2) \\ (3) \\ (4) \\ (5) \end{array} \quad \begin{bmatrix} a_{11} & a_{12} & a_{13} & a_{15} & a_{16} \\ a_{21} & a_{22} & a_{23} & a_{24} & a_{25} \\ a_{31} & a_{32} & a_{33} & a_{34} & a_{35} \\ a_{41} & a_{42} & a_{43} & a_{44} & a_{45} \\ a_{51} & a_{52} & a_{53} & a_{54} & a_{55} \end{bmatrix}$$

The successive minors are the determinants of the series of submatrices shown.

Figure 8 Successive Minors of a Matrix.

Once the eigenvalues have been found the corresponding eigenvectors of the tridiagonal matrix can be found by inverse iteration (Wielandt iteration) [8]. Two iterations are usually sufficient. The eigenvectors of the standard equation (22) are then calculated by premultiplying the eigenvectors of the tridiagonal matrix by the $\underset{\sim}{P}_i$. Finally the eigenvectors of the original system (14) are determined by equation (22a).

## 7. Solution of the Original Equations using Stürm Sequence Technique

It can be shown that the sign of each of the tridiagonal matrix is the same as the sign of the corresponding minor of $(\underset{\sim}{K} - \lambda \underset{\sim}{M})$ [9, 10]. Thus instead of reducing the equations to standard form and then finding the equivalent tridiagonal matrix, it is possible to use the Stürm sequence property of the minors of the original matrices.

However, unless the bandwidth of the matrices is very small, this process can require excessive computer time. To overcome this, it is possible to use the Stürm sequence method applied to the original matrices simply to find a series of intervals each containing only one eigenvalue. This avoids the possibility that a root might be missed. The roots are then determined accurately by evaluating the numerical value of the complete determinant and using interpolation techniques to accelerate the process of searching for a zero.

## 8. Simultaneous Iteration

The method of simultaneous iteration (or direct iteration) [11] is a method of finding the roots of the standard eigenvalue equation (22). A set of mutually orthogonal trial vectors are selected. If possible these should be realistic estimates of the eigenvectors corresponding to the lowest eigenvalues but in a general scheme may be, for instance, the columns of the identity matrix. Let us assume we have chosen 3 mutually orthogonal vectors $\underset{\sim}{V}_i$. These may be written as linear combinations of the (unknown) eigenvectors of the matrix $\underset{\sim}{B}$.

$$\begin{aligned} \underset{\sim}{V}_1 &= \underset{\sim}{Z}_1 + C_{12}\underset{\sim}{Z}_2 + C_{13}\underset{\sim}{Z}_3 + \ldots \\ \underset{\sim}{V}_2 &= C_{21}\underset{\sim}{Z}_1 + \underset{\sim}{Z}_2 + C_{23}\underset{\sim}{Z}_3 + \ldots \\ \underset{\sim}{V}_3 &= C_{31}\underset{\sim}{Z}_1 + C_{32}\underset{\sim}{Z}_2 + \underset{\sim}{Z}_3 + \ldots \end{aligned} \tag{28}$$

Now if the $C_{ij}$ are less than 1 then, since the $\underset{\sim}{V}_i$ are mutually orthogonal, we have that the equations can approximately be written as

$$\begin{aligned} \underset{\sim}{V}_1 &= \underset{\sim}{Z}_1 + C_{12}\underset{\sim}{Z}_2 + C_{13}\underset{\sim}{Z}_3 + \ldots \\ \underset{\sim}{V}_2 &= -C_{12}\underset{\sim}{Z}_1 + \underset{\sim}{Z}_2 + C_{23}\underset{\sim}{Z}_3 + \ldots \\ \underset{\sim}{V}_3 &= -C_{13}\underset{\sim}{Z}_1 - C_{23}\underset{\sim}{Z}_2 + \underset{\sim}{Z}_3 + \ldots \end{aligned} \tag{29}$$

A new set of vectors $\underset{\sim}{V}_i'$ can be defined by

$$\underset{\sim}{V}' = \underset{\sim}{B}\,\underset{\sim}{V} \tag{30}$$

where $\underset{\sim}{V}' = [\underset{\sim}{V}_1'\ \underset{\sim}{V}_2'\ \underset{\sim}{V}_3']$ and $\underset{\sim}{V} = [\underset{\sim}{V}_1\ \underset{\sim}{V}_2\ \underset{\sim}{V}_3]$.

Now since

$$\underset{\sim}{B}\,\underset{\sim}{Z}_i = \underset{\sim}{\lambda}_i\,\underset{\sim}{Z}_i \,, \tag{31}$$

$$\begin{aligned} \underset{\sim}{V}_1' &= \lambda_1\underset{\sim}{Z}_1 + \lambda_2 C_{12}\underset{\sim}{Z}_2 + \lambda_3 C_{13}\underset{\sim}{Z}_3 + \ldots \\ \underset{\sim}{V}_2' &= -\lambda_1 C_{12}\underset{\sim}{Z}_1 + \lambda_2\underset{\sim}{Z}_2 + \lambda_3 C_{23}\underset{\sim}{Z}_3 + \ldots \\ \underset{\sim}{V}_3' &= -\lambda_1 C_{13}\underset{\sim}{Z}_1 - \lambda_2 C_{23}\underset{\sim}{Z}_2 + \lambda_3\underset{\sim}{Z}_3 + \ldots \end{aligned} \tag{32}$$

We can now define the matrix $\underset{\sim}{D}$ by

$$\underset{\sim}{D} = \underset{\sim}{V}^T\underset{\sim}{V}' = \underset{\sim}{V}^T\,\underset{\sim}{B}\,\underset{\sim}{V} \tag{33}$$

Thus

$$\underset{\sim}{D} = \begin{bmatrix} \lambda_1 & -C_{12}(\lambda_1-\lambda_2) & -C_{13}(\lambda_1-\lambda_3) \\ -C_{12}(\lambda_1-\lambda_2) & \lambda_2 & -C_{23}(\lambda_2-\lambda_3) \\ -C_{13}(\lambda_1-\lambda_3) & -C_{23}(\lambda_2-\lambda_3) & \lambda_3 \end{bmatrix} \tag{34}$$

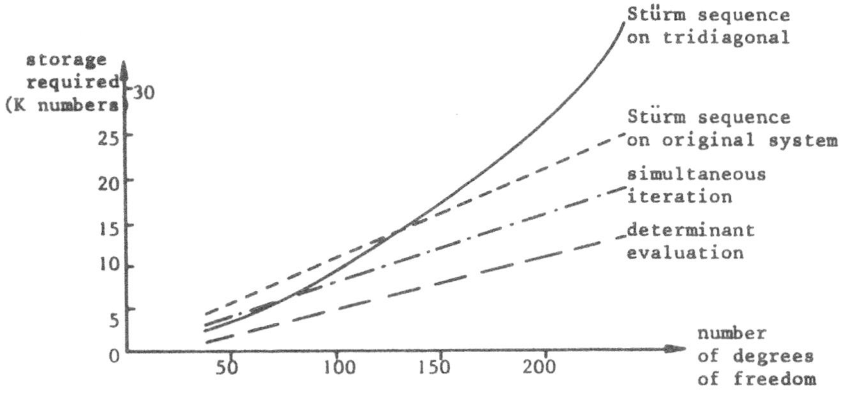


**Figure 9** **Store Required for Eigenvalue Solution**

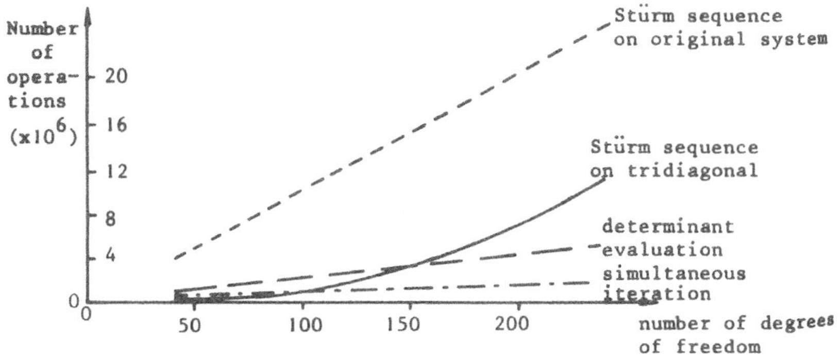


**Figure 10** **Number of Operations During Eigenvalue Solution**

If the trial vectors we selected had been the true eigenvectors then D would be a diagonal matrix with the entries on the diagonal equal to the eigenvalues. In general the off-diagonal terms will be non-zero because of the coupling between the trial vectors. The entries of $\underset{\sim}{D}$ can then be used to suggest a second trial set of vectors. After they have been made mutually orthogonal, this second set of vectors can be used to repeat the cycle. This process is continued until the required accuracy is obtained.

## 9. Comparison of Eigenvalue Solution Methods

We have now considered four methods of solution and it is useful to compare the different approaches. The methods were

1) direct evaluation of the zeros of the determinant
2) application of Stürm sequence technique to the original equations perhaps using determinant evaluation for final root determination
3) reduction to tridiagonal form and then application of Stürm sequence technique.
4) simultaneous iterations.

The first two methods are applied to the original system of equations, the second two to the equation after it has been reduced to standard form. As mentioned previously, direct evaluation of the determinant can fail to find close or identical roots. Because of this, determinant evaluation on its own is not suitable for a general solution scheme. Simultaneous iteration suffers from the disadvantage that its rate of convergence is dependent on the problem and in particular on how well separated the roots are. Both the methods using a Stürm sequence technique have no difficulty in finding pairs of identical eigenvalues. It should be noted however that they do not produce the correct eigenvectors in this case.

The storage required and the number of operations performed during the calculations by each of the methods is very much dependent on the type and size of problem. Estimated formula for these computational parameters based on references [12] and [1] are given in Table 1. For the simultaneous iteration method, estimates have been made of the number of iterations and number of trial vectors required.

Figures 9 and 10 show plots based on these formulae for the storage required and the number of operations for a typical problem. It has been assumed that 10 eigenvalues are required and that the semi-bandwidth of the matrices is 20. In general there is no real best buy. All the methods discussed (and others [1,4]) are used. Perhaps method 3, the reduction to tridiagonal form and application of the Stürm sequence technique is the most widely used in this country. Because of this, it

| Method | Storage required | Number of operations |
|---|---|---|
| 1 determinant evaluation | $2n(b+1) + 9n$ | $(3nb^2 + 39nb + 114n)m$ |
| 2 Sturm sequence on original system | $5nb$ | $25nb^2 m$ |
| 3 Sturm sequence on tridiagonal | $\frac{1}{2}n^2 + 2nb$ | $3/2n^2 + \frac{2}{3}n^3 + mn^2$ |
| 4 simultaneous iteration | $2nb + 3n(m+3) + \frac{1}{2}(m+3)^2$ | $\frac{1}{2}nb^2 + (m+3)^3 + +15n(m+3) + 18nb(m+3)$ |

n = number of degrees of freedom; b = semi-bandwidth; m = number of eigenvalues required.

Table 1 Storage and Number of Operations for Eigenvalue Solution

has the advantage that there is a number of well tried algorithms available which are based on it.

## 10. Node Condensation

For large systems, the size of the matrices is such that it may not be possible to solve the eigenvalue equation in the computer store available. One technique which is used widely to overcome this difficulty in the finite element method is node condensation [13, 14] . Consider a structure which is represented by n degrees of freedom but with forces applied only to the first r of them. The resulting deflections are given by

$$\begin{bmatrix} \underset{\sim}{P}_1 \\ \underset{\sim}{0} \end{bmatrix} = \begin{bmatrix} \underset{\sim}{k}_{11} & \underset{\sim}{k}_{12} \\ \underset{\sim}{k}_{12}^T & \underset{\sim}{k}_{22} \end{bmatrix} \begin{bmatrix} \underset{\sim}{u}_1 \\ \underset{\sim}{u}_2 \end{bmatrix} \tag{35}$$

i.e. $$\underset{\sim}{P}_1 = \underset{\sim}{k}_{11}\underset{\sim}{u}_1 + \underset{\sim}{k}_{12}\underset{\sim}{u}_2 \tag{36}$$

and $$\underset{\sim}{0} = \underset{\sim}{k}_{12}^T\underset{\sim}{u}_1 + \underset{\sim}{k}_{22}\underset{\sim}{u}_2 \tag{37}$$

Substituting for $\underset{\sim}{u}_2$ from (37) into equation (36) gives

$$\underset{\sim}{P}_1 = (\underset{\sim}{k}_{11} - \underset{\sim}{k}_{12}\underset{\sim}{k}_{22}\underset{\sim}{k}_{12}^T)\ \underset{\sim}{u}_1 \tag{38}$$

Thus the system is described by the equation

$$\underset{\sim}{P}_1 = \underset{\sim}{k}^*\ \underset{\sim}{u}_1\ , \tag{39}$$

where $$\underset{\sim}{k}^* = \underset{\sim}{k}_{11} - \underset{\sim}{k}_{12}\underset{\sim}{k}_{22}\underset{\sim}{k}_{12}^T$$

This is equivalent to a coordinate transformation from the original coordinates $\underset{\sim}{u}$ to a new coordinate system $\underset{\sim}{y}(= \underset{\sim}{u}_1)$. The transformation matrix $\underset{\sim}{\Lambda}$ is given by

$$\underset{\sim}{u} = \underset{\sim}{\Lambda}\ \underset{\sim}{y} \tag{40}$$

Since

$$\underset{\sim}{u} = \begin{bmatrix} \underset{\sim}{u}_1 \\ \underset{\sim}{u}_2 \end{bmatrix} = \begin{bmatrix} \underset{\sim}{I} \\ -\underset{\sim}{k}_{22}\ \underset{\sim}{k}_{12}^T \end{bmatrix} \begin{bmatrix} \underset{\sim}{u}_1 \end{bmatrix} \tag{41}$$

we have $$\underset{\sim}{\Lambda} = \begin{bmatrix} \underset{\sim}{I} \\ -\underset{\sim}{k}_{22}\ \underset{\sim}{k}_{12}^T \end{bmatrix} \tag{42}$$

The potential energy of the system is given in the original coordinate system by

$$V = \tfrac{1}{2} \underset{\sim}{u}^T \underset{\sim}{K} \underset{\sim}{u} \qquad (43)$$

$$V = \tfrac{1}{2}(\underset{\sim}{\Lambda} \underset{\sim}{y})^T \underset{\sim}{K}(\underset{\sim}{\Lambda} \underset{\sim}{y}) = \tfrac{1}{2}\underset{\sim}{y}^T (\underset{\sim}{\Lambda}^T \underset{\sim}{K} \underset{\sim}{\Lambda})\underset{\sim}{y}$$

$$V = \tfrac{1}{2}\underset{\sim}{y}^T \underset{\sim}{k}^* \underset{\sim}{y} , \qquad (44)$$

where $\underset{\sim}{k}^* = \underset{\sim}{\Lambda}^T \underset{\sim}{K} \underset{\sim}{\Lambda}$

The stiffness matrix calculated in this way is identical to that given by equation (39).

In a similar manner we can introduce a mass matrix defined by

$$\underset{\sim}{m}^* = \underset{\sim}{\Lambda}^T \underset{\sim}{M} \underset{\sim}{\Lambda} . \qquad (45)$$

These reduced matrices can now form an eigenvalue equation which is able to be solved in the available computer store.

For the vibration of a practical structure, there will be forces applied at every degree of freedom having a mass associated with it. (By D'Alembert's principle equivalent force = $-m\ddot{u}$.) However we can decide to retain only the more important freedoms, the *master* degrees of freedom, and condense out the others, the *slave* degrees of freedom. The eigenvalue equation can then be solved in terms of the masters. The complete mode shape can be found by substituting in equation (40) for the slave freedoms.

The choice of master freedoms is largely a matter of experience. It is relatively simple if one already knows roughly what the mode shapes will be. Perhaps the only general guideline is that it is usually better to eliminate rotational freedoms and retain only translation freedoms. This method does however allow reasonably accurate predictions to be made of the vibrational behaviour of very large systems.

## 11. Substructure Analysis

The technique of substructure analysis [15] can be used to reduce the size of eigenvalue equations instead of node condensation. For a very complicated structure rather than solving in terms of all the freedoms, only those freedoms which couple the main sections of the structure are used. For example, an aeroplane could be considered as an assemblage of fuselage, wings and tail. The effect of each sub-structure is represented in terms of its normal modes calculated with constraints applied to the coupling freedoms. Modes of the complete structure are represented as a synthesis of the modes of the substructures.

This method has two advantages over node condensation. Firstly it is often important to know the local resonances of a structure and this information is given directly by a substructure analysis. Secondly, substructure analysis allows a great reduction in computation where the structure is repetitive, having a number of identical components. On the other hand, it is not as suitable as node condensation for use in a general solution scheme.

## 12. Rate of Change of Eigenvalues

For large systems, it requires a great deal of computer time to solve the eigenvalue equation. It is not feasible to solve for many design variations. One way round this is to calculate the rate of change of the eigenvalues [16, 17] with change in design parameter.

When a structure is vibrating in its ith mode, the maximum potential and kinetic energies are equal.

i.e.
$$\tfrac{1}{2}\,\boldsymbol{\phi}_i^T \mathbf{K}\,\boldsymbol{\phi}_i = \tfrac{1}{2}\,\lambda_i\,\boldsymbol{\phi}_i^T \mathbf{M}\,\boldsymbol{\phi}_i \tag{46}$$

If we now consider a small change in design, then there will be new stiffness and mass matrices $\mathbf{K} + \Delta\mathbf{K}$ and $\mathbf{M} + \Delta\mathbf{M}$. Thus if the corresponding changes to the ith eigenvector and eigenvalue are $\Delta\boldsymbol{\phi}_i$ and $\Delta\boldsymbol{\phi}_i$, we have

$$\tfrac{1}{2}(\boldsymbol{\phi}_i + \Delta\boldsymbol{\phi}_i)^T(\mathbf{K} + \Delta\mathbf{K})(\boldsymbol{\phi}_i + \Delta\boldsymbol{\phi}_i)$$

$$= \tfrac{1}{2}(\lambda_i + \Delta\lambda_i)(\boldsymbol{\phi}_i + \Delta\boldsymbol{\phi}_i)(\mathbf{M} + \Delta\mathbf{M})(\boldsymbol{\phi}_i + \Delta\boldsymbol{\phi}_i) \tag{47}$$

Expanding and retaining only the first order terms we have,

$$\Delta\lambda_i = \frac{\boldsymbol{\phi}_i^T(\Delta\mathbf{K} - \lambda\,\Delta\mathbf{M})\boldsymbol{\phi}_i}{\boldsymbol{\phi}_i^T\,\mathbf{M}\,\boldsymbol{\phi}_i} \tag{48}$$

Thus once an eigenvalue equation has been solved, an estimate of the effect of a small change can be made without having to solve a new set of equations.

References

1. Bathe, K.J. and E.L. Wilson, Solution methods for eigenvalue problems in structural mechanics, *International Journal of Numerical Methods in Engineering*, Vol.6, 213-226, 1973.

2. Akyuz, F.A. and S. Utku, An automatic scheme for bandwidth minimization of stiffness matrices, *AIAA Journal*, Vol.6, No.4, 728-730, 1968.

3. Barlow, J. and C.G. Marples, Comment on reference 2, *AIAA Journal*, Vol.7, No.2, 380-382, 1969.

4. Wilkinson, J.H. *The Algebraic Eigenvalue Problem*, Clarendon Press, Oxford,65.

5. Martin, R.S. and J.H. Wilkinson, Reduction of the symmetric eigen-problem $Ax = \lambda Bx$ and related problems to standard form, *Numerische Mathematik*, Vol. 11, 99-110, 1968.

6. Wilkinson, J.H. Householder's method for symmetric matrices, *Numerische Mathematik*, Vol.4, 354-361, 1962.

7. Barth, W., R.S. Martin and J.H. Wilkinson, Calculation of the eigenvalues of a symmetric tridiagonal matrix by the method of bisection, *Numerische Mathematik*, Vol.9, 386-393, 1967.

8. Wilkinson, J.H. Calculation of the eigenvectors of a symmetric tridiagonal matrix by inverse iteration, *Numerische Mathematik*, Vol.4, 368-376, 1962.

9. Gupta, K.K. Solution of eigenvalue problems by Stürm sequence method, *International Journal for Numerical Methods in Engineering*, Vol.4, 379-404,72.

10. Peters, G. and J.H. Wilkinson, Eigenvalues of $Ax = \lambda Bx$ with band symmetric A and B, *Computer Journal*, Vol.12, 398-404, 1969.

11. Jennings, A. A direct iteration method of obtaining latent roots and vectors of a symmetric matrix, *Proceedings of the Cambridge Philosophic Society*, Vol.63, 755-765, 1967.

12. Jennings, A. and D.R.L. Orr, Application of the simultaneous iteration method to undamped vibration problems, *International Journal for Numerical Methods in Engineering*, Vol.3, 13-24, 1971.

13. Guyan, R.J. Reduction of stiffness and mass matrices, *AIAA Journal*, Vol.3, No.2, 380, 1965.

14. Irons, B. Structural eigenvalue problems: elimination of unwanted variables, *AIAA Journal*, Vol.3, No.5, 961-962, 1965.

15. Craig, R.R. and M.C.C. Bampton, Coupling of substructure for dynamic analysis. *AIAA Journal*, Vol.6, No. 7, 1313-1319, 1968.

16. Fox, R.L. and M.P. Kapoor, Rates of change of eigenvalues and eigenvectors, *AIAA Journal*, Vol. 6, No.2, 2426-2429, 1968.

17. Wilson, R.R. and C.A. Brebbia, Rate of change of dynamic response, 10/67-10/77, *Variational Methods in Engineering*, (editors C. Brebbia and H. Tottenham) Southampton University Press, 1973.

CHAPTER 5

# APPROXIMATE METHODS FOR CALCULATING NATURAL FREQUENCIES AND DYNAMIC RESPONSE OF ELASTIC SYSTEMS

by

H. Tottenham

Since the dynamic analysis of any but the simplest structure is a complicated business it is useful to have some approximate methods which will at least indicate whether a more detailed analysis is necessary. The purpose of this chapter is to outline a few simple devices which can, in many cases, help us to determine the natural frequencies and dynamic response of elastic structures.

## 1. Equivalent One Degree of Freedom Systems

The equation of motion of a single degree of freedom system is

$$M\frac{d^2w}{dt^2} = -kw + p(t) \tag{1}$$

where M is the mass of the system, k the elastic stiffness, w(t) the displacement of the mass and p(t) is the applied load. The natural frequency is given by

$$\omega = \sqrt{\frac{k}{M}} \qquad n = \frac{1}{2\pi}\sqrt{\frac{k}{M}} \tag{2}$$

and thus depends only upon the ratio k/M. The actual response however depends upon both quantities separately, or what is equivalent, upon one of these and upon the ratio. If then we wish to replace some system by an equivalent one degree of freedom system we must make sure that we have appropriate values for both k and M.

In passing we note a useful result. If the mass M is acting under the influence of gravity as a static load Mg the displacement $w_s$ would be

$$w_s = \frac{Mg}{k} \qquad \text{or} \qquad k = \frac{Mg}{w_s}$$

Substituting this into (2) gives

$$\omega = \sqrt{\frac{g}{w_s}} \qquad n = \frac{1}{2\pi}\sqrt{\frac{g}{w_s}}$$

If $w_s$ is measured in mm, we have g = 9810 and hence

$$\omega = 99 / \sqrt{w_s(\text{mm})} \tag{3}$$

or if $w_s$ is measured in cm, $g = 981$ (cms$^{-2}$) and

$$n = \frac{1}{2\pi}\sqrt{\frac{981}{w_s}} = \sqrt{\frac{981}{4\pi^2 w_s}} = \frac{5}{\sqrt{w_s(cm)}} \tag{3a}$$

Simple Beams

We consider now a simply supported beam for which we know

$$w_s = \frac{5}{384} mgL \cdot \frac{L^3}{EI} \qquad \omega = \pi^2\sqrt{\frac{EI}{mL^4}}$$

The total mass is $M = mL$ and the 'stiffness' is thus

$$k = \frac{384\ EI}{5L^3}$$

We wish to replace the beam by an equivalent one degree of freedom system, with some mass $M_1$ and spring stiffness $k_1$. If we consider the beam as a weightless elastic spring we have

$$w = \frac{1}{48}\ P\frac{L^3}{EI_1}$$

and its stiffness $k_1$ is $\frac{48EI_1}{L^3}$. For the same deflection we must have

$$\frac{5}{384}\ \frac{PL^3}{EI} = \frac{1}{48}\ \frac{PL^3}{EI_1}$$

and hence $k_1 = \frac{5}{8}k$ .

We put $\qquad \alpha_2 = \frac{k_1}{k} = \frac{5}{8} \qquad (4)$

To get the correct natural frequency we must have

$$\omega^2 = \frac{k_1}{M_1} = \pi^4\ \frac{EI}{ML^3}$$

and hence

$$\frac{M_1}{M} = \frac{384}{8\pi^4} = 0.493 \approx 0.5 = \alpha_1 \tag{5}$$

If for a central load $P(t)$ the amplitication factor is $\eta$ then

$$w_{max} = \frac{P(t)}{k_1}\ \eta \tag{6}$$

and the moments and shear forces can be estimated by the usual methods of structural mechanics. A distributed load $p(x)$ must be replaced by a concentrated equivalent

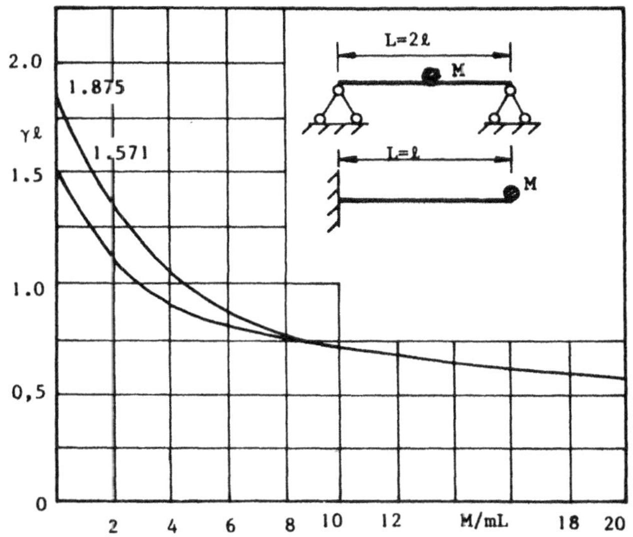


Figure 1

force and for a uniform load $P_{eff} = \alpha_2 p(x)L$. Considering other cases similarly we get

| Beam | $\alpha_1$ | $\alpha_2$ |
|---|---|---|
| | .5 | .625 |
| | .4 | .57 |
| | .40 | .53 |
| | .24 | .25 |

Beams with concentrated masses

Although Durkerly's formula is intended to be applied to the coupling of elastic systems we can also use it for superposition in one system. For example if we have a concentrated mass M at midspan in a beam whose mass is m per unit length we have

$$\omega_1 = 2\sqrt{\frac{EI}{mL^4}} \qquad \omega_2 = 4\sqrt{\frac{48EI}{mL^3}}$$

and can estimate the natural frequency from

$$\frac{1}{\omega^2} = \frac{1}{\omega_1^{\,2}} + \frac{1}{\omega_2^{\,2}} \tag{7}$$

The graph (Fig. 1) shows the accurate values of $\omega$ for a simply supported beam and a cantilever loaded at the tip. For convenience the factor plotted is $\gamma$ which is such that

$$\omega = \gamma^2 \sqrt{\frac{EI}{mL^4}} \tag{8}$$

2. Continuous Beams

If we have two separate lumped mass systems with natural frequencies $\omega_1$ and $\omega_2$ then the natural frequencies of these when they are coupled, $\bar{\omega}_1$ and $\bar{\omega}_2$ are such that

$$\bar{\omega}_1 < \omega_1\,, \qquad \omega_2 > \bar{\omega}_2 \tag{9}$$

However when we have continuous systems this is not so. The fundamental frequency of a coupled system is between the fundamental frequencies of the separate systems. This arises from the fact that we have two degrees of freedom, rotation and displacement, in the continuous system. A good first estimate of the natural frequency can be found from

$$\omega = n\sqrt{\omega_1 \omega_2 \omega_2 \dots \omega_n} \tag{10}$$

where $\omega_i$ are the natural frequencies of the i separate spans, provided that the values are not too disparate, i.e. all the values are in the range $\omega_{max} \leqslant 2\omega_{min}$. Figure 2 shows values of $\gamma$ for two and three span beams of different effective spans the effective span being the actual span multiplied by $k_1$

$$k_1 = \sqrt{\frac{m}{m_s} \frac{E_s I_s}{EI}} \tag{11}$$

where the suffix s denotes the quantities of the span taken as the standard.

## 3. Distribution Methods

If we have a continuous beam, each span of uniform section and mass, we can derive dynamic "three moment" equations

$$\alpha_n M_{n-1} + (\beta_n + \beta_{n+1}) M_n + \alpha_{n+1} M_{n+1} = 0 \tag{12}$$

where M are now moments

$$\alpha_n = \frac{L_n}{E_n I_n} \left[ \frac{\sinh \gamma_n - \sin \gamma_n}{2\sinh \gamma_n \sin \gamma_n} \right] \tag{13a}$$

$$\beta_n = \frac{L_n}{E_n I_n} \left[ \frac{\cosh \gamma_n \sin \gamma_n - \sinh \gamma_n \cos \gamma_n}{2 \sinh \gamma_n \sin \gamma_n} \right] \tag{13b}$$

and

$$\gamma_n^4 = \frac{\omega^2 m_n L_n^4}{E_n I_n} \tag{14}$$

We can approximate to these coefficients by

$$\alpha_n = \frac{E_n I_n}{6L_n} \left(1 + \frac{\gamma_n^4}{20}\right) / \left(1 - \frac{\gamma_n^4}{98}\right) \tag{15a}$$

$$\beta_n = \frac{E_n I_n}{3L_n} \left(1 - \frac{\gamma_n^4}{210}\right) / \left(1 - \frac{\gamma_n^4}{98}\right) \tag{15b}$$

We first of all take a value for $\omega$ and evaluate the coefficients. Starting at one end of the continuous beam we can take $M_1 = 1$, say, and calculate $M_2$, $M_3$, etc. until we reach the other end of the beam, and find $M_m$, say. We then use the last equation to find $M_{m-1}$ from $M_m$, it will generally differ from that obtained initially.

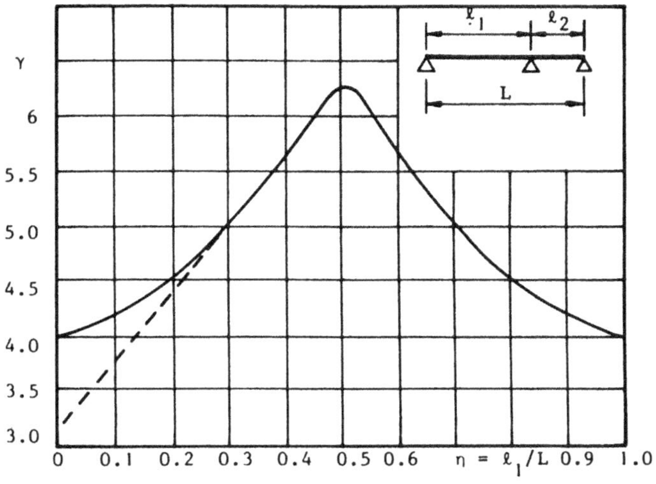
γ
6
5.5
5.0
4.5
4.0
3.5
3.0
0 0.1 0.2 0.3 0.4 0.5 0.6 η = ℓ1/L 0.9 1.0
ℓ1 ℓ2
L

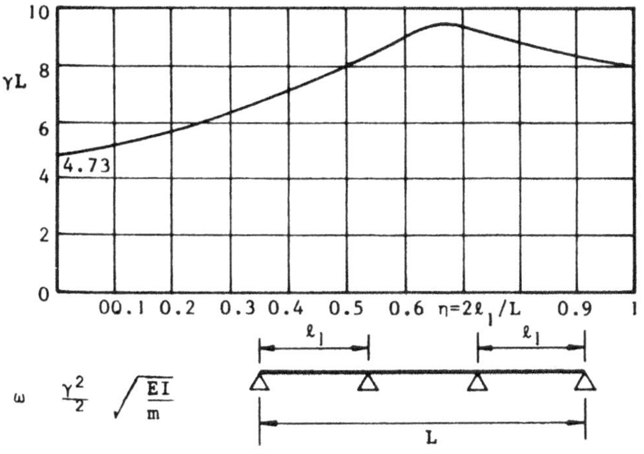
10
γL 8
6
4.73
4
2
0
00.1 0.2 0.3 0.4 0.5 0.6 η=2ℓ1/L 0.9 1
ℓ1 ℓ1
L
ω γ²/2 √(EI/m)

Figure 2

We then take a different value for $\omega$ and repeat the process. Plotting the difference in value found for $M_{m-1}$ in each case against $\omega$ will generally lead to a value of $\omega$ sufficiently accurate for general use after two such calculations. Nowadays with calculators which readily produce hyperbolic functions the more accurate formulae may be used. The quantities (13) are also readily tabulated against $\gamma_n$.

## 4. Multi-storey Frames

Let us now consider rigid jointed frameworks consisting of uniform beams and columns. The two limiting cases for the natural frequencies are (a) the beams are very stiff compared with the columns and (b) the columns are very stiff compared with the beams.

For the first case let us look at a frame of r columns each having storey heights h. The stiffness of a single storey against lateral motion is

$$k = \frac{12rEI}{h^3}$$

and hence if the mass of each floor is M we have for a single storey

$$\omega_1 = 2\sqrt{3}\sqrt{\frac{rEI}{Mh^3}} \tag{16}$$

For a two storey building we can easily find the lowest frequency as

$$\omega_2 = \sqrt{\left(\frac{3-\sqrt{5}}{2}\right)}\, 2\sqrt{3}\sqrt{\frac{rEI}{Mh^3}} \tag{17}$$

and for n storeys (n large)

$$\omega_n = \frac{\pi}{2n+1}\; 2\sqrt{3}\sqrt{\frac{rEI}{Mh^3}} \tag{18}$$

It will be seen that even with n = 2 we have factors 0.618 and 0.626 from (17) and (18).

At the other extreme the system acts as a cantilever with lumped masses. The single storey system gives

$$\omega_1 = \sqrt{3}\,.\sqrt{\frac{rEI}{Mh^3}} \tag{19}$$

For a large number of storeys we can take a uniform mass/unit height M/h, and thus find

$$\omega_n = 3.516 \sqrt{\frac{rEI}{Mh^3n^4}}$$

$$= \frac{3.516}{n^2} \sqrt{\frac{rEI}{Mh}} \approx \frac{\omega_1}{2n^2} \qquad (20)$$

Using formula (20) gives an error of only 4% when $n = 2$ and less than 1% for higher values of n.

Since the frequency (19) for the 'rigid columns' case is one half of that given for the 'rigid beam case' we can write

$$\omega_i = \frac{\pi}{2i+1} \omega^*, \quad (a) \quad \text{or} \quad \frac{1}{4i^2} \omega^*, \quad (b) \qquad (21)$$

or the two cases, $\omega^*$ being given by (6). If the 'dynamic' stiffness $\gamma$ of the floor beam is more than three times that of the column the formula (21a) may be used.

CHAPTER 6

# DETERMINATION OF RESPONSE

by

G.B. Warburton

## 1. Introductory Remarks

In addition to the outline in the introduction, methods of determining response have been given in Chapter 5. The purpose of this chapter is not to give further methods, but to consider in greater depth the normal mode method. It will be recalled that there are three main types of excitation, namely harmonic, transient and random. It has been shown that the normal mode method can be used to determine the response to harmonic and transient excitations; also this method yields the complex frequency response function, which is required in the determination of random response. For comparison the frequency response method yields directly steady-state solutions for harmonic problems and has similar applicability to the determination of random response as the normal mode method. The direct or numerical integration methods are applied in practice only to transient problems, but have added importance because of their applicability to non-linear problems. Thus the normal mode method is considered in greater depth here because of its versatility, but it is also the method which provides some physical understanding of the vibration of complex structures.

The linear vibrations of any structure can be represented by the matrix equation

$$\underset{\sim}{M}\,\ddot{\underset{\sim}{U}} + \underset{\sim}{C}\,\dot{\underset{\sim}{U}} + \underset{\sim}{K}\,\underset{\sim}{U} = \underset{\sim}{P}\,(t) \qquad (1)$$

where $\underset{\sim}{M}$, $\underset{\sim}{C}$ and $\underset{\sim}{K}$ are the symmetric mass, damping and stiffness matrices of order n x n; $\underset{\sim}{U}$, $\dot{\underset{\sim}{U}}$ and $\ddot{\underset{\sim}{U}}$ are the vectors containing the displacements, velocities and accelerations respectively, and $\underset{\sim}{P}(t)$ is the vector of excitation forces. Consistent definitions must be used for $\underset{\sim}{U}$ and $\underset{\sim}{P}(t)$; thus if $u_j$ is the displacement at coordinate j, $p_j(t)$ is the force applied at j and acts in the direction of $u_j$ where $u_j$ and $p_j(t)$ are the $j^{th}$ entries in the column matrices $\underset{\sim}{U}$ and $\underset{\sim}{P}(t)$ respectively. As engineering structures are usually complex, their response to specified excitations can be determined only by approximate methods. The finite element method is the most general and powerful method available; when it is applied to a structure, the resulting approximate mathematical model is a multi degree-of-freedom system, which can be represented by equation (1). In some examples in earlier chapters that related to equation (1) a diagonal mass matrix was obtained for multi degree-of-freedom systems, which consist of a chain of springs and masses. In conventional applications of the finite element method consistent mass matrices, which are symmetric, are generated; thus $\underset{\sim}{M}$ in equation (1) is assumed to be symmetric. Equation (1) is obtained also if the variational form of the finite difference method or the Rayleigh-Ritz method is used to generate the approximate model of the structure.

As has been shown in the chapter on transient response, a transformation from the original displacement vector $\underset{\sim}{U}$ to the generalized or normal coordinate vector $\underset{\sim}{q}$ is made through the relation

$$\underset{\sim}{U} = \underset{\sim}{Z}\,\underset{\sim}{q} \tag{2}$$

where the modal matrix Z consists of the normalized modal vectors $\underset{\sim}{z}_1, \underset{\sim}{z}_2, \dots \underset{\sim}{z}_n$ with the $r^{th}$ vector $\underset{\sim}{z}_r$ in the $r^{th}$ column of $\underset{\sim}{Z}$. This leads to the matrix equation

$$\ddot{\underset{\sim}{q}} + \underset{\sim}{B}\,\dot{\underset{\sim}{q}} + \underset{\sim}{\Omega}\,\underset{\sim}{q} = \underset{\sim}{Z}^T\,\underset{\sim}{P}(t) \tag{3}$$

where $\underset{\sim}{\Omega}$ is a diagonal matrix, containing the squares of the natural frequencies, i.e. $\omega_1{}^2, \omega_2{}^2, \dots \omega_n{}^2$, and

$$\underset{\sim}{B} = \underset{\sim}{Z}^T\,\underset{\sim}{C}\,\underset{\sim}{Z} \tag{4}$$

If we have proportional damping (also known as classical or Rayleigh damping), i.e.

$$\underset{\sim}{C} = \lambda_k\,\underset{\sim}{K} + \lambda_m\,\underset{\sim}{M} \tag{5}$$

the matrix $\underset{\sim}{B}$ is diagonal and equation (3) consists of a set of uncoupled equations; a typical equation can be written

$$\ddot{q}_r + 2\gamma_r\omega_r\dot{q}_r + \omega_r{}^2 q_r = f_r(t), \quad r=1,2, \dots n \tag{6}$$

where $2\gamma_r\omega_r = \lambda_m + \lambda_k\,\omega_r{}^2$ and $\gamma_r$ is the non-dimensional modal damping parameter for mode r, and $f_r(t) = \sum_{j=1}^{n} z_{jr}\,p_j(t)$.

Equation (6) is of similar form to that for a single degree-of-freedom system and can be solved by identical methods; when the coordinates $q_r$ have been determined, the displacement can be obtained from equation (2).

## 2. Steady-state Response

The particular case, when the excitation is a harmonic function of time and the steady-state response is required, will be discussed in more detail. The excitation force at coordinate j is assumed to be $p_j \sin(\omega t + \beta_j)$, where $p_j$ is not a function of time; inertial loading can be allowed for by taking

$p_j = d_j \omega^2$ where $d_j$ is a constant. As shown in a previous chapter, we consider a force component $p_j e^{i\omega t} e^{i\beta j}$, put $\dot{q}_r = i\omega q_r$ and $\ddot{q}_r = -\omega^2 q_r$, obtain a complex expression for $q_r$ and take the imaginary part of the complete solution, as sin $(\omega t + \beta_j)$ is the imaginary part of $e^{i\omega t} e^{i\beta j}$. Thus from equation (6) the complex form of $q_r$ is

$$q_r = \frac{e^{i\omega t} \sum_{j=1}^{n} z_{jr} p_j e^{i\beta j}}{\omega_r^2 - \omega^2 + 2i\gamma_r \omega_r \omega}$$

$$= \frac{e^{i\omega t} \sum_{j=1}^{n} [z_{jr} p_j e^{i\beta j}][\omega_r^2 - \omega^2 - 2i\gamma_r \omega_r \omega]}{(\omega_r^2 - \omega^2)^2 + 4\gamma_r^2 \omega_r^2 \omega^2}$$

Using equation (2) to determine a typical displacement $u_s$ and taking the imaginary part of the complete complex expression

$$u_s = \underline{\text{Im}} \sum_{r=1}^{n} \left[ \frac{z_{sr} e^{i\omega t} \sum_{j=1}^{n} \{z_{jr} p_j e^{i\beta j}\} (\omega_r^2 - \omega^2 - 2i\gamma_r \omega_r \omega)}{(\omega_r^2 - \omega^2)^2 + 4\gamma_r^2 \omega_r^2 \omega^2} \right]$$

Thus for $\gamma_r$ small $q_r$ is large when $\omega \approx \omega_r$; if in addition $\omega_r$ is well separated from the adjacent natural frequencies $\omega_{r-1}$ and $\omega_{r+1}$, $q_k$ with $k \neq r$ is not large when $\omega \approx \omega_r$. Considering the displacement $u_s$ (and assuming that $z_{sr}$ is not small), we have the $r^{th}$ resonant peak at an excitation frequency in the vicinity of $\omega = \omega_r$ and the magnitude of this peak is dominated by the contribution from mode r, i.e. by $q_r$. Thus for a single excitation $p_j \sin \omega t$ the resonant amplitude may be approximated (provided that the above conditions hold) as

$$U_s \approx \frac{z_{sr} z_{jr} p_j}{2\gamma_r \omega_r^2}$$

Considering this single excitation the response at any excitation frequency $\omega$ can be written in complex form as

$$u_s = H_{sj}(\omega)\, p_j e^{i\omega t}$$

where the complex frequency response function (or receptance)

$$H_{sj}(\omega) = \sum_{r=1}^{n} \left[ \frac{z_{sr} z_{jr} (\omega_r^2 - \omega^2 - 2i\gamma_r \omega_r \omega)}{(\omega_r^2 - \omega^2)^2 + 4\gamma_r^2 \omega_r^2 \omega^2} \right]$$

The response function $H_{sj}(\omega)$ is required when evaluating the response of a multi-degree-of-freedom system to a random excitation.

Damping

The following points will be discussed: the significance of equation (5); more general conditions than equation (5) for which $\underset{\sim}{B}$ is diagonal; and procedures when $\underset{\sim}{B}$ is not diagonal.

Equation (5) is important and not as restrictive as it appears. In order to allow for internal damping in the members of a structure it is conventional to replace Hooke's law, i.e.

$$\sigma_x = E\,\varepsilon_x$$

where $\sigma_x$ and $\varepsilon_x$ are the stress and strain in the X-direction and E is Young's modulus, by

$$\sigma_x = E(\varepsilon_x + \lambda_k \frac{\partial \varepsilon_x}{\partial t})$$

The change in relationship implies replacing $\underset{\sim}{K}\,\underset{\sim}{U}$ for an undamped system by $\underset{\sim}{K}\,\underset{\sim}{U} + \lambda_k \underset{\sim}{K}\,\dot{\underset{\sim}{U}}$ for the damped system; thus we have satisfied equation (5) and $\lambda_m = 0$. This assumes that $\lambda_k$ is a constant for the whole structure. From the general relation

$$2\gamma_r\,\omega_r = \lambda_m + \lambda_k\,\omega_r^2 \qquad (7)$$

it follows that for this special case

$$\gamma_r = \gamma_1 \frac{\omega_r}{\omega_1}$$

i.e. the modal damping parameter $\gamma_r$ increases as the mode number increases. Available experimental evidence suggests that this relation may overestimate the damping in higher modes.

If the damping in two modes is prescribed, for example $\gamma_1$ and $\gamma_2$, $\lambda_m$ and $\lambda_k$ can be found to satisfy equation (5), but the damping of the higher modes, $\gamma_r$ with $r \geqslant 3$, will have to satisfy equation (7). If the damping in more than two modes is prescribed, it will be impossible, in general, to satisfy equation (5). However, additional terms can be added to the right-hand side of equation (5); for example, $\lambda_1 \underset{\sim}{K}\,\underset{\sim}{M}^{-1}\,\underset{\sim}{K}$ and $\lambda_2 \underset{\sim}{M}\,\underset{\sim}{K}^{-1}\,\underset{\sim}{M}$ are allowable additional terms. With four terms the coefficients $\lambda$ can be selected to satisfy four modal damping parameters $\gamma_r$. Further terms can be added if required; the uncoupling condition is that each term $\underset{\sim}{C}_j$ (j=1, 2, ...) on the right-hand side of equation (5) must satisfy

$$\underset{\sim}{C}_j\,\underset{\sim}{M}^{-1}\,\underset{\sim}{K} = \underset{\sim}{K}\,\underset{\sim}{M}^{-1}\,\underset{\sim}{C}_j$$

If all modal damping parameters $\gamma_r$ can be estimated, it is not necessary to form the damping matrix $\underset{\sim}{C}$, but merely add the appropriate term, $2\gamma_r \omega_r \dot{q}_r$, in each uncoupled equation (6) at the last stage of the modal analysis.

There are problems for which proportional damping does not exist; for instance, in most soil-structure interaction problems the level of damping in the soil, which is mainly due to radiation, is higher than that in the structure. Then for a system with $n$ degrees of freedom equation (3) consists of $n$ equations, which are coupled through their damping terms. These equations can be solved by one of the methods of numerical integration. At first sight this procedure is not advantageous, as the original $n$ coupled equations in the physical coordinates, equation (1), have been exchanged for $n$ coupled equations in terms of $q_r$. However, as only $i$ modes make a significant contribution to the response and $i << n$, working in terms of the normal coordinates $q_r$ requires numerical integration of significantly smaller matrix equations. This procedure is advocated and illustrated by Clough and Mojtahedi [1].

Approximations that allow the normal mode method to be used when the matrix $\underset{\sim}{B}$ is not diagonal are important in practice. Thomson et al [2] suggest that the non-diagonal matrix $\underset{\sim}{B}$ should be replaced by a diagonal matrix with the same diagonal terms as the original matrix, (i.e. the off-diagonal terms of $\underset{\sim}{B}$ are replaced by zeros); then the standard normal mode procedure is followed. Although the literature contains some conflicting numerical evidence, it appears that, provided damping is light and natural frequencies are reasonably well spaced, this approximation introduces acceptable errors allowing for the uncertainties regarding damping values in real structures. The author [3] has given a criterion, which should be satisfied in order that neglecting the off-diagonal terms in $\underset{\sim}{B}$ does not lead to excessive errors in major response quantities. The criterion is:

$$\gamma_r \leqslant 0.05 \left| \frac{b_{rr}\,\omega_s^2}{2b_{rs}\,\omega_r^2} - 1 \right|_{\min s} \tag{8}$$

where $\gamma_r$ is the damping ratio for the $r^{th}$ mode, $\omega_r$ and $\omega_s$ are natural frequencies, $b_{rr}$ and $b_{rs}$ are elements from the matrix $\underset{\sim}{B}$ and the minimum of the expression $|.....|$ with respect to $s$ is taken; $s$ may take any integer value between 1 and $n$ other than $r$. In practice, equation (8) is applied only to the lower values of $r$, where significant resonant response may occur. For each of these values of $r$ the right-hand side is obtained by considering a few values of $s$ on either side of $r$. Although the criterion was developed from a study of test problems, further numerical evidence, which relates to large practical systems, would be beneficial.

Uncoupled equations can be obtained for any damping distribution by working in terms of damped normal modes (instead of the classical undamped modes), but the $n$ second order differential equations (1) have to be replaced by $2n$ first order equations and analysis is in terms of complex eigenvalues and eigenvectors. Traill-Nash [4] has contributed recently to this method and also surveyed earlier applications by others.

## 4. Truncation of Series Solution

Considering the determination of maximum response (displacement, acceleration, stress etc) from equations (2) and (3), only a fraction $i$ of the total modes $n$, where in general $i << n$, will make a significant contribution. These $i$ modes must be identified; $i$, or the ratio $i/n$, depends upon the response quantity of interest and the time history of the excitation, as well as on the system, and general rules cannot be formulated, although previous experience of similar situations is valuable. Assuming that equation (1) represents a finite element idealization of an elastic structure, it should be checked that these $i$ modes are reasonable approximations to modes of the true structure. This can be achieved by varying the idealization and demonstrating that these modes are unchanged within a specified error criterion.

The effect of the time history of the excitation and the response quantity of interest on convergence will be illustrated by a simple example. We consider the two-storey frame of Fig. 1; the base EF is subjected to a horizontal displacement

$$x_o = 0.01 \sin \pi t/t_o \ (m) \qquad 0 \leqslant t \leqslant t_o$$

$$x_o = 0 \qquad t \geqslant t_o$$

where (i) $t_o = 0.2s$ and (ii) $t_o = 0.075s$. For free vibrations $\omega_1 = 19.54$ rad/s and $\omega_2 = 51.17$ rad/s. The modal matrix

$$\underset{\sim}{Z} = \begin{bmatrix} 0.5257 & 0.8506 \\ & \\ 0.8506 & -0.5257 \end{bmatrix} \times 10^{-2}$$

For response $\qquad \underset{\sim}{x} = \underset{\sim}{Z}\,\underset{\sim}{q}$

The response is determined for the excitation era, $0 \leqslant t \leqslant t_o$, and for the ensuing free vibrations ($t \geqslant t_o$). For simplicity damping is neglected. For brevity expressions for $q_r$ will be given only for $t > t_o$, as maximum displacements in the free vibration era are greater if $t_o = 0.075s$ and comparable to maxima in

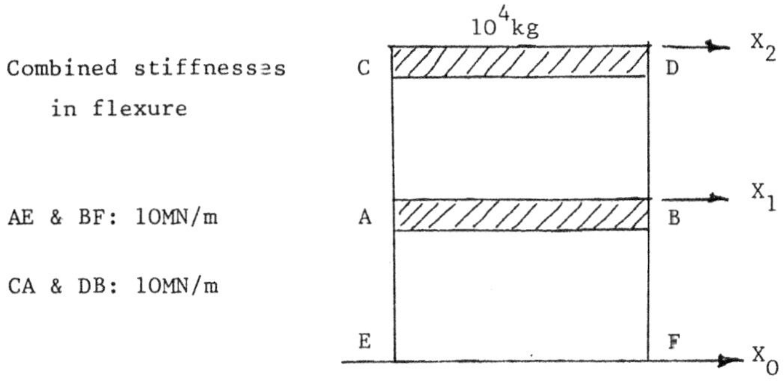


2 DOF frame. Excitation - base displacement

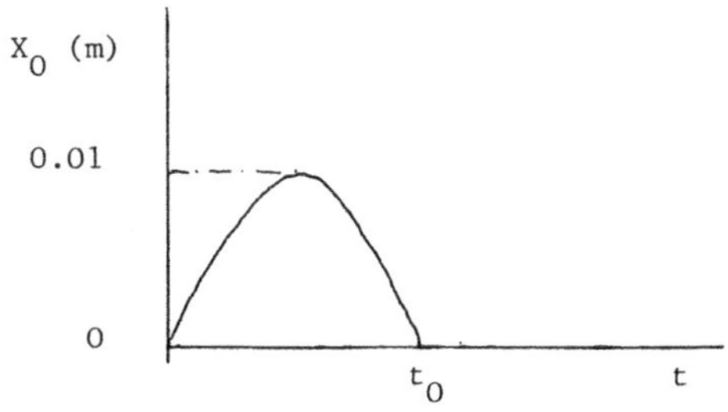


Excitation

Fig. 1

the excitation era if $t_o = 0.2s$. (Further details are given in Reference [5]).

For (i) $t_o = 0.2s$, $q_1 = 2.339 \sin(\omega_1 t + \alpha_1)$

$q_2 = 0.0866 \sin(\omega_2 t + \alpha_2)$

for (ii) $t_o = 0.075s$, $q_1 = 1.220 \sin(\omega_1 t + \alpha_1)$

$q_2 = 0.550 \sin(\omega_2 t + \alpha_2)$

For case (i) the response is dominated by the first modal contribution $q_1$ and only a small error will occur if the contribution from $q_2$ to $x_1$ and $x_2$ is neglected. For case (ii) the contributions from the two modes are of comparable magnitude. If now the accelerations of the two masses, $\ddot{x}_1$ and $\ddot{x}_2$, are required, then from

$$\ddot{\underset{\sim}{x}} = \underset{\sim}{Z}\, \ddot{\underset{\sim}{q}}$$

and

$$\ddot{q}_r = -\omega_r^2 q_r ,$$

we have for case (i)

$$\ddot{q}_1 = -893.1 \sin(\omega_1 t + \alpha_1)$$

$$\ddot{q}_2 = -226.7 \sin(\omega_2 t + \alpha_2)$$

Thus the contribution of the second mode to the acceleration response is now significant.

When truncation is applied, the response is found by summing contributions from i modes with $i << n$. Thus, instead of using equation (2), we evaluate

$$\underset{\sim}{U} = \underset{\sim}{Z}^* \; \underset{\sim}{q}^* \qquad (9)$$

where

$$\underset{\sim}{Z}^* = \begin{bmatrix} z_{11} & z_{12} ---- & z_{1i} \\ z_{21} & z_{22} ---- & z_{2i} \\ - - - & - - - & - - - \\ z_{n1} & z_{n2} & z_{ni} \end{bmatrix} \quad \text{and} \quad q^* = \begin{bmatrix} q_1 \\ q_2 \\ \vdots \\ q_i \end{bmatrix}$$

An improvement in the accuracy of this evaluation can be obtained by adding the 'static' solutions, which correspond to modes $i + 1, i + 2, ---- n$. Alternatively, for given accuracy a smaller number of modes i has to be retained in the above equation if the 'static' contributions from the higher modes are included. In theory 'static' contributions are determined by solving equation (6) for $r = i + 1$, i + 2, ---- with the inertia and damping terms neglected. In practice, after some matrix manipulation the improved solution can be expressed as

$$\underset{\sim}{U} = \underset{\sim}{Z}^* \underset{\sim}{q}^* + \underset{\sim}{K}^{-1}\underset{\sim}{P} - \underset{\sim}{Z}^* \; [\underset{\sim}{\Omega}^*]^{-1} \; \underset{\sim}{Z}^{*T} \underset{\sim}{P} \tag{10}$$

$$\text{where} \quad [\underset{\sim}{\Omega}^*]^{-1} = \begin{bmatrix} 1/\omega_1^2 & 0 & 0 - - - 0 \\ 0 & 1/\omega_2^2 & 0 - - - 0 \\ - - - & - - - & - - - - - \\ 0 & 0 & 0 - - -1/\omega_i^2 \end{bmatrix}$$

The second and third terms in equation (10) represent the 'static' contributions from modes 1, 2, --- n and from modes 1, 2, --- i respectively. The expression requires that only the natural frequencies and mode shapes for modes 1, 2, --- i are evaluated. It has been known for many years that inclusion of the 'static' contributions associated with higher modes improved accuracy, but this modern form is based on the work of Hansteen and Bell [6].

In an earlier chapter the frequency response method of determining the steady-state response to harmonic excitation has been given. If the excitation vector is $\underset{\sim}{P}$ sin $\omega t$ [replacing $\underset{\sim}{P}(t)$ on the right-hand side of equation (1)], the response is given by

$$\underset{\sim}{U} = \underline{\text{Im}} \; [\underset{\sim}{J}^{-1} \underset{\sim}{P} \, e^{i\omega t}]$$

$$\text{where} \qquad \underset{\sim}{J} = \underset{\sim}{K} - \omega^2 \underset{\sim}{M} + i\omega \underset{\sim}{C}$$

This method can be used whether or not the modal method gives uncoupled equations, but requires the inversion of the complex matrix $\underset{\sim}{J}$ at each excitation frequency $\omega$ of interest. In practical problems the order of this matrix n may be large, so that considerable computation is required. If the modal method gives uncoupled equations series expressions can be used to determine steady-state response; in practice, considerable truncation of the response should be possible. If the modal method gives coupled equations, i.e. $\underset{\sim}{B}$ is non-diagonal, a combination of modal truncation and the frequency response method requires the inversion of a complex matrix of order j, where j is the number of retained modes and in general $j \ll n$, but it may be difficult to decide upon an optimum value of j. Using the truncation defined by equation (9), the truncated form of equation (3) is

$$\ddot{\underset{\sim}{q}}^* + \underset{\sim}{B}^* \dot{\underset{\sim}{q}}^* + \underset{\sim}{\Omega}^* \underset{\sim}{q}^* = \underset{\sim}{Z}^{*T} \underset{\sim}{P} \, e^{i\omega t}$$

where $\underset{\sim}{B}^* = \underset{\sim}{Z}^{*T} \underset{\sim}{C}\, \underset{\sim}{Z}^*$ and $\underset{\sim}{\Omega}^*$ is the diagonal matrix containing $\omega_1{}^2$, $\omega_2{}^2$, .... $\omega_j{}^2$. From the frequency response method

$$\underset{\sim}{U} = \underline{\underline{\mathrm{Im}}}\, [\underset{\sim}{Z}^* \underset{\sim}{N}^{-1} \underset{\sim}{Z}^{*T} \underset{\sim}{P}\, e^{i\omega t}]$$

where $\underset{\sim}{N} = \underset{\sim}{\Omega}^* - \omega^2 \underset{\sim}{I} + i\omega \underset{\sim}{B}^*$, $\underset{\sim}{I}$ is the identity matrix and $\underset{\sim}{N}$, $\underset{\sim}{\Omega}^*$ and $\underset{\sim}{B}^*$ are of order $j \times j$.

## 5. Response Spectrum Methods

The displacement at degree of freedom $s$ is [from equation (2)].

$$u_s = \sum_r z_{sr}\, q_r(t) \,. \tag{11}$$

As each coordinate $q_r$ is a different function of time, determination of the maximum value of $u_s$ requires evaluation of each significant coordinate $q_r$ for a large number of values of $t$. Now

$$u_s(\max) \leq \sum_r | \; z_{sr}\, q_r\, (\max) \; | \tag{12}$$

where $q_r(\max)$ is the maximum value of $q_r(t)$. A simple upper bound is obtained if the equals sign in equation (12) is used. The value of $q_r(\max)$ can be determined from a response spectrum . For a single applied force $p_j(t)$ applied at coordinate $j$, the equation for the $r^{th}$ mode is

$$\ddot{q}_r + 2\gamma_r\, \omega_r\, \dot{q}_r + \omega_r{}^2\, q_r = z_{jr}\, p_j(t) \tag{13}$$

The equation of motion for a single-degree-of-freedom system of mass $m$, which is subjected to a force $P_o f(t)$, is

$$\ddot{x} + 2\gamma\omega_n\, \dot{x} + \omega_n{}^2 x = P_o f(t)/m \tag{14}$$

A response spectrum shows the variation of the dynamic magnification factor (DMF), $k\, x(\max)/P_o$, where $x(\max)$ is the maximum displacement of the mass $m$, $P_o$ is the maximum value of the force $P_o f(t)$, $k$ is the stiffness and $\omega_n{}^2 = k/m$, with a period or frequency ratio, e.g. the ratio of some characteristic time (duration or rise time) of $f(t)$, $T_o$, to the period $T_n$ $(= 2\pi/\omega_n)$ for a specified value of the damping ratio $\gamma$. Comparing equations (13) and (14) and assuming that $p_j(t) = P_o f(t)$, $q_r(\max)$ can be found from the response spectrum, associated with $f(t)$ and $\gamma_r$, by multiplying the DMF, corresponding to $T_o/T_r$, where $T_r = 2\pi/\omega_r$, by $P_o z_{jr}/\omega_r{}^2$. Thus using the upper bound as an approximation,

$$u_s(\max) = P_o \sum_r \left| \frac{z_{sr} z_{jr}}{\omega_r^2} (DMF)_r \right| \qquad (15)$$

where $(DMF)_r$ is obtained from a response spectrum for force f(t) and damping ratio $\gamma_r$ at $T_o/T_r$.

The upper bound approximation may seriously overestimate the maximum response for complex excitations, particularly earthquakes, as the underlying assumption - that all coordinates $q_r$ reach their maximum at the same time - is not true. An alternative, empirical expression, which is based on the square root of the sum of the squares (SRSS), is

$$u_s(\max) = [ \sum_r \{z_{sr} \, q_r(\max)\}^2 ]^{1/2} \qquad (16)$$

Equation (16) is not a bound and thus may underestimate the maximum response.

Other approximations are based on combinations of the upper bound expression (12) and the SRSS value (16). When two natural frequencies of a structure are close together, the response may exhibit beating phenomena; in such cases maxima are very sensitive to the level of damping and use of equations (12) or (16) seriously overestimates the combined response from these two closely coupled modes. Expressions exist which allow for modal damping values when combining contributions from two closely spaced modes and can be used in conjunction with equation (16) for the remaining modes [7,8]. Anagnostopoulos [9] has tested the various methods of combining response spectrum values. He subjected finite element models of three offshore platforms to each of thirty earthquake acceleration records and determined various response quantities. For each method he obtained mean errors in maximum response and their standard deviations (these are means of several hundred values). For horizontal excitations for which the natural frequencies of the responding modes are well separated the SRSS method, equation (16), gives the best results, but for vertical excitations some closely spaced frequencies exist and this should be allowed for by using a modified form to combine response spectrum values.

## References


1. Clough, R.W. and S. Mojtahedi, "Earthquake response analysis considering non-proportional damping", Earthquake Engineering and Structural Dynamics, 4, 1976, 489-496.

2. Thomson, W.T., T. Calkins and P. Caravani, "A numerical study of damping", Earthquake Engineering and Structural Dynamics, 3, 1974, 97-103.

3. Warburton, G.B. and S.R. Soni, "Errors in response calculations for non-clasically damped structures", Earthquake Engineering and Structural Dynamics, 5, 1977, 365-376.

4. Traill-Nash, R.W., "Modal methods in the dynamics of systems with non-classical damping", Earthquake Engineering and Structural Dynamics, 9, 1981, 153-169.

5. Warburton, G.B. "The Dynamical Behaviour of Structures" 2nd. edition, Pergamon, Oxford, 1976.

6. Hansteen, O.E. and K. Bell, "On the accuracy of mode superposition analysis in structural dynamics", Earthquake Engineering and Structural Dynamics, 7, 1979, 405-411.

7. Kelly, J.M. and J.L. Sackman, "Conservatism in summation rules for closely spaced modes", Earthquake Engineering and Structural Dynamics, 8, 1980, 63-74.

8. Wilson, E.L., A.Der Kiureghian and E.P. Bayo, "A replacement for the SRSS method in seismic analysis". Earthquake Engineering and Structural Dynamics, 9, 1981, 187-192.

9. Anagnostopoulos, S.A., "Response spectrum techniques for three-component earthquake design", Earthquake Engineering and Structural Dynamics, 9, 1981, 459-476.

CHAPTER 7

# THE FINITE ELEMENT TECHNIQUE

by

C.A. Brebbia

## 1. Introduction

The finite element method is an approximate method of analysis which can be used to solve complex structural problems.

In the analysis of structures the method can be applied in terms of displacements, forces or both. In what follows we will only refer to the displacement finite element technique which is the most popular for dynamic problems.

The method consists in taking the displacement measures at discrete points in the body as the unknowns and defining the displacement field in terms of these *discrete* variables. Once the discrete displacements are known, the strains are evaluated from the strain-displacement relations and, finally, the stresses are determined from the stress-strain relations.

Contrary to Rayleigh-Ritz and similar methods where the expressions for displacement are applicable to the complete domain, finite element expressions only apply on a part of the domain or 'element'.

In the displacement method, the application of the principle of virtual displacements results in a set of simultaneous algebraic equations for the unknown nodal displacements. Because of the large number of variables, the analysis is most conveniently formulated in terms of matrix algebra.

The selected displacement fields satisfy the admissibility and completeness conditions for the problem. As the number of elements increases we can obtain convergence of certain parameters.

The stiffness and mass matrices for the structure are obtained by superposing the contribution of the element stiffness and mass matrices at each node and the system load vector is generated in a similar way, i.e. by superposing the element force vectors. The displacement boundary conditions are then enforced. These steps result in a set of algebraic equations relating the displacement measures.

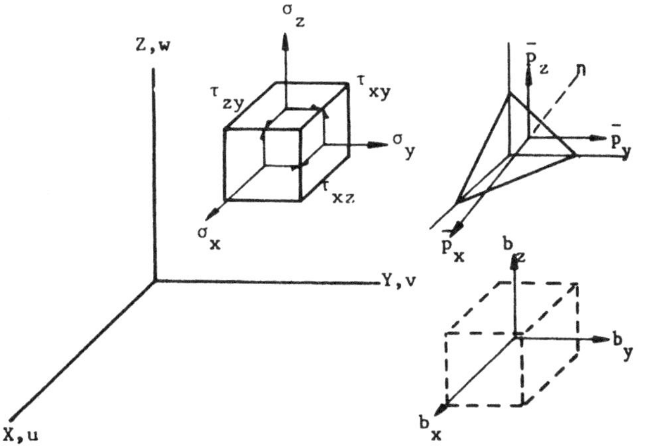


**Figure 1  Notation for force, stress and displacement.**

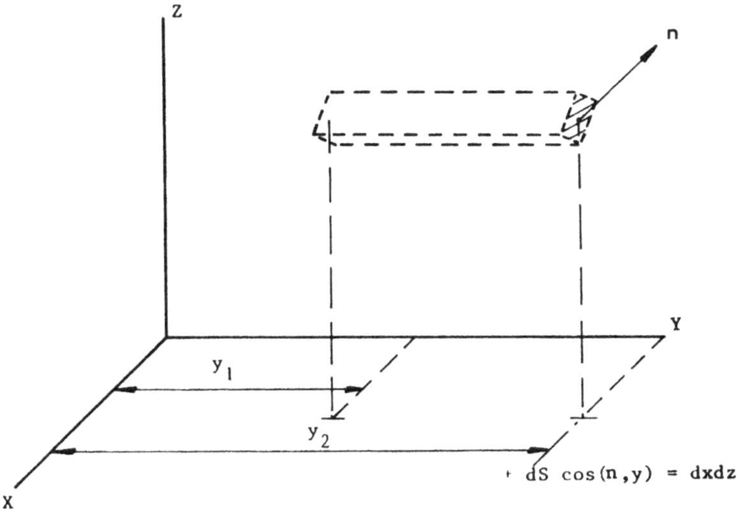


$$\left.\begin{aligned} \ell &= \cos(n,x) \\ m &= \cos(n,y) \\ n &= \cos(n,z) \end{aligned}\right\} \quad \text{direction cosines}$$

**Figure 2  Integration Limits**

The various phases of the method are

1. Discretization of the body, i.e. selection of elements interconnected at certain nodal points.
2. Evaluation of the element stiffness, mass and force matrices.
3. Assemblage of the stiffness mass and force matrices for the system of elements and nodes (system equations) and introduction of displacement boundary conditions.
4. Solution of the resulting system equations and calculation of strains and stresses based on the nodal displacements.

## 2. The Principle of Virtual Displacements

Consider a body in *equilibrium* under loading $b_x$, ... $\bar{p}_z$ and internal forces $\sigma_x$, ... $\tau_{zx}$. Now, visualize the body displaced from the equilibrium position and let $\delta u$, $\delta v$, $\delta w$ define the *virtual* displacements.

If the initial position is an *equilibrium* position, the first-order-work $\delta W_E$ done by the external forces, is equal to the first order work $\delta W_D$ done by the internal forces (stresses) during the virtual displacement:

$$\delta W_E = \delta W_D \quad \text{for arbitrary } \delta u,\ \delta v,\ \delta w. \tag{1}$$

This equation is called the Principle of Virtual Displacement. It is an alternate statement of the equilibrium conditions and it is independent of material behaviour and magnitude of displacement, i.e. it is valid for non linear geometry and arbitrary material behaviour.

The three-dimensional form of (1) is,

$$\iiint\limits_{\text{volume}} (\sigma_x \delta\varepsilon_x + \sigma_y \delta\varepsilon_y + \sigma_z \delta\varepsilon_z + \tau_{xy}\delta\gamma_{xy} + \tau_{yz}\delta\gamma_{yz} + \tau_{xz}\delta\gamma_{xz})\,dx\,dy\,dz$$

$$= \iiint\limits_{\text{volume}} (b_x \delta u + b_y \delta v + b_z \delta w)\,dx\,dy\,dz + \iint\limits_{S_\sigma} (\bar{p}_x \delta u + \bar{p}_y \delta v + \bar{p}_z \delta w)\,dS \tag{2}$$

where $S_\sigma$ is the part of the boundary where forces are prescribed and $S_u$ where displacements are given. $(S = S_u + S_\sigma)$. We will assume that the displacements identically satisfy the conditions on $S_u$ hence $\delta u = \delta v = \delta w = 0$. $\delta\varepsilon$ and $\delta\gamma$ are the first-order strain increments due to the virtual displacements. They reduce to,

$$\delta\varepsilon_x = \frac{\partial\delta u}{\partial x}, \quad \delta\varepsilon_y = \frac{\partial\delta v}{\partial y}, \quad \delta\varepsilon_z = \frac{\partial\delta w}{\partial z}$$

$$\delta\gamma_{xy} = \frac{\partial\delta u}{\partial y} + \frac{\partial\delta v}{\partial x}, \quad \delta\gamma_{yz} = \frac{\partial\delta v}{\partial z} + \frac{\partial\delta w}{\partial y}, \quad \delta\gamma_{zx} = \frac{\partial\delta u}{\partial z} + \frac{\partial\delta w}{\partial x} \tag{3}$$

for the geometrically linear case. We allow for dynamic behaviour by expressing the body forces as

$$b_x \rightarrow b_y - \rho\ddot{u}, \quad b_y \rightarrow b_y - \rho\ddot{v}, \quad b_z \rightarrow b_z - \rho\ddot{w} \tag{4}$$

where $b_x$, $b_y$ ... are prescribed, $\rho$ is the mass density and the dot notation indicates differentiation with respect to time,

$$(\ddot{\ }) = \frac{\partial^2}{\partial t^2}(\ ) \tag{5}$$

Equation (4) expresses D'Alembert's principle.

The Principle of Virtual Displacements now takes the form,

$$\iiint (\sigma_x\delta\varepsilon_x + \ldots + \tau_{xz}\delta\gamma_{xz})\, d(\text{vol.}) + \iiint \rho(\ddot{u}\delta u + \ddot{v}\delta v + \ddot{w}\delta w)\, d(\text{vol.})$$

$$= \iiint (b_x\delta u + \ldots)\, d(\text{vol.}) + \iint (\bar{p}_x\delta u + \ldots)\, dS \tag{6}$$

and is valid for a given time 't'.

The principle takes a compact form when matrix notation is utilized. We let

$$\begin{aligned} \underset{\sim}{\sigma} &= \{\sigma_x, \sigma_y \ldots \tau_{zx}\} \\ \delta\underset{\sim}{\varepsilon} &= \{\delta\varepsilon_x \ldots\ldots \delta\gamma_{zx}\} \\ \delta\underset{\sim}{u} &= \{\delta u, \delta v, \delta w\}, \qquad \ddot{\underset{\sim}{u}} = \{\ddot{u}, \ddot{v}, \ddot{w}\} \\ \bar{\underset{\sim}{p}} &= \{\bar{p}_x, \bar{p}_y, \bar{p}_z\}, \qquad \underset{\sim}{b} = \{b_x, b_y, b_z\}. \end{aligned} \tag{7}$$

This reduces equation (6) to,

$$\int \underset{\sim}{\sigma}^T \delta\underset{\sim}{\varepsilon}\, d(\text{vol.}) + \int \rho\, \ddot{\underset{\sim}{u}}^T \delta\underset{\sim}{u}\, d(\text{vol.}) = \int \underset{\sim}{b}^T \delta\underset{\sim}{u}\, d(\text{vol.}) + \int_{S_\sigma} \underset{\sim}{p}^T \delta\underset{\sim}{u}\, dS \tag{8}$$

To show that the principle of virtual displacements is equivalent to the equilibrium equations, we integrate the 1st term on the left hand side of (6) using Gauss theorem.

Typical terms are (figure 2),

$$\int \sigma_y \delta\varepsilon_y \, d(\text{vol.}) = \iint \left[\sigma_y \delta v\right]_{y_1}^{y_2} dx\, dz - \int \delta v \frac{\partial \sigma_y}{\partial y} d(\text{vol.}) =$$

$$= \oint \sigma_y \, \delta v \cos(n,y)\, dS - \int \delta v \frac{\partial \sigma_y}{\partial y} d(\text{vol.})$$

$$\int \tau_{xy} \delta\gamma_{xy} \, d(\text{vol.}) = \oint \tau_{xy} \, (\delta v \cos(n,x) + \delta u \cos(n,y))dS$$

$$- \int \left(\frac{\partial \tau_{xy}}{\partial x} \delta v + \frac{\partial \tau_{xy}}{\partial y} \delta u\right) d(\text{vol.}) \qquad (9)$$

Repeating for the other terms we obtain

$$\int (\sigma_x \delta\varepsilon_x + \ldots + \tau_{zx} \delta\gamma_{zx}) \, d(\text{vol.}) =$$

$$= \oint \Big[ (\sigma_x \ell + \tau_{xy} m + \tau_{zx} n)\delta u + (\tau_{xy} \ell + \sigma_y m + \tau_{yz} n)\delta v$$

$$+ (\tau_{zx} \ell + \tau_{yz} m + \sigma_z n)\, \delta w \Big] \, dS - \qquad (10)$$

$$- \int \Big[ \left( \frac{\partial \sigma_x}{\partial x} + \frac{\partial \tau_{xy}}{\partial y} + \frac{\partial \tau_{zx}}{\partial z} \right) \delta u + \left( \frac{\partial \tau_{xy}}{\partial x} + \frac{\partial \sigma_y}{\partial y} + \frac{\partial \tau_{yz}}{\partial z} \right) \delta v$$

$$+ \left( \frac{\partial \tau_{zx}}{\partial x} + \frac{\partial \tau_{yz}}{\partial y} + \frac{\partial \sigma_z}{\partial z} \right) \delta w \Big] \, d(\text{vol.})$$

Finally, after substituting (10) into (6) we obtain

$$\int \Big[ \left( \frac{\partial \sigma_x}{\partial x} + \frac{\partial \tau_{xy}}{\partial y} + \frac{\partial \tau_{zx}}{\partial z} + b_x - \rho \ddot{u} \right) \delta u + (\ldots)\delta v + (\ldots)\delta w \Big] \, d(\text{vol.})$$

$$= \int_{S_\sigma} \Big[ (\sigma_x \ell + \tau_{xy} m + \tau_{zx} n - \bar{p}_x)\, \delta u + (\ldots)\, \delta v + (\ldots)\, \delta w \Big] \, dS \qquad (11)$$

The terms on the left hand side are the equilibrium equations for a three dimensional body and those on the right the stress boundary conditions on $S_\sigma$ (on $S_u$ part of the boundary we do not apply forces).

In this proof it was required that the displacements and stresses are *continuous* and that the stresses are *in equilibrium* in the interior and on the surface of the body.

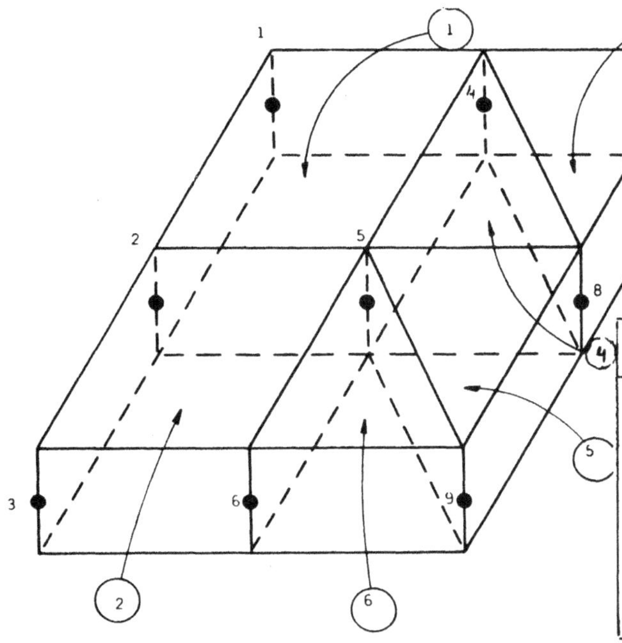


| Elements | Nodes | | | |
|---|---|---|---|---|
| | $n_1$ | $n_2$ | $n_3$ | $n_4$ |
| 1 | 1 | 2 | 5 | 4 |
| 2 | 2 | 3 | 6 | 5 |
| 3 | 4 | 8 | 7 | - |
| 4 | 4 | 5 | 8 | - |
| 5 | 5 | 9 | 8 | - |
| 6 | 5 | 6 | 9 | - |

Figure 3 Plate divided into 6 elements

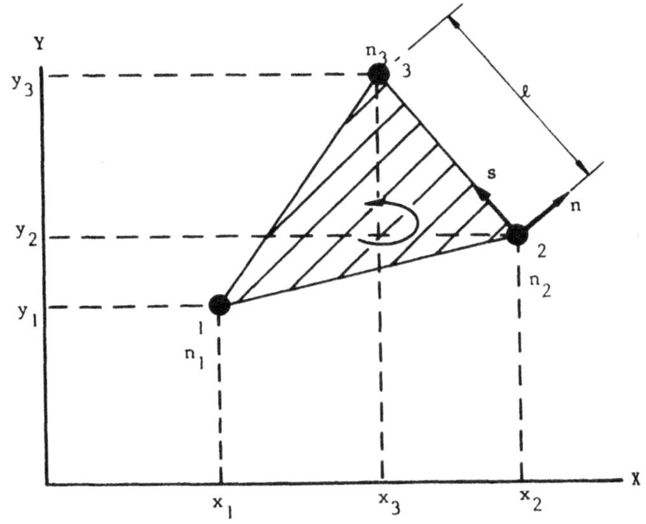


Figure 4 Triangular Element

Note that when the strain-displacement relations and geometrical boundary conditions are specified, the equations of equilibrium and the mechanical boundary conditions can be deduced from (6). Conversely when the stress field equations are defined we can deduce the geometric boundary conditions and strain-displacement equations starting with (11).

## 3. Finite Element Discretization and Element Matrices

In the finite element method we consider the body to be divided into volume elements having finite dimensions and we select certain points on the interior and exterior boundary surfaces. The volume elements are referred to as 'finite' elements since their dimensions are finite: the boundary points are called nodal points or nodes. We number the elements and nodes and specify the element-node connectivity by listing, for each element, the nodes associated with that element. A typical discretization for a flat plate, either in plane stress or bending is shown in Figure 3. We take the nodes at the corner of the elements on the middle surface of the plate. One could also select additional nodes along the element boundaries.

The element connectivity table is shown in Figure 3. Note that the nodes have to be listed in the same direction (clockwise or anticlockwise). It is irrelevant which is the starting node.

The subdivision of the continuum into elements is the most critical step of the method. A general procedure for sub-dividing the continuum does not exist. However, there are some guidelines which have evolved from experience with the method. The most important are:

a) - Irregularly shaped elements, such as long thin rectangular and flat triangles, should be avoided. Equilateral triangles and 'square' rectangles give the most accurate results.

b) - More nodes are required for stress concentration zones (high stress gradients) than for regions where the stresses vary smoothly).

c) - To evaluate the accuracy of the results it is advisable to solve the same example with a finer grid. This will provide a measure of convergence. In addition, one should always check statics.

Next, we define nodal displacement quantities. The number and choice of displacement quantities is problem-dependent but they have to be at least the displacements which satisfy the boundary conditions on $S_u$. For plane stress we take the two in-plane displacement components. In plate bending, we can work with the transverse displacement and the two rotations of the normal to middle surface. For three dimensional analysis we take as nodal displacement quantities the three displacement

We will use two reference systems. When discussing a single element we express nodal variables with reference to a *local* numbering system but when discussing the assembly of elements we shall express them with reference to a *global* system.

For a triangular element as in figure 4, the numbers 1,2,3 refer to the *local* system; $n_1$ $n_2$ $n_3$ instead refer to the *global* one (e.g. $n_1 = 94$, $n_2 = 96$, $n_3 = 92$).

The variable vector for a node i can then be written as

$$\underset{\sim}{U}_i = \begin{Bmatrix} u \\ v \\ w \end{Bmatrix}_i \quad \text{or} \quad \underset{\sim}{U}_{n_i} = \begin{Bmatrix} u \\ v \\ w \end{Bmatrix}_{n_i} \tag{12}$$

*local*          *global*

The elements of $\underset{\sim}{U}_i$ or $\underset{\sim}{U}_{n_i}$ in (12) are the nodal unknowns.

The vector formed by the vectors of unknowns at the elements nodes will be called

$$\underset{\sim}{U}_e = \begin{Bmatrix} \underset{\sim}{U}_1 \\ \underset{\sim}{U}_2 \\ \cdot \\ \cdot \\ \cdot \\ \underset{\sim}{U}_s \end{Bmatrix} \quad \text{or} \quad \begin{Bmatrix} u_1 \\ v_1 \\ w_1 \\ u_2 \\ v_2 \\ w_2 \\ \cdot \\ \cdot \\ \cdot \\ u_s \\ v_s \\ w_s \end{Bmatrix} \tag{13}$$

The superscript n denotes that the vector extends over all nodes of the element, s is the number of nodes in the elements, $\underset{\sim}{U}_e$ is called the *element* nodal unknowns vector.

Now we introduce expansions for the displacement over the element domain in terms of a set of parameters,

$$\underset{\sim}{u} = \begin{Bmatrix} u \\ v \\ w \end{Bmatrix} = \underset{\sim}{A}\,\underset{\sim}{\alpha} \tag{14}$$

where $\underset{\sim}{A}$ contains prescribed functions (of x,y,z) and $\underset{\sim}{\alpha}$ contains the displacement parameters for the element. $\underset{\sim}{\alpha}$ and $\underset{\sim}{U}_e$ are related by evaluating (14) at the nodes for the element. This leads to,

$$\underset{\sim}{U}_e = \underset{\sim}{C}\,\underset{\sim}{\alpha} \tag{15}$$

The order of $\underset{\sim}{\alpha}$ must be equal to or greater than the order of $\underset{\sim}{U}_e$. This discussion is restricted to the case where $\underset{\sim}{C}$ is square and non singular. Finally, inverting (15) and substituting in (14) we obtain,

$$\underset{\sim}{u} = \underset{\sim}{A} (\underset{\sim}{C}^{-1}) \underset{\sim}{U}_e = \underset{\sim}{G} \underset{\sim}{U}_e \tag{16}$$

Instead of starting with (14), one could establish (16) directly by using interpolation functions. The procedure outlined above is the 'original' approach. Employing interpolation functions is a subsequent innovation and is much more convenient.

We now introduce the displacement expansions in the Principle of Virtual Displacements (equation (8)), and obtain a set of algebraic equations relating the nodal displacements and generalized nodal forces. In what follows, the steps are outlined and the expressions for the various element matrices are developed.

Let us restrict this discussion to linear elastic behaviour for which the stress-strain relations can be expressed as,

$$\underset{\sim}{\sigma} = \underset{\sim}{D} \underset{\sim}{\varepsilon} \tag{17}$$

where D is a matrix of elastic constants. $\underset{\sim}{D}$ is symmetric and positive definite for a real material ($\underset{\sim}{D}$ degenerates to a positive semi-definite matrix if the material is assumed to be incompressible). The form of $\underset{\sim}{D}$ depends on the material, i.e., whether it is isotropic, orthotropic or anisotropic. Substituting for $\underset{\sim}{\sigma}$, into (8) we obtain

$$\int \delta\underset{\sim}{\varepsilon}^T \underset{\sim}{D} \underset{\sim}{\varepsilon} \, d(\text{vol.}) + \int \rho \, \delta\underset{\sim}{u}^T \ddot{\underset{\sim}{u}} \, d(\text{vol.}) = \int \delta\underset{\sim}{u}^T \underset{\sim}{b} \, d(\text{vol.}) + \int \delta\underset{\sim}{u}^T \bar{\underset{\sim}{p}} \, dS \tag{18}$$

Since we are expanding the displacements only over one element domain, we must write the Principle of Virtual displacements for the whole structure as,

$$\sum_{n_e} \left\{ \int \delta\underset{\sim}{\varepsilon}^T \underset{\sim}{D} \underset{\sim}{\varepsilon} \, d(\text{vol.}) + \int \rho \, \delta\underset{\sim}{u}^T \ddot{\underset{\sim}{u}} \, d(\text{vol.}) \right\}_{\text{element } n} = \sum_{n_e} \left\{ \int \delta\underset{\sim}{u}^T \underset{\sim}{b} \, d(\text{vol.}) + \int \delta\underset{\sim}{u}^T \bar{\underset{\sim}{p}} \, dS \right\}_{\text{element } n} \tag{19}$$

$n_e$ denotes the total number of elements.

Consider now the strains. Applying the strain-displacement relations to (16) results in

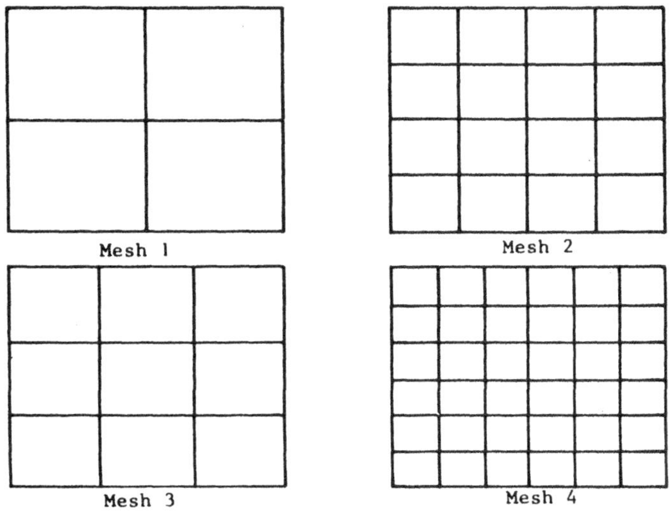


Figure 5  Different Types of Mesh

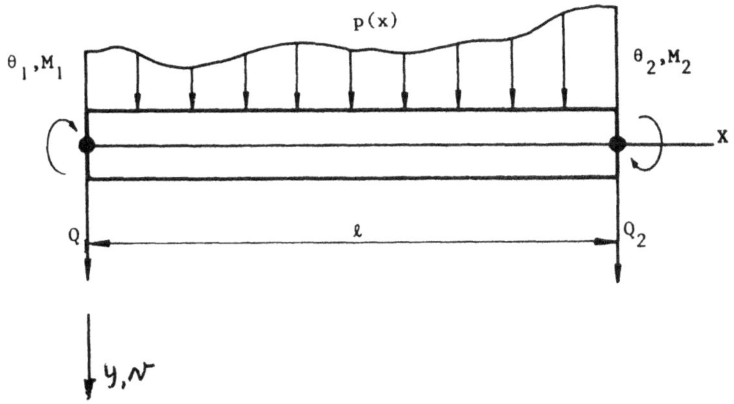


Figure 6  Beam Element

$$\underset{\sim}{\varepsilon} = \underset{\sim}{B}\,\underset{\sim e}{U} \tag{20}$$

where $\underset{\sim}{B}$ contains prescribed functions of the position coordinates. The stresses are given by

$$\underset{\sim}{\sigma} = \underset{\sim}{D}\,\underset{\sim}{B}\,\underset{\sim e}{U} \tag{21}$$

Since $\underset{\sim}{B}$ is independent of displacements,

$$\delta\underset{\sim}{\varepsilon} = \underset{\sim}{B}\,\delta\underset{\sim e}{U} \tag{22}$$

Substituting the above relationships into (18) we obtain for an element

$$\begin{aligned} \delta\underset{\sim e}{U}^{T} \left\{ \int \underset{\sim}{B}^{T}\,\underset{\sim}{D}\,\underset{\sim}{B}\, d(\text{vol.})\, \underset{\sim e}{U} + \int \rho\, \underset{\sim}{G}^{T}\,\underset{\sim}{G}\, d(\text{vol.})\, \ddot{U}_{e} \right\} &= \\ = \delta\underset{\sim e}{U}^{T} \left\{ \int \underset{\sim}{G}^{T}\,\underset{\sim}{b}\, d(\text{vol.}) + \int_{S_\sigma} \underset{\sim}{G}^{T}\,\underset{\sim}{p}\, dS \right\} & \end{aligned} \tag{23}$$

or,

$$\delta\underset{\sim e}{U}^{T} \{ \underset{\sim}{k}\,\underset{\sim e}{U} + \underset{\sim}{m}\,\underset{\sim e}{\ddot{U}} \} = \delta\underset{\sim e}{U}^{T}\,\underset{\sim}{f} \tag{24}$$

where

$\underset{\sim}{k}$ = stiffness matrix = $\int \underset{\sim}{B}^{T}\,\underset{\sim}{D}\,\underset{\sim}{B}\, d(\text{vol.})$

$\underset{\sim}{m}$ = mass matrix = $\int \underset{\sim}{G}^{T}\,\rho\,\underset{\sim}{G}\, d(\text{vol.})$

$\underset{\sim}{f}$ = consistent element force matrix = $\int \underset{\sim}{G}^{T}\,\underset{\sim}{b}\, d(\text{vol.}) + \int_{S_\sigma} \underset{\sim}{G}^{T}\,\bar{\underset{\sim}{p}}\, dS$

Note that $\underset{\sim}{p}$ contains the prescribed external surface forces. Hence the area integral involves only the exterior position of the surface area for the element, i.e. the interior surface area which is common for adjacent elements is not considered.

With this notation the Principle of Virtual displacements takes the following discretized form,

$$\sum_{n_e} \delta\underset{\sim e}{U}^{T} (\underset{\sim}{k}\,\underset{\sim e}{U} + \underset{\sim}{m}\,\underset{\sim e}{\ddot{U}}) = \sum_{n_e} \delta\underset{\sim e}{U}^{T}\,\underset{\sim}{f} \tag{25}$$

The critical step in the finite element displacement method is the selection of displacement expansions, i.e., the form of $\underset{\sim}{G}$. If the displacement expansion includes all possible rigid body displacements, all uniform strain states and if displacement compatibility along the boundaries between elements is satisfied. The finite element solution represents an upper bound on the total potential energy and the solution will converge to the true solution as the mesh size is decreased.

Inter-element compatibility requires that the assumed displacement field be continuous up to the derivative of one order lower than the highest derivative appearing in the strain-displacement relations. For example for plane stress - plane stress the strain displacements relations involves first order derivatives, hence the functions themselves must be continuous on the inter-element boundaries. For plate bending instead, the strain-displacement equations are second order, therefore the first derivatives as well as the functions must be continuous between element.

Most of the element formulations that have been developed are based on polynomial expansions. To satisfy the requirements of rigid modes and constant strain states, the expansion must be at least a complete polynomial of order equal to the highest derivative occurring in the strain-displacement relations. For plane stress this requires a first-order polynomial, for beam or plate bending, a complete second-order polynomial is required. Additional terms are included to complete the expansion, i.e. to obtain the necessary number of displacement parameters.

Convergence of certain parameters (for instance potential energy to static cases, eigenvalues in free vibration) can be ensured when inter-element compatibility is satisfied. As the discretization is refined the parameter (potential energy or eigenvalue) will converge to the true solution, provided that all uniform strain states can be represented by the expansion. However, the convergence will be monotonic only if the discretization comprises a *minimising sequence*. This means that by suitably specializing the nodal displacements for the n'th discretization we must be able to reproduce the displacement patterns corresponding to the n-1 previous discretization. In order to satisfy this requirement, the n'th discretization must contain all the previous nodes and element sides and the element displacement expansions must be invariant (i.e. their form must not depend on the orientation or dimensions of the element). As an illustration consider figure 5. Patterns 1 and 2 comprise one minimising sequence while 3 and 4 comprise another minimising sequence since they contain a different set of nodes.

In some cases, particularly for plate bending and shell elements, it is quite difficult to satisfy inter-element compatibility without resorting to rather complex displacement expansions. Formulations which violate inter-element compatibility are used and they exhibit good convergence in comparison with compatible models. A non-compatible element may converge to the true solution if all the body modes and uniform strain states are included. However, non-compatible elements do not provide bounds, i.e. we do not know whether the potential energy is below or above.

Example 1

To illustrate the generation of the displacement expansion and element matrices consider the prismatic beam element shown in Figure 6. The choice of nodal variables is v and $\theta$, and it follows that $\underset{\sim}{U}_e$ is at least 1 x 4.

$$\underset{\sim}{U}_e = \{ v_1 \; \theta_1 \; v_2 \; \theta_2 \; . \; . \; . \} \tag{a}$$

This discussion is restricted to negligible transverse shear deformation. The extensional strain varies linearly through the depth and $\theta$ is equal to the rotation of the tangent.

$$\varepsilon = - z \frac{d^2w}{dx^2} \quad , \qquad \theta = \frac{dw}{dx} \tag{b}$$

Since $\varepsilon$ involves the second derivative, the displacement expansion must contain a complete quadratic in order to be able to represent rigid body motion and constant strain. We write

$$v = \alpha_1 + \alpha_2 x + \alpha_3 x^2 + \alpha_4 x^3 =$$

$$= [1 \quad x \quad x^2 \quad x^3] \begin{Bmatrix} \alpha_1 \\ \alpha_2 \\ \alpha_3 \\ \alpha_4 \end{Bmatrix} = \underset{\sim}{A} \, \underset{\sim}{\alpha} \tag{c}$$

The $\underset{\sim}{\alpha}$ may be expressed in function of the generalized displacements $v_i$. Thus,

$$\begin{Bmatrix} v_1 \\ \theta_1 \\ v_2 \\ \theta_2 \end{Bmatrix} = \begin{bmatrix} 1 & 0 & 0 & 0 \\ 0 & 1 & 0 & 0 \\ 1 & \ell & \ell^2 & \ell^3 \\ 0 & 1 & 2\ell & 3\ell^2 \end{bmatrix} \begin{Bmatrix} \alpha_1 \\ \alpha_2 \\ \alpha_3 \\ \alpha_4 \end{Bmatrix} \tag{d}$$

of $\underset{\sim}{U}_e = \underset{\sim}{C} \, \underset{\sim}{\alpha}$. Next we invert (d)

$$\begin{aligned} \underset{\sim}{\alpha} &= \underset{\sim}{C}^{-1} \underset{\sim}{U}_e \\ &= \begin{bmatrix} 1 & 0 & 0 & 0 \\ 0 & 1 & 0 & 0 \\ -\dfrac{3}{\ell^2} & -\dfrac{2}{\ell} & \dfrac{3}{\ell^2} & -\dfrac{1}{\ell} \\ \dfrac{2}{\ell^3} & \dfrac{1}{\ell^2} & -\dfrac{2}{\ell^3} & \dfrac{1}{\ell^2} \end{bmatrix} \begin{Bmatrix} v_1 \\ \theta_1 \\ v_2 \\ \theta_2 \end{Bmatrix} \end{aligned} \tag{e}$$

Substituting $\underset{\sim}{\alpha}$ in (c) we obtain the final result.

$$v = \underset{\sim}{G}\,\underset{\sim}{U}_e = g_1 v_1 + g_2 \theta_1 + g_3 v_2 + g_4 \theta_2 \qquad \text{(f)}$$

where $g_1 = [\, 1 - 3(\frac{x}{\ell})^2 + 2(\frac{x}{\ell})^3 ]$, $\quad g_2 = [\, x - 2\frac{x^2}{\ell} + \frac{x^3}{\ell^2}\, ]$

$g_3 = [\, 3(\frac{x}{\ell})^2 - 2(\frac{x}{\ell})^3 ]$, $\quad g_4 = [\, -\frac{x^2}{\ell} + \frac{x^3}{\ell^2}\, ]$

$g_i$ are the interpolation functions.

The strain expansion is obtained by substituting for w in (b).

$$\varepsilon = -\,y\,\frac{d^2 v}{dx^2} = -\,z\,\frac{d^2}{dx^2}\,(\underset{\sim}{G})\,\underset{\sim}{U}_e = \underset{\sim}{B}\,\underset{\sim}{U}_e$$

$$\underset{\sim}{B} = -\,z\,\{-\frac{6}{\ell^2} + \frac{12x}{\ell^3}\,,\; -\frac{4}{\ell} + \frac{6x}{\ell^2}\,,\; \frac{6}{\ell^2} - \frac{12x}{\ell^3}\,,\; -\frac{2}{\ell} + \frac{6}{\ell^2}\,\} \qquad \text{(g)}$$

Considering only normal stresses, the stress-strain relations $\underset{\sim}{\sigma} = \underset{\sim}{D}\,\underset{\sim}{\varepsilon}$, reduce to,

$$\sigma = E\,\varepsilon \qquad \text{(h)}$$

We can now generate the *element matrices*.

(i) Stiffness Matrix

$$\underset{\sim}{k} = \int \underset{\sim}{B}^T\,\underset{\sim}{D}\,\underset{\sim}{B}\;d(\text{vol.}) = \frac{EI}{\ell^3}\begin{bmatrix} 12 & 6\ell & -12 & 6\ell \\ & 4\ell^2 & -6\ell & 2\ell^2 \\ & & 12 & -6\ell \\ \text{sym.} & & & 4\ell^2 \end{bmatrix} \qquad \text{(i)}$$

This result is the 'exact' stiffness matrix for a prismatic beam. The agreement is due to the expansion employed. One can readily show that the exact homogeneous solution of the governing equations for a prismatic element is a cubic polynomial.

(ii) Mass Matrix

$$\underset{\sim}{m} = \int \rho\,\underset{\sim}{G}^T\,\underset{\sim}{G}\;d(\text{vol.}) = \frac{\rho A\ell}{420}\begin{bmatrix} 156 & 22\ell & 54 & -13\ell \\ & 4\ell^2 & 13\ell & -3\ell^2 \\ & & 156 & -22\ell \\ \text{sym.} & & & 4\ell^2 \end{bmatrix} \qquad \text{(j)}$$

(iii) Consistent Element Nodal Force Matrix for the case of only distributed transverse loading p(x) is

$$\underset{\sim}{f} = \int_0^{\ell} \underset{\sim}{G}^T p(x)\, dx \qquad \text{(k)}$$

or

$$\begin{Bmatrix} p_{1v} \\ p_{1\theta} \\ p_{2v} \\ p_{2\theta} \end{Bmatrix} = \ell \int_0^{\ell} \begin{Bmatrix} p(1-3\bar{x}^2 + 2\bar{x}^3) \\ p\ell(1-2\bar{x}^2 + \bar{x}^3) \\ p(3\bar{x}^2 - 2\bar{x}^3) \\ p\ell(-\bar{x}^2 + \bar{x}^3) \end{Bmatrix} d\bar{x}$$

where $\bar{x} = x/\ell$.

The form of $\underset{\sim}{f}$ will depend on how p(x) varies. If p is constant we obtain, after integration,

$$\underset{\sim}{f} = \begin{Bmatrix} \frac{1}{2} \\ \frac{1}{12}\ell \\ \frac{1}{2} \\ -\frac{1}{12}\ell \end{Bmatrix} p\ell$$

Application Let us consider a simply supported beam represented by only one element. Hence $v_1 = v_2 = 0$ are the displacement boundary conditions to be satisfied and the element matrices in the absence of external forces become,

$$\underset{\sim}{K}\,\underset{\sim}{U} + \underset{\sim}{M}\,\ddot{\underset{\sim}{U}} = 0$$

or,

$$\frac{EI}{\ell}\begin{bmatrix} 4 & 2 \\ 2 & 4 \end{bmatrix}\begin{Bmatrix} \theta_1 \\ \theta_2 \end{Bmatrix} + \frac{\rho A \ell^3}{420}\begin{bmatrix} 4 & -3 \\ -3 & 4 \end{bmatrix}\begin{Bmatrix} \ddot{\theta}_1 \\ \ddot{\theta}_2 \end{Bmatrix} = \begin{Bmatrix} 0 \\ 0 \end{Bmatrix} \qquad (\ell)$$

The solution of an homogeneous second order system of equations as this is

$$\theta_j = \bar{\theta}_j\, e^{i\omega t}$$

where $\omega$ is the circular frequency and $i = \sqrt{-1}$. Hence ($\ell$) becomes,

$$\left\{ \frac{EI}{\ell}\begin{bmatrix} 4 & 2 \\ 2 & 4 \end{bmatrix} - \frac{\rho A \omega^2 \ell^3}{420}\begin{bmatrix} 4 & -3 \\ -3 & 4 \end{bmatrix} \right\} \begin{Bmatrix} \bar{\theta}_1 \\ \bar{\theta}_2 \end{Bmatrix} = \begin{Bmatrix} 0 \\ 0 \end{Bmatrix} \qquad \text{(m)}$$

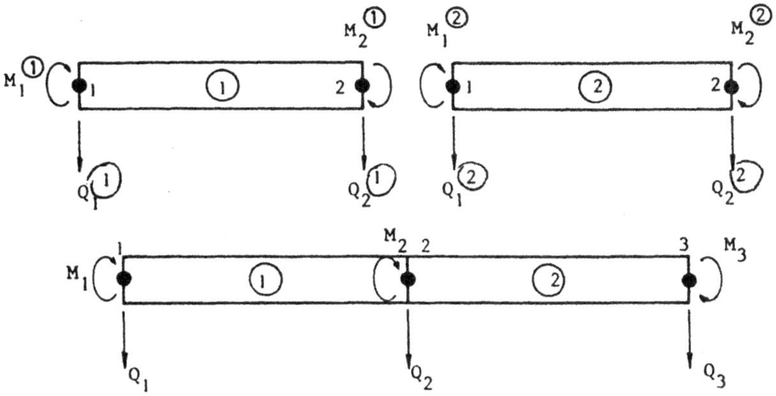


Figure 7 Beam Composed of Two Elements

| Element | Nodes | | |
|---|---|---|---|
| | $n_1$ | $n_2$ | $n_3$ |
| ① | 1 | 4 | 2 |
| ② | 3 | 4 | 1 |
| ③ | 3 | 6 | 4 |
| ④ | 5 | 6 | 3 |

Figure 8 Four Elements Body

and defining $\lambda = \frac{\rho A \ell^4 \omega^2}{420\ EI}$, the system of equation (m) has a solution if the determinant is zero, i.e.

$$\begin{vmatrix} 4-3\lambda & 2+3\lambda \\ 2+3\lambda & 4-4\lambda \end{vmatrix} = 0 \qquad \text{(n)}$$

Thus $\lambda = \frac{2}{7}$ or 6 and the value of $\frac{\rho A \ell^4 \omega^2}{EI}$ = 120 cr 2520, which compare reasonably well with the exact values of 97.41 and 1559. In order to improve on our approximation we can take two elements and assemble them together. This is done in the next example to give some background of how to obtain the system equations of a structure.

Example 2.

Consider now the beam represented by two elements of equal length (Figure 7). For element $\textcircled{1}$ we have

$$\underset{\sim}{k}^{\textcircled{1}}\, \underset{\sim}{U}^{\textcircled{1}} + \underset{\sim}{m}^{\textcircled{1}}\, \underset{\sim}{\ddot{U}}^{\textcircled{1}} = \underset{\sim}{f}^{\textcircled{1}} \qquad \text{(a)}$$

and for element $\textcircled{2}$

$$\underset{\sim}{k}^{\textcircled{2}}\, \underset{\sim}{U}^{\textcircled{2}} + \underset{\sim}{m}^{\textcircled{2}}\, \underset{\sim}{\ddot{U}}^{\textcircled{2}} = \underset{\sim}{f}^{\textcircled{2}} \qquad \text{(b)}$$

These equations can be assembled by using the compatibility (local node 2 of beam $\textcircled{1}$ is the same as node 1 of beam $\textcircled{2}$ ) and equilibrium conditions,

$$\begin{aligned} Q_1 &= Q_1^{\textcircled{1}}, & Q_2 &= Q_2^{\textcircled{1}} + Q_1^{\textcircled{2}}, & Q_3 &= Q_2^{\textcircled{2}} \\ M_1 &= M_1^{\textcircled{1}}, & M_2 &= M_2^{\textcircled{1}} + M_1^{\textcircled{2}}, & M_3 &= M_2^{\textcircled{2}} \end{aligned} \qquad \text{(c)}$$

The above relationships imply that the coefficients of the mass and stiffness element matrices can now be written as,

equilibrium equation which corresponds to

$$\begin{matrix} Q_1 \rightarrow \\ M_1 \rightarrow \\ Q_2 \rightarrow \\ M_2 \rightarrow \\ Q_3 \rightarrow \\ M_3 \rightarrow \end{matrix} \begin{bmatrix} o & o & o & o & & \\ o & o & o & o & & \\ o & o & o & o & o & o \\ o & o & o & o & o & o \\ & & o & o & o & o \\ & & o & o & o & o \end{bmatrix} \text{ multiplied by } \begin{Bmatrix} v_1 \\ \theta_1 \\ v_2 \\ \theta_2 \\ v_3 \\ \theta_3 \end{Bmatrix} \text{ or their acceleration} \qquad \text{(d)}$$

This gives,

$$\frac{EI}{\ell^3}\begin{bmatrix} 12 & 6\ell & -12 & 6\ell & 0 & 0 \\ & 4\ell^2 & -6\ell & 2\ell^2 & 0 & 0 \\ & & 24 & 0 & -12 & 6\ell \\ & & & 8\ell^2 & -6\ell & 2\ell^2 \\ \text{sym.} & & & & 12 & -6\ell \\ & & & & & 4\ell^2 \end{bmatrix}\begin{Bmatrix} v_1 \\ \theta_1 \\ v_2 \\ \theta_2 \\ v_3 \\ \theta_3 \end{Bmatrix} + \qquad \text{(e)}$$

$$\frac{\rho A\ell}{420}\begin{bmatrix} 156 & 22\ell & 54 & -13\ell & 0 & 0 \\ & 4\ell^2 & 13 & -3\ell^2 & 0 & 0 \\ & & 312 & 0 & 54 & -13\ell \\ & & & 8\ell^2 & 13\ell & -3\ell^2 \\ & & & & 156 & -22\ell \\ & & & & & 4\ell^2 \end{bmatrix}\begin{Bmatrix} \ddot{v}_1 \\ \ddot{\theta}_1 \\ \ddot{v}_2 \\ \ddot{\theta}_2 \\ \ddot{v}_3 \\ \ddot{\theta}_3 \end{Bmatrix} = \begin{Bmatrix} Q_1 \\ M_1 \\ Q_2 \\ M_2 \\ Q_3 \\ M_3 \end{Bmatrix}$$

Finally we can impose the displacement boundary conditions $v_1 = v_3 = 0$ and assume the right hand side of (e) is zero, i.e. the vibrations are free. We obtain the following system of equations

$$\begin{bmatrix} (4-4\lambda)\ell^2 & (-6-12\lambda)\ell & (2+3\lambda)\ell^2 & 0 \\ & (24-312\lambda)\ell & 0 & (6+13\lambda)\ell \\ & & (8-8\lambda)\ell^2 & (2+3\lambda)\ell^2 \\ & \text{sym.} & & (4-4\lambda)\ell^2 \end{bmatrix}\begin{Bmatrix} \theta_1 \\ v_2 \\ \theta_2 \\ \theta_3 \end{Bmatrix} = \begin{Bmatrix} 0 \\ 0 \\ 0 \\ 0 \end{Bmatrix} \qquad \text{(f)}$$

where $\lambda = \dfrac{\rho A\ell^4\omega^2}{420\ EI}$ .

The solution of the system of equations (f) gives the following eigenvalues,

$$\bar{\lambda} = \frac{\rho AL^4\omega^2}{EI} = 98.18,\quad 1920,\quad 12130,\quad 40320$$

($L = 2\ell$, total length of the beam). The exact solution for the beam gives,

$$\bar{\lambda} = 97.41,\quad 1559,\quad 7890,\quad 24940 .$$

## 4. System Equations

In order to obtain the global equilibrium equations for a body formed by different elements we have to assemble the element matrices and apply the boundary conditions. One first expands the element nodal unknowns vector $\underset{\sim}{U}_e$ in terms of the vector of nodal displacements $\underset{\sim}{U}_{n_i}$, now deferred to the *global* numbering

$$\underset{\sim}{U}_e = \{ \underset{\sim}{U}_{n_i} \} = \begin{Bmatrix} \underset{\sim}{U}_{n_1} \\ \underset{\sim}{U}_{n_2} \\ \vdots \\ \end{Bmatrix} \qquad i = 1,2, \ldots s \tag{26}$$

where s is the number of nodes in the element and $n_i$ their number referred to the complete system. We partition $\underset{\sim}{k}$, $\underset{\sim}{m}$ and $\underset{\sim}{f}$ for the element consistently with the partitioning of $\underset{\sim}{U}_e$. they are understood to be resulting from a derivation based on several unknowns for node, i.e. based on $\underset{\sim}{U}_e$.

$$\begin{aligned} \underset{\sim}{k} &= \{ k_{ij} \} \\ \underset{\sim}{m} &= \{ m_{ij} \} \\ \underset{\sim}{f} &= \{ f_i \} \quad i,j = 1,2 \ldots 5 \end{aligned} \tag{27}$$

With this notation the terms in (25) take the form

$$\begin{aligned} \delta\underset{\sim}{U}_e^T \underset{\sim}{k} \underset{\sim}{U}_e &= \sum_{i=1}^{s} \delta \underset{\sim}{U}_{n_i}^T \{ \sum_{j=1}^{s} \underset{\sim}{k}_{ij} \underset{\sim}{U}_{n_j} \} \\ \delta\underset{\sim}{U}_e^T \underset{\sim}{m} \underset{\sim}{U}_e &= \sum_{i=1}^{s} \delta \underset{\sim}{U}_{n_i}^T \{ \sum_{j=1}^{s} \underset{\sim}{m}_{ij} \underset{\sim}{U}_{n_j} \} \\ \delta\underset{\sim}{U}_e^T \underset{\sim}{f} &= \sum_{i=1}^{s} \delta \underset{\sim}{U}_{n_i}^T \underset{\sim}{f}_i \end{aligned} \tag{28}$$

If, in the derivation of the element matrices, the unknowns are referred to the local frame rather than the basic frame, it is necessary to transform the nodal unknowns, $\underset{\sim}{U}_{n_i}$, $\delta\underset{\sim}{U}_{n_i}$ in (28) from the element frame to the basic frame. We use an asterisk to indicate the global frame and obtain

$$\begin{aligned} U_{n_i} &= R\, U^*_{n_i} \\ \delta\underset{\sim}{U}_{n_i} &= \underset{\sim}{R}\, \delta\underset{\sim}{U}^*_{n_i} \end{aligned} \tag{29}$$

where $\underset{\sim}{R}$ contains the direction cosines for the local direction with respect to the global matrices. Equation (28) can now be written,

$$
\begin{aligned}
\delta U_e^T \, k \, U_e &= \sum_{i=1}^{s} \delta U^{*}{}'^{T}_{n_i} \left\{ \sum_{j=1}^{s} k^{*}_{ij} \, U^{*}_{nj} \right\} \\
\delta U_e^T \, m \, U_e &= \sum_{i=1}^{s} \delta U^{*}{}'^{T}_{n_i} \left\{ \sum_{j=1}^{s} m^{*}_{ij} \, U^{*}_{nj} \right\} \qquad (30) \\
\delta U_e^T \, f &= \sum_{i=1}^{s} \delta U^{*}{}'^{T}_{n_i} \, f_i
\end{aligned}
$$

where

$$
\begin{aligned}
k^{*}_{ij} &= R^T \, k_{ij} \, R \\
m^{*}_{ij} &= R^T \, m_{ij} \, R \qquad (31) \\
f^{*}_{i} &= R^T \, f_i
\end{aligned}
$$

The governing equations for the *whole* body can be written as,

$$
\sum_{n_e} \delta U_e^T \left( k \, U_e + m \, \ddot{U}_e \right) = \sum_{n_e} \delta U_e^T \, f \qquad (32)
$$

If N denotes the total number of nodes we can define a system nodal unknown vector. (In what follows we assume $U_{n_i}$ are referred to the global frame, that is we drop the asterisk for simplicity).

$$
U = \{ U_1, \; U_2 \; \ldots \; U_N \} \qquad (33)
$$

Expanding (32) by summing the contribution of the elements incident on each node, we have,

$$
\delta U^T \{ K \, U + M \, \ddot{U} \} = \delta U^T \, F \qquad (34)
$$

or for arbitrary $\delta U$,

$$
K \, U + M \, \ddot{U} = F \qquad (35)
$$

The partitioned form of (35) is

$$
\begin{bmatrix} K_{11} K_{12} & \cdots & K_{1N} \\ K_{21} K_{22} & \cdots & K_{2N} \\ \cdots & & \\ K_{N1} K_{N2} & \cdots & K_{NN} \end{bmatrix}
\begin{Bmatrix} U_1 \\ U_3 \\ \vdots \\ U_N \end{Bmatrix}
+
\begin{bmatrix} M_{11} M_{12} & \cdots & M_{1N} \\ M_{21} M_{22} & \cdots & M_{2N} \\ & & \\ M_{N1} M_{N2} & \cdots & M_{NN} \end{bmatrix}
\begin{Bmatrix} U_1 \\ U_2 \\ \vdots \\ U_N \end{Bmatrix}
=
\begin{Bmatrix} F_1 \\ F_2 \\ \vdots \\ F_N \end{Bmatrix} \qquad (36)
$$

We assemble $\underset{\sim}{K}$, $\underset{\sim}{M}$ and $\underset{\sim}{F}$ in *partitioned* form working with successive elements. The contribution for an element is listed below.

$$
\begin{array}{lll}
\underline{\text{In}}\ \underset{\sim}{F} & & \\
 & \underset{\sim}{f}_i \quad \text{in row } i & \\
 & \qquad\qquad i = 1,2 \ldots s. & \\
\underline{\text{In}}\ \underset{\sim}{K} & & \\
 & \underset{\sim}{k}_{ij} \quad \text{in row } i, \text{ column } j & \\
 & \qquad\qquad i,j = 1,2 \ldots s. & \\
\underline{\text{In}}\ \underset{\sim}{M} & & \\
 & \underset{\sim}{m}_{ij} \quad \text{in row } i, \text{ column } j & \\
 & \qquad\qquad i,j = 1,2 \ldots s. &
\end{array}
\tag{37}
$$

These operations are carried out for all the elements. Since $\underset{\sim}{m}$ and $\underset{\sim}{k}$ are symmetrical for our example, the $\underset{\sim}{M}$ and $\underset{\sim}{K}$ matrices of (35) are going to be symmetrical and only the coefficients on and above the diagonal need to be stored.

Example 3

Let us consider a body ccmposed of only 4 triangular elements (Figure 8). We assume to know the element matrices $\underset{\sim}{k}$, $\underset{\sim}{m}$, $\underset{\sim}{f}$. For instance for the element (2) the matrices can be written

| Element | Nodes $n_1$ | $n_2$ | $n_3$ |
|---|---|---|---|
| 1 | 1 | 4 | 2 |
| 2 | 3 | 4 | 1 |
| 3 | 3 | 6 | 4 |
| 4 | 5 | 6 | 3 |

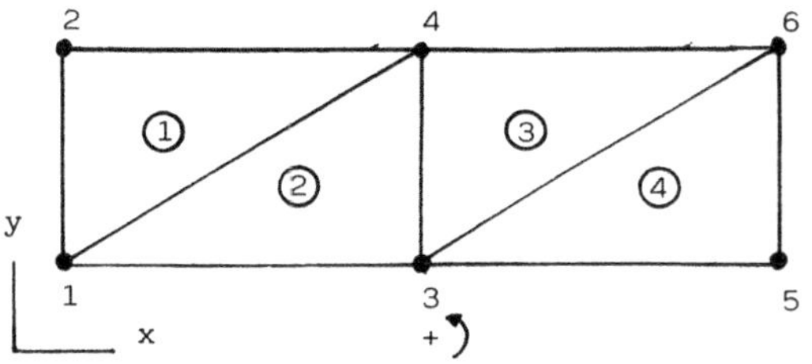


Figure 8 Four Elements Body

$$
\begin{bmatrix}
\underset{\sim}{k}_{11}^{(2)} & \underset{\sim}{k}_{12}^{(2)} & \underset{\sim}{k}_{13}^{(2)} \\
\underset{\sim}{k}_{21}^{(2)} & \underset{\sim}{k}_{22}^{(2)} & \underset{\sim}{k}_{23}^{(2)} \\
\underset{\sim}{k}_{31}^{(2)} & \underset{\sim}{k}_{32}^{(2)} & \underset{\sim}{k}_{33}^{(2)}
\end{bmatrix}
\begin{Bmatrix}
\underset{\sim}{U}_{1}^{(2)} \\
\underset{\sim}{U}_{2}^{(2)} \\
\underset{\sim}{U}_{3}^{(2)}
\end{Bmatrix}
+
\begin{bmatrix}
\underset{\sim}{m}_{11}^{(2)} & \underset{\sim}{m}_{12}^{(2)} & \underset{\sim}{m}_{13}^{(2)} \\
\underset{\sim}{m}_{21}^{(2)} & \underset{\sim}{m}_{22}^{(2)} & \underset{\sim}{m}_{23}^{(2)} \\
\underset{\sim}{m}_{31}^{(2)} & \underset{\sim}{m}_{32}^{(2)} & \underset{\sim}{m}_{33}^{(2)}
\end{bmatrix}
\begin{Bmatrix}
\ddot{\underset{\sim}{U}}_{1}^{(2)} \\
\ddot{\underset{\sim}{U}}_{2}^{(2)} \\
\ddot{\underset{\sim}{U}}_{3}^{(2)}
\end{Bmatrix}
=
\begin{Bmatrix}
\underset{\sim}{f}_{1}^{(2)} \\
\underset{\sim}{f}_{2}^{(2)} \\
\underset{\sim}{f}_{3}^{(2)}
\end{Bmatrix}
\tag{a}
$$

$\underset{\sim}{k}\,\underset{\sim}{U}^n + \underset{\sim}{m}\,\ddot{\underset{\sim}{U}}^n = \underset{\sim}{f}$ for element 2, where k and m are symmetric.

The unknowns in the above equation are referred to the local numbering system. For the *global* nodal unknowns numbering system we can write,

$$\begin{bmatrix} \mathbf{k}_{11}^{(2)} & \mathbf{k}_{12}^{(2)} & \mathbf{k}_{13}^{(2)} \\ \mathbf{k}_{21}^{(2)} & \mathbf{k}_{22}^{(2)} & \mathbf{k}_{23}^{(2)} \\ \mathbf{k}_{31}^{(2)} & \mathbf{k}_{32}^{(2)} & \mathbf{k}_{33}^{(2)} \end{bmatrix} \begin{Bmatrix} \mathbf{U}_3 \\ \mathbf{U}_4 \\ \mathbf{U}_1 \end{Bmatrix} + \begin{bmatrix} \mathbf{m}_{11}^{(2)} & \mathbf{m}_{12}^{(2)} & \mathbf{m}_{13}^{(2)} \\ \mathbf{m}_{21}^{(2)} & \mathbf{m}_{22}^{(2)} & \mathbf{m}_{23}^{(2)} \\ \mathbf{m}_{31}^{(2)} & \mathbf{m}_{32}^{(2)} & \mathbf{m}_{33}^{(2)} \end{bmatrix} \begin{Bmatrix} \ddot{\mathbf{U}}_3 \\ \ddot{\mathbf{U}}_4 \\ \ddot{\mathbf{U}}_1 \end{Bmatrix} = \begin{Bmatrix} \mathbf{f}_1^{(2)} \\ \mathbf{f}_2^{(2)} \\ \mathbf{f}_3^{(2)} \end{Bmatrix} \qquad \text{(b)}$$

We are interested in superimposing the effects of all the elements in order to form the system matrices for the whole body. If the structure has 6 nodes we will finally obtain a (6d x 6d) matrix where d are the number of degrees of freedom per node. A typical element like ② will have the $k_{ij}$, $m_{ij}$ coefficients plus the 3 right hand side terms distributed in the global matrix, as follows.

$$\mathbf{K} = \begin{bmatrix} \mathbf{k}_{33}^{(2)} & \cdot & \mathbf{k}_{31}^{(2)} & \mathbf{k}_{32}^{(2)} & \cdot & \cdot \\ \cdot & \cdot & \cdot & \cdot & \cdot & \cdot \\ \mathbf{k}_{13}^{(2)} & \cdot & \mathbf{k}_{11}^{(2)} & \mathbf{k}_{12}^{(2)} & \cdot & \cdot \\ \mathbf{k}_{23}^{(2)} & \cdot & \mathbf{k}_{21}^{(2)} & \mathbf{k}_{22}^{(2)} & \cdot & \cdot \\ \cdot & \cdot & \cdot & \cdot & \cdot & \cdot \\ \cdot & \cdot & \cdot & \cdot & \cdot & \cdot \end{bmatrix} \quad \text{and } \mathbf{F} = \begin{bmatrix} \mathbf{f}_3^{(2)} \\ \cdot \\ \mathbf{f}_1^{(2)} \\ \mathbf{f}_2^{(2)} \\ \cdot \\ \cdot \end{bmatrix} \qquad \text{(c)}$$

and similarly for $\mathbf{M}_{ij}$ coefficients.

Once all the elements have been superimposed we will find that the global matrices for $\mathbf{M}$ and $\mathbf{K}$ look like

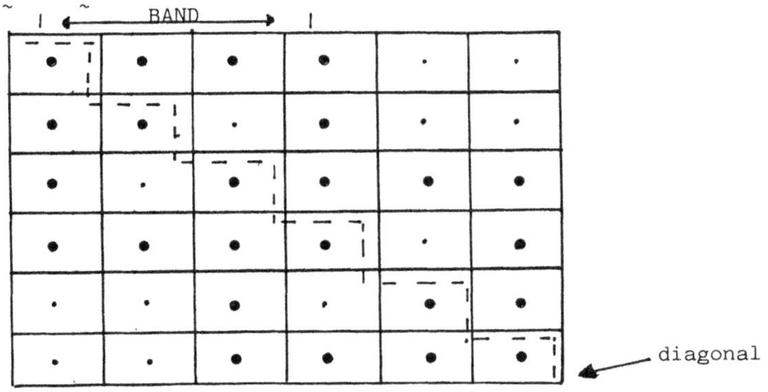


(d)

where • represents a filled store area and · an empty one. This matrix is symmetric and banded, the band equal to 4 x ·. Note that the band is proportional to the largest difference between nodes in the same element. Because of symmetry one needs only to store the diagonal and upper diagonal elements.

## 5. Solution

The governing equations for the unrestrained case are given by

$$\underset{\sim}{M}\,\ddot{\underset{\sim}{U}} + \underset{\sim}{K}\,\underset{\sim}{U} = \underset{\sim}{F} \tag{38}$$

The system mass matrix $\underset{\sim}{M}$ is positive definite but $\underset{\sim}{K}$ is singular due to the rigid body terms. If rigid body motion of the system is suppressed, $\underset{\sim}{K}$ becomes positive definite. In what follows we consider only the restrained case and rewrite the modified (i.e. with the displacement boundary conditions imposed) equations in the same way as (38)* to avoid proliferation of notation.

Let us first consider the free vibration problems, i.e. when $\underset{\sim}{F} = 0$. Equation (38) reduces to,

$$\underset{\sim}{M}\,\ddot{\underset{\sim}{U}} + \underset{\sim}{K}\,\underset{\sim}{U} = \underset{\sim}{0} \tag{39}$$

The form of this equation suggests that we express the solutions

$$\underset{\sim}{U} = e^{i\omega t}\,\underset{\sim}{z} \tag{39}$$

where $\omega$ is the circular frequency and $\underset{\sim}{z}$ defines the displacement pattern. Substituting for $\underset{\sim}{U}$ transforms (39) to

$$(\underset{\sim}{K} - \lambda\,\underset{\sim}{M})\,\underset{\sim}{z} = \underset{\sim}{0} \qquad \lambda = \omega^2 \tag{41}$$

The determination of the values of $\lambda$ gives the eigenvalues of the system. In general, there are n values of $\lambda$ which satisfy (41). Also, all n values are positive when both $\underset{\sim}{K}$ and $\underset{\sim}{M}$ are positive definite, which is the case here.

Let $\lambda_i = \omega_i^2$ denote an eigenvalue of (41) and $\underset{\sim}{z}_i$ the corresponding non-trivial solution. By definition,

$$\underset{\sim}{K}\,\underset{\sim}{z}_i = \lambda_i\,\underset{\sim}{M}\,\underset{\sim}{z}_i \tag{42}$$

---

* although their dimensions are now different as the number of degrees of freedom has been reduced after applying the known displacement boundary conditions.

Each eigenvector contains an arbitrary constant which is generally evaluated by normalizing $\underset{\sim}{\phi}$ with respect to $\underset{\sim}{M}$, i.e. by requiring

$$\underset{\sim}{z}_i^T \underset{\sim}{M} \underset{\sim}{z}_i = 1 \tag{43}$$

where $\underset{\sim}{z}_i$ is the normalized value of $\underset{\sim}{\phi}_i$.

Premultiplying the normalized version of (42) by $\underset{\sim}{z}_i^T$ we have,

$$\underset{\sim}{z}_i^T \underset{\sim}{K} \underset{\sim}{z}_i = \lambda_i \tag{44}$$

Due to orthogonality we have

$$\begin{aligned} \underset{\sim}{z}_j^T \underset{\sim}{M} \underset{\sim}{z}_i &= 0 \qquad \text{for } i \neq j \\ \underset{\sim}{z}_j^T \underset{\sim}{K} \underset{\sim}{z}_i &= 0 \qquad \text{for } i \neq j \end{aligned} \tag{45}$$

Equations (45) express the orthogonality relationship among the natural modes.

The results for the free vibration case are utilised to generate the solution for applied loading. We express the solution as a linear combination of s eigenvectors.

$$\underset{\sim}{U} = \sum_{i=1}^{s} q_i \underset{\sim}{z}_i \tag{46}$$

where $s \lesssim n$ and $q_i = q_i(t)$ can be interpreted as generalized coordinates.

Substituting $\underset{\sim}{U}$ into (38) we have,

$$\sum_{i=1}^{s} (\ddot{q}_i \underset{\sim}{M} \underset{\sim}{z}_i + q_i \underset{\sim}{K} \underset{\sim}{z}_i) = \underset{\sim}{F} \tag{47}$$

Premultiplying (50) by $\underset{\sim}{z}_j^T$ and noting the orthogonality relations gives s uncoupled differential equations

$$\ddot{q}_j + \omega_j^2 q_j = F_j \qquad j = 1,2 \dots s$$

$$F_j = \underset{\sim}{z}_j^T \underset{\sim}{F} \tag{48}$$

One can interpret $F_j$ as the generalized force corresponding to $q_j$, i.e. to the j'th mode. the static solution is

$$q_j = \frac{1}{\omega_j^2} (F_j)_{static} \tag{49}$$

We number the eigenvalues according to the increase magnitude $\omega_1 \lesssim \omega_2 \lesssim \ldots \lesssim \omega_n{}^2$. Equation (49) shows that the contribution of the higher modes decreases with increasing mode number, assuming the generalized modal forces are of the same order of magnitude. The ratio $(\underset{\sim}{F}_j)_{static}/\omega^2$ is called the 'participation factor' for the j'th mode.

The results for the static case suggest that we express $q_j$ and $F_j$ as,

$$\begin{aligned} F_j &= F_j^{(m)} f(t) \\ q_j &= \frac{F_j^{(m)}}{\omega_j{}^2} \bar{q}_j \end{aligned} \tag{50}$$

where $F_j^{(m)}$ represents the maximum values of $F_j$ and $F_j^{(m)}/\omega^2$ can be interpreted as a 'quasi-static' participation factor. We can now write (48) as

$$\frac{d^2}{dt^2}(\bar{q}_j) + \omega_j{}^2\bar{q}_j = \omega_j{}^2 f(t) \tag{51}$$

and the general solution is

$$\begin{aligned} \bar{q}_j = \left| q_j \right|_{t=o} \cos \omega_j t + \frac{1}{\omega_j} \left|\frac{d\bar{q}_j}{dt}\right|_{t=o} \sin \omega_j t \, + \\ + \, \omega_j \int_o^t f(\xi) \sin \omega_j (t - \xi) d\xi \end{aligned} \tag{52}$$

where $\xi$ is a dummy integration variable. The first two terms, which vanish if the system is initially at rest, represent the free vibration solution.

We can now evaluate $\underset{\sim}{U}$ for any given time from (46) which gives

$$\underset{\sim}{U}(t) = \sum_{j=1}^{s} \left(\frac{F_j^{(m)}}{\omega_j{}^2}\right) \underset{\sim}{z}_j \, \bar{q}_j(t) \tag{53}$$

The essential difficulty with model superposition is the decision of how many generalized coordinates - s - one should take. The computation required for s = n is usually prohibitive when n is large and the interpretation of the higher modes is difficult since the higher freqencies tend to be closely spaced. In selecting which modes to include for a particular loading, one should compare the modal participation factors.

References

Brebbia, C.A. and J.J. Connor, *Fundamentals of Finite Element Techniques for Structural Engineers*, Butterworths, 1973.

McCallion, H., *Vibration of Linear Mechanical Systems*, Longman, 1973.

Meirovitch, L., *Analytical Methods in Vibrations*, MacMillan Co., 1967.

Hurty, W.C. and M. Rubinstein, *Dynamics of Structures*, Prentice-Hall, 1964.

CHAPTER 8

# TWO DIMENSIONAL AND PLATE BENDING APPLICATIONS

by

R.R. Wilson

## 1. Introduction

In the previous chapter the principles of the finite element displacement method as applied to vibration problems were presented. This discussion is now extended to an examination of the factors which influence the choice of a displacement function for an element, with the particular cases of in-plane and transverse vibration of plates being considered. Different displacement functions are used to derive a number of elements, and the results which these different elements give when applied to sample problems compared. Finally, it is shown how in-plane plate elements, transverse plate elements and beam elements can be combined together to analyse a composite beam and plate structure.

The basis of the finite element displacement method is the representation of the displacement field throughout an element in terms of the value of the displacement at a discrete number of nodal points. The number of parameters used in specifying the displacement field must equal the total number of nodal variables.

Consider the situation where, because of limited computer storage, a fixed number of degrees of freedom is available for an analysis. We are then faced with the choice of using a large number of elements each with a displacement field based on a small number of parameters, or using a smaller number of elements each with displacement fields involving more unknowns. It is usually desirable to select a compromise between the two extremes of a very few high order elements and a large number of low order elements.

When very high order elements are used some of the advantages of the finite element method are lost. For a complex structure a large number of elements is required to represent the geometry, with the different structural components represented by elements with different lengths, thicknesses, material properties etc. It can often be the need to represent the geometry of a structure accurately that determines the number of elements used rather than a requirement for a high order displacement field to match the stress variation. In these cases when high order elements are used, some of the degrees of freedom are "wasted".

On the other hand, if the displacement field is represented by a very low order function, then the conditions required for convergence to the true solution as the mesh is refined may not be satisfied. Possibly the most suitable elements for inclusion in a general analysis package are those which have displacement functions which just satisfy the convergence criteria.

Once the displacement field for an element has been selected, there is a further choice. The unknowns determining the displacement function can be evaluated in terms of a large number of variables at each of a small number of nodes, or more nodes can be used, each with fewer freedoms.

A difficulty can arise when a high order displacement function is used with a small number of nodes. In order that the parameters defining the displacement fields can be evaluated, high order derivatives of the displacement are used as nodal variables.

This situation occurs most frequently in the analysis of shells, but consider for example a beam element, in which the second derivative of the transverse displacement is used as a nodal variable. If we use this element to analyse a stepped beam then the second derivative nodal variable will enforce continuity of the bending stress across the change in section. This imposes a theoretical constraint on the beam which is not there in practice.

To overcome this it is possible to *uncouple* this freedom and solve for the two values of the derivative on either side of the discontinuity. However this complicates the analysis and it is probably preferable to avoid the difficulty by forming an element with additional nodes when a high order displacement function is to be used.

The shape of an element also determines the number of nodes. With 3 nodes for a triangular element and 4 nodes for a rectangular element, the element shape can be defined in terms of the nodal coordinates. Similarly, curved elements are more easily defined when they have at least one node in the middle of each side of the element in addition to the corner nodes.

The shape of a structure often determines the elements to be used. If an irregularly shaped plate is to be analysed, it is probably easiest to use triangular plate elements, but if a rectangular plate is being considered the data preparation is simplest if rectangular elements are used. In the following sections different formulations for rectangular plate elements will be examined.

2. In-plane Plate Elements

The strains $\underset{\sim}{\varepsilon}$ produced by in-plane stretching of a plate are given by

$$\underset{\sim}{\varepsilon} = \begin{bmatrix} \varepsilon_x \\ \varepsilon_y \\ \varepsilon_{xy} \end{bmatrix} = \begin{bmatrix} \partial u/\partial x \\ \partial v/\partial y \\ \frac{\partial u}{\partial y} + \frac{\partial v}{\partial x} \end{bmatrix} , \quad (1)$$

where u and v are the displacements parallel to the x- and y-axes respectively. The corresponding stresses $\underset{\sim}{\sigma}$ are given by

$$\underset{\sim}{\sigma} = \underset{\sim}{D}\,\underset{\sim}{\varepsilon} , \quad (2)$$

where $$\underset{\sim}{D} = \frac{E}{1-\nu^2} \begin{bmatrix} 1 & \nu & 0 \\ \nu & 1 & 0 \\ 0 & 0 & \frac{1-\nu}{2} \end{bmatrix} , \quad (3)$$

where E is Young's modulus and $\nu$ is Poisson's ratio.

Since the strains consist only of the first order derivatives, to ensure inter-element compatibility it is necessary only that the displacements u and v are continuous between elements. We can define a possible displacement field throughout an element by the expression

$$\underset{\sim}{u} = \begin{bmatrix} u \\ v \end{bmatrix} \quad (4)$$

where

$$u = \alpha_1 + \alpha_2 x + \alpha_3 y + \alpha_4 xy$$

$$(5)$$

and $$v = \alpha_5 + \alpha_6 x + \alpha_7 y + \alpha_8 xy$$

i.e. $$\underset{\sim}{u} = \underset{\sim}{A}\,\underset{\sim}{\alpha} \quad (6)$$

where $$\underset{\sim}{A} = \begin{bmatrix} 1 & x & y & xy & 0 & 0 & 0 & 0 \\ 0 & 0 & 0 & 0 & 1 & x & y & xy \end{bmatrix} \quad (7)$$

and

$$\underset{\sim}{\alpha} = \begin{bmatrix} \alpha_1 \\ \vdots \\ \alpha_8 \end{bmatrix} \quad (8)$$

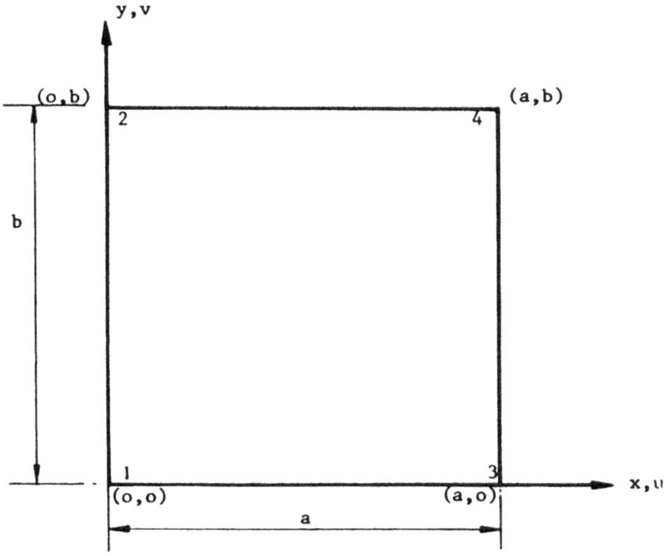


Figure 1 Plate Element

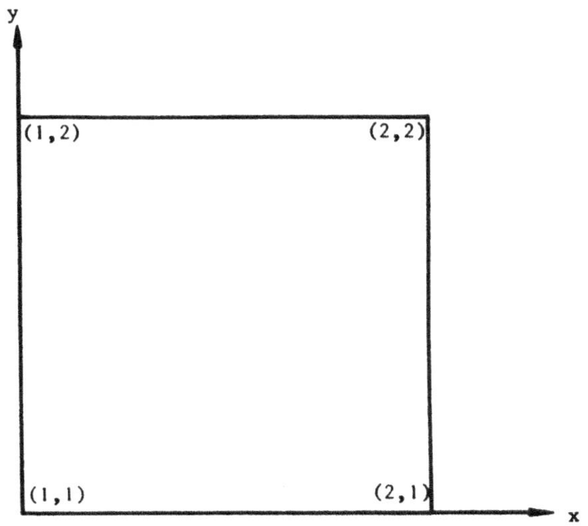


Figure 2 Node Numbering for use with Interpolation Polynomials.

If we consider a four noded rectangular element as shown in Figure 1, then the eight parameters $\alpha_i$ can be evaluated in terms of the value of the two displacements u and v at each of the four nodes.

We have

$$\underset{\sim}{U}_e = \begin{bmatrix} u_1 \\ v_1 \\ u_2 \\ v_2 \\ u_3 \\ v_3 \\ u_4 \\ v_4 \end{bmatrix} = \begin{bmatrix} 1 & 0 & 0 & 0 & 0 & 0 & 0 & 0 \\ 0 & 0 & 0 & 0 & 1 & 0 & 0 & 0 \\ 1 & 0 & b & 0 & 0 & 0 & 0 & 0 \\ 0 & 0 & 0 & 0 & 1 & 0 & b & 0 \\ 1 & a & 0 & 0 & 0 & 0 & 0 & 0 \\ 0 & 0 & 0 & 0 & 1 & a & 0 & 0 \\ 1 & a & b & ab & 0 & 0 & 0 & 0 \\ 0 & 0 & 0 & 0 & 1 & a & b & ab \end{bmatrix} \begin{bmatrix} \alpha_1 \\ \alpha_2 \\ \alpha_3 \\ \alpha_4 \\ \alpha_5 \\ \alpha_6 \\ \alpha_7 \\ \alpha_8 \end{bmatrix} \tag{9}$$

where a and b are the length and breadth of the plate element.

i.e.

$$\underset{\sim}{U}_e = \underset{\sim}{C}\,\underset{\sim}{\alpha} \tag{10}$$

$$\therefore \quad \underset{\sim}{\alpha} = \underset{\sim}{C}^{-1}\,\underset{\sim}{U}_e \tag{11}$$

Substituting in equation (6) gives

$$\underset{\sim}{u} = \underset{\sim}{G}\,\underset{\sim}{U}_e$$

where

$$\underset{\sim}{G} = \underset{\sim}{A}\,\underset{\sim}{C}^{-1}\ . \tag{13}$$

Writing equation (12) in full,

$$u = (1-\bar{x})(1-\bar{y})u_1 + (1-\bar{x})\bar{y}u_2 + \bar{x}(1-\bar{y})u_3 + \bar{x}\bar{y}u_4 \tag{14}$$

$$v = (1-\bar{x})(1-\bar{y})v_1 + (1-\bar{x})\bar{y}v_2 + \bar{x}(1-\bar{y})v_3 + \bar{x}\bar{y}v_4 \tag{15}$$

where $\bar{x} = \frac{x}{a}$ and $\bar{y} = \frac{y}{b}$ .

If we number the nodes as shown in Figure 2, then equation (14) may be written as

$$u = H_1(\bar{x})H_1(\bar{y})u_{11} + H_1(\bar{x})H_2(\bar{y})u_{12} + H_2(\bar{x})H_1(\bar{y})u_{21} + H_2(\bar{x})H_2(\bar{y})u_{22} \tag{16}$$

i.e.
$$u = \sum_{i=1}^{2} \sum_{j=1}^{2} H_i(\bar{x})H_j(\bar{y})u_{ij} \tag{17}$$

where
$$H_1(\bar{x}) = 1-\bar{x} \quad \text{and} \quad H_2(\bar{x}) = \bar{x} \tag{18}$$

The $H_1(\bar{x})$, $H_2(\bar{x})$ are examples of interpolation polynomials. They have the properties that

$$H_1(0) = H_2(1) = 1 \tag{19}$$

and
$$H_1(1) = H_2(0) = 0 \tag{20}$$

Once equation (12) has been formed the matrix $\underset{\sim}{B}$ can be calculated as described in the previous chapter from

$$\underset{\sim}{k} = \int \underset{\sim}{B}^T \underset{\sim}{D} \underset{\sim}{B} \, d(vol) \tag{21}$$

$$\underset{\sim}{m} = \int \underset{\sim}{G}^T \underset{\sim}{\rho} \underset{\sim}{G} \, d(vol) \tag{22}$$

The resultant element matrices are given in Appendix 1.

We shall now consider a second element, based on a higher order displacement function. If we have nodal variables, $u$, $\frac{\partial u}{\partial x}$, $\frac{\partial u}{\partial y}$, $v$, $\frac{\partial v}{\partial x}$, $\frac{\partial v}{\partial y}$ then there are a total of 24 nodal variables in the element. The displacement field throughout the element could be defined in terms of polynomials as in equation (5) but with 24 parameters $\alpha_i$. Instead of this, let us express the displacement throughout the element directly in terms of the nodal variables by using a new set of interpolation polynomials.

We have
$$u = \sum_{i=1}^{2} \sum_{j=1}^{2} [H_{0i}(\bar{x}) H_{0j}(\bar{y})u_{ij} + H_{1i}(\bar{x}) H_{0j}(\bar{y})(\frac{\partial u}{\partial x})_{ij} + H_{0i}(\bar{x}) H_{ij}(\bar{y})(\frac{\partial u}{\partial y})_{ij} ] \tag{23}$$

and similarly for v,

with
$$H_{01}(\bar{x}) = \frac{1}{a^3} (a^3 - 3ax^2 + 2x^3)$$

$$H_{02}(\bar{x}) = \frac{1}{a^3} (3ax^2 - 2x^3)$$

$$H_{11}(\bar{x}) = \frac{1}{a^2} (a^2x - 2ax^2 + x^3)$$

$$H_{12}(\bar{x}) = \frac{1}{a^2}(-ax^2 + x^3)$$

The properties possessed by these polynomials are summarized in Table 1.

| | $\bar{x} = 0$ | $\bar{x} = 1$ | | $\bar{x} = 0$ | $\bar{x} = 1$ |
|---|---|---|---|---|---|
| $H_{01}(\bar{x})$ | 1 | 0 | $\frac{\partial H_{01}}{\partial x}(\bar{x})$ | 0 | 0 |
| $H_{02}(\bar{x})$ | 0 | 1 | $\frac{\partial H_{02}}{\partial \bar{x}}(\bar{x})$ | 0 | 0 |
| $H_{11}(\bar{x})$ | 0 | 0 | $\frac{\partial H_{11}}{\partial \bar{x}}(\bar{x})$ | 1 | 0 |
| $H_{12}(\bar{x})$ | 0 | 0 | $\frac{\partial H_{12}}{\partial \bar{x}}(\bar{x})$ | 0 | 1 |

Table 1 Interpolation Polynomials

The element matrices can now be calculated as before.

3. <u>In-plane Vibration of Plates</u>

We have derived two finite elements for the in-plane vibration of plates; the first has 2 variables at each node and element matrices of order 8 x 8, and the second has 6 variables at each node and element matrices of order 24 x 24.

The equations of motion are [1],

$$\frac{\partial^2 u}{\partial x^2} + \tfrac{1}{2}(1-\nu)\frac{\partial^2 u}{\partial y^2} + \tfrac{1}{2}(1+\nu)\frac{\partial^2 v}{\partial x \partial y} = \rho\frac{(1-\nu^2)}{E}\frac{\partial^2 u}{\partial t^2} \tag{24}$$

$$\tfrac{1}{2}(1-\nu)\frac{\partial^2 v}{\partial x^2} + \frac{\partial^2 v}{\partial y^2} + \tfrac{1}{2}(1+\nu)\frac{\partial^2 u}{\partial x \partial y} = \rho\frac{(1-\nu^2)}{E}\frac{\partial^2 v}{\partial t^2}$$

For the boundary conditions that $u = 0$ on the sides parallel to the x-axis and $v = 0$ on the sides parallel to the y-axis, the equations are satisfied by

$$u = A\cos\frac{m\pi}{c}x\sin\frac{n\pi}{d}y\,e^{i\omega t}$$

$$v = B \sin \frac{m\pi}{c} x \cos \frac{n\pi}{d} y \; e^{i\omega t} \qquad (25)$$

where c and d are the length and breadth of the plate and m and n are integers. The value of $\omega$ is found by equating the determinant of the coefficients A and B to zero and is given by

$$\omega^2 = \frac{E\pi^2}{4\rho(1-\nu^2)} \left( \frac{m^2}{c^2} + \frac{n^2}{d^2} \right) [(3-\nu) \pm (1+\nu)] \qquad (26)$$

Because it has many fewer variables at each node, it is possible to use a much finer mesh with the element based on the 1st order polynomial than with the element based on the second order polynomial.

The problem was solved with the 1st element using meshes a - f, as shown in Figure 3, and using meshes a - c with the 2nd element. Figure 4 shows a comparison of the values calculated for the lowest five natural frequencies of a plate using each of the elements. The values are plotted against the final number of degrees of freedom since this is equal to the size of the system matrices and determines the time taken to solve the equations.

It can be seen that both elements give results which converge towards the exact solution from above. This is to be expected since they both satisfy the criteria for uniform convergence. It should be noted however that although the convergence was monotonic we were not guaranteed this since, for example, mesh b is not a refinement of mesh a. The element with the higher order displacement function gives more accurate answers for a given number of degrees of freedom. Thus if we have a simple in-plane plate vibration problem, this would appear to be the better element to use.

## 4. Plate Bending Elements

We shall now look at some of the elements which have been derived for the analysis of transverse vibration of plates. A total of seven elements will be introduced to show the range of formulations possible.

The first element (element a) we shall consider is based on a 12 term polynomial
2 for the transverse displacement

$$w = \alpha_1 + \alpha_2 x + \ldots + \alpha_{12} x y^3 . \qquad (27)$$

The terms included in this expression are

$$\begin{array}{ccccccc} & & & 1 & & & \\ & & x & & y & & \\ & x^2 & & xy & & y^2 & \\ x^3 & & x^2y & & xy^2 & & y^3 \\ & x^3y & & & & xy^3 & \end{array} \qquad (28)$$

The 12 unknowns are evaluated in terms of three nodal variables w, $\frac{\partial w}{\partial x}$ and $\frac{\partial w}{\partial y}$ at each of the four nodes at the corners of the rectangle.

Table 2 summarizes the number of elements, number of nodes and number of degrees of freedom for each mesh. The initial and final degrees of freedom refer to the number of degrees of freedom before and after the boundary conditions are applied .

| | Mesh | Number of elements | Number of nodes | Initial no. of degrees of freedom | Final number of degrees of freedom |
|---|---|---|---|---|---|
| First element | a | 4 | 9 | 18 | 6 |
| | b | 9 | 16 | 32 | 16 |
| | c | 16 | 25 | 50 | 30 |
| | d | 25 | 36 | 72 | 48 |
| | e | 36 | 49 | 98 | 70 |
| | f | 49 | 64 | 128 | 96 |
| Second element | a | 4 | 9 | 54 | 30 |
| | b | 9 | 16 | 96 | 64 |
| | c | 16 | 25 | 150 | 110 |

Table 2 Finite Element Meshes

It is possible to formulate an element with w, $\frac{\partial w}{\partial x}$ and $\frac{\partial w}{\partial y}$ as nodal variables by using interpolation polynomials, as in equation (23). The displacement function for this element, however, does not include the term xy, and so the element is not able to represent a state of constant twist $\frac{\partial^2 w}{\partial x \partial y}$ . It thus does not satisfy the conditions required to ensure convergence as the mesh is refined.

An element (element b) which does satisfy the convergence criteria can be derived by introducing an additional nodal variable $\frac{\partial^2 w}{\partial x \partial y}$ [3] . The displacement function can be written in terms of the interpolation polynomials,

$$w(x,y) = \sum \sum [H_{0i}(x)\, H_{0j}(y) w_{ij} + H_{1i}(x)\, H_{0j}(y) \left(\frac{\partial w}{\partial x}\right)_{ij}$$

$$+ H_{0i}(x)\, H_{1j}(y) \left(\frac{\partial w}{\partial y}\right)_{ij} + H_{1i}(x)\, H_{1j}(y) \left[\frac{\partial^2 w}{\partial x \partial y}\right]_{ij} ] \tag{29}$$

This element has 4 nodal variables at each of the 4 nodes, and the displacement function can be written as a polynomial with the following 16 terms

$$\begin{array}{c} 1 \\ x \quad y \\ x^2 \quad xy \quad y^2 \\ x^3 \quad x^2y \quad xy^2 \quad y^3 \\ x^3y \quad x^2y^2 \quad xy^3 \\ x^3y^2 \quad x^2y^3 \\ x^3y^3 \end{array} \tag{30}$$

$$\begin{array}{c} 1 \\ x \quad y \\ x^2 \quad xy \quad y^2 \\ x^3 \quad x^2y \quad xy^2 \quad y^3 \\ x^4 \quad x^3y \qquad xy^3 \quad y^4 \\ x^5 \quad x^4y \quad x^3y^2 \quad x^2y^3 \quad xy^4 \quad y^5 \\ x^6 \quad x^5y \quad x^4y^2 \qquad x^2y^4 \quad xy^5 \quad y^6 \\ x^7 \qquad x^4y^3 \quad x^3y^4 \qquad y^7 \\ x^5y^3 \qquad x^3y^5 \\ x^6y^3 \qquad x^3y^6 \end{array}$$

Finally it is possible to derive an element based on smooth surface interpolation rather than the linear interpolation polynomials. This element [5] has the nodal variables $w$, $\frac{\partial w}{\partial x}$ and $\frac{\partial w}{\partial y}$ giving a total of 12 variables in the displacement function. The displacement field is obtained by dividing the rectangle into four triangular areas and using a separate expansion for each.

It will have become apparent that there are many ways of deriving a rectangular plate element. We have considered seven and have by no means exhausted the possibilities. Four of the elements we have mentioned have nodal variable $w$, $\frac{\partial w}{\partial x}$ and $\frac{\partial w}{\partial y}$ and a 12 term displacement expansion. These all give different results for the same mesh and cne of the elements does not even converge to the true solution as the mesh is refined. Clearly, therefore, the choice of a displacement function for an element requires some care.

## 5. Transverse Vibration of Plates

We shall use three of the elements to calculate the natural frequencies of two plates, a simply supported square plate and a rectangular clamped plate. The elements used are element a which is based on the simple 12 term polynomial, element b which has the 16 term displacement field and element c which is based on an approximation to the 16 term polynomial.

A variation of this element (element c) can be obtained by expressing the twist variable $\frac{\partial^2 w}{\partial x \partial y}$ at each node in terms of the slopes at the adjacent nodes. This gives 4 constraint equations reducing the original 16 degrees of freedom to 12. It is not possible to write the displacement field explicitly as a simple polynomial but it does include the term xy and so the element still can represent constant twist conditions. The nodal variables in this simplified element are w, $\frac{\partial w}{\partial x}$, $\frac{\partial w}{\partial y}$ .

A fourth element can be obtained by including the additional nodal variables $\frac{\partial^2 w}{\partial x^2}$ and $\frac{\partial^2 w}{\partial y^2}$ as well as $\frac{\partial^2 w}{\partial x \partial y}$ and using higher order interpolation polynomials [3]. We now have 6 variables at each node and the equivalent of a 24 degree of freedom polynomial.

Instead of increasing the number of nodal variables, the number of nodes can be increased to allow higher order displacement functions to be used [4], with the nodal variables, w, $\frac{\partial w}{\partial x}$ and $\frac{\partial w}{\partial y}$ . The first of these elements has one mid-side node giving a total of 24 degrees of freedom with a polynomial expansion having the terms

$$\begin{array}{ccccccccccc}
 & & & & & 1 & & & & & \\
 & & & & x & & y & & & & \\
 & & & x^2 & & xy & & y^2 & & & \\
 & & x^3 & & x^2y & & xy^2 & & y^3 & & \\
 & x^4 & & x^3y & & & & xy^3 & & y^4 & \\
x^5 & & x^4y & & x^3y^2 & & x^2y^3 & & xy^4 & & y^5 \\
 & x^5y & & x^4y^2 & & & & x^2y^4 & & xy^5 &
\end{array} \qquad (31)$$

With 2 additional nodes on each side, the element has 36 degrees of freedom. In this case the terms in the polynomial are obtained from element b by removing the twist term.

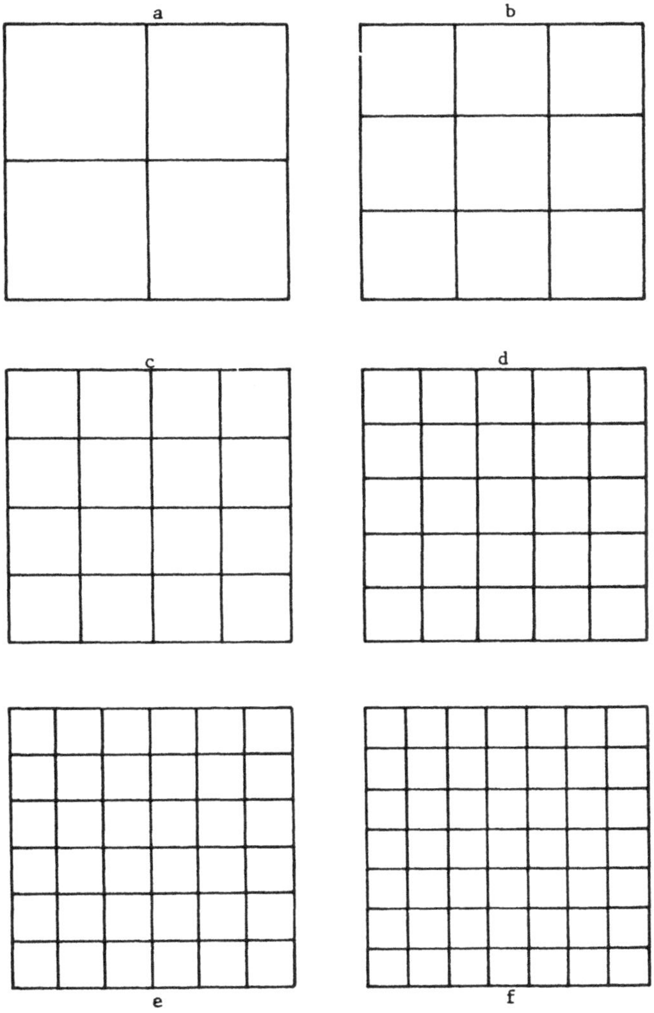


Figure 3 Plate Element Meshes

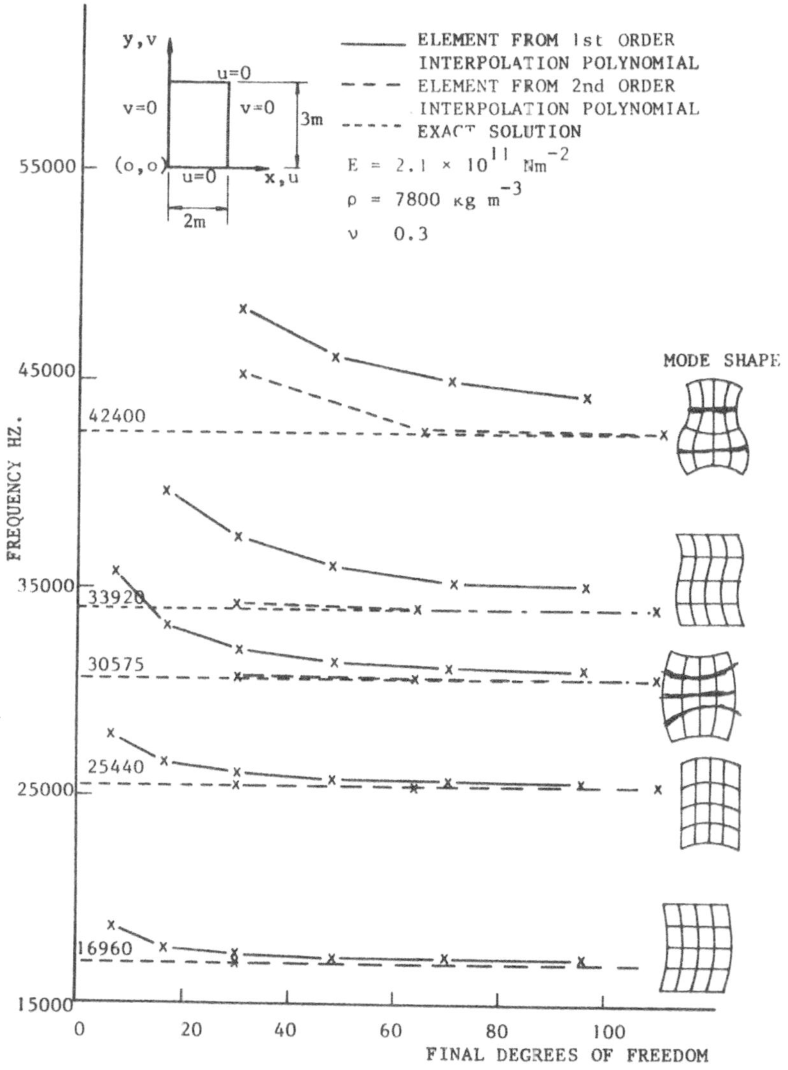


Figure 4 Natural Frequencies of In-plane Vibration of a Rectangular Plate

All three elements satisfy the conditions which ensure convergence as the mesh is refined. Only element b however satisfies the additional condition which ensures that the convergence is uniform; it is the only element of the three for which the normal slope between elements is continuous.

Figure 5 shows the results obtained for the natural frequencies of a simply supported plate using the different elements. It can be seen that for all three elements, the results tend to converge towards the exact values [6] as the mesh is refined.

Figure 6 shows the results obtained for the natural frequencies of a clamped rectangular plate. These are compared with the values obtained using a Rayleigh-Ritz solution [7] as described in Chapter 6.

In both examples all the elements converge as the mesh is refined. However only the fully-conforming element, element b converges monotonically, consistently giving an upper bound to the frequencies. It can be seen that, although they have the same order of displacement function, element a and element c give different answers for the same mesh.

## 6. Combination of Plate and Beam Elements

When it is necessary to select elements for the analysis of a complex structure, then considerations other than the behaviour in test examples such as those discussed in the previous sections are important.

For instance even though it has a better rate of convergence in the examples considered, the in-plane element based on the 2nd order interpolation polynomial is not suitable for most applications. It can represent a rapid variation of the in-plane displacement accurately with few degrees of freedom. However in practice the in-plane displacements do not usually vary in this complex way.

The in-plane element based on the 1st order interpolation polynomial will usually give satisfactory answers for the type of problem encountered. In addition since it is possible to use very many more elements, cut-outs and projections in the plates can be more accurately represented with this element.

Similar considerations favour choice of one of the lower order plate bending elements. Thus if we wish a plate element for combined bending and stretching then one possibility would be to have 20 x 20 element matrices with nodal variables u,

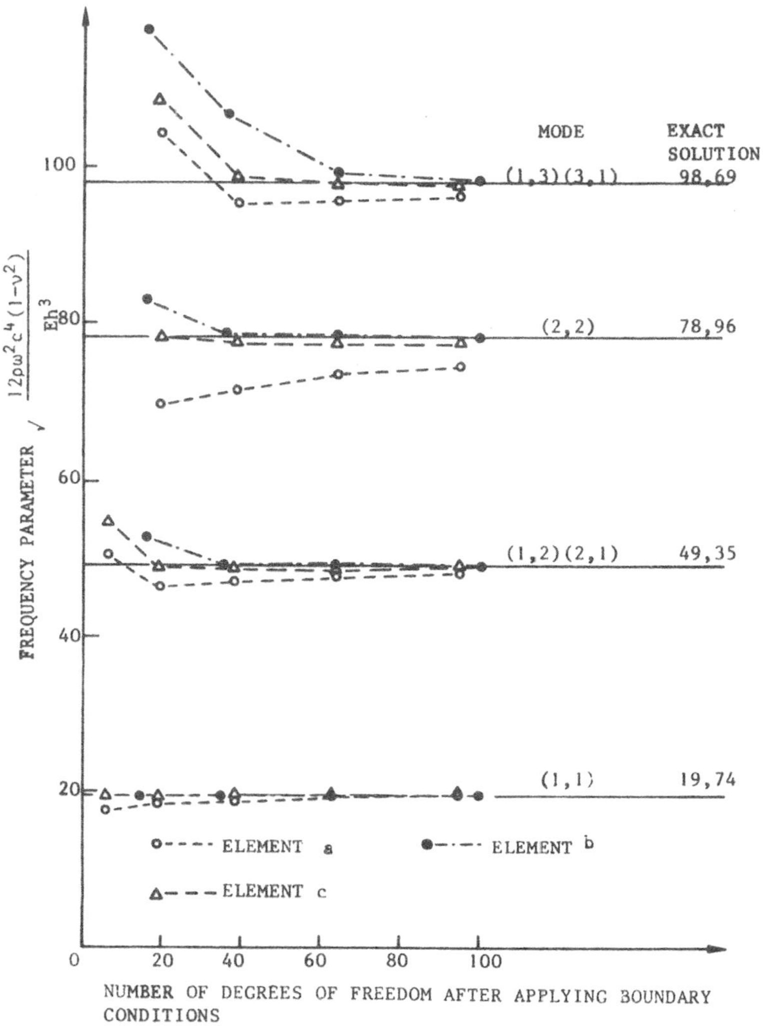


Figure 5 Convergence for Simply Supported Square Plate.

$v$, $w$, $\frac{\partial w}{\partial x}$ and $\frac{\partial w}{\partial y}$ . The in-plane components of the elements could be based on the 1st order interpolation polynomial with the transverse components corresponding to element a, element c or the element based on smooth surface interpolation.

This element has no coupling between the in-plane and transverse displacements and the corresponding portion of the element matrices is filled with zeros. Where there is coupling, for example in a curved plate, then the process can be repeated as before but with the assumed fields for the two displacements being used together in deriving the element matrices.

A suitable beam element for use with the above plate element as described can be obtained by assuming the following displacement functions for an element parallel to the x-axis.

$$
\begin{aligned}
w &= \alpha_1 + \alpha_2 x + \alpha_3 x^2 + \alpha_4 x^3 \\
v &= \alpha_5 + \alpha_6 x + \alpha_7 x^2 + \alpha_8 x^3 \\
u &= \alpha_9 + \alpha_{10} x \\
\theta &= \alpha_{11} + \alpha_{12} x
\end{aligned}
\qquad (33)
$$

We have 12 parameters which can be evaluated in terms of the six nodal variables, $w$, $\frac{\partial w}{\partial x}$ , $v$, $\frac{\partial v}{\partial x}$ , $u$, $\theta$ at each end of the element. Cubic displacement functions are used for the two transverse displacements. These parts of the element matrices are derived as described in Chapter 7. Linear functions are used to represent the extension and twisting of the beams. The strain energy expression corresponding to each of these two displacements involves only the first order derivatives and so a linear displacement function is sufficient to give convergence.

The complete element mass and stiffness matrices for this beam element are given in Appendix 2. The transverse displacement part of the matrices is based on Euler beam theory. If the beams being considered were deep, Timoshenko beam theory could be used [8].

Similarly if in a particular problem it was important to represent accurately the taper of a beam then a special element could be formulated. Expressions for say the variation of the cross-sectional area and the second moment of area along the length of an element would be assumed. The element matrices would be derived as before but with these expressions included under the integration. The formulation of plate and beam elements described can be varied to include additional effects as a particular problem demands.

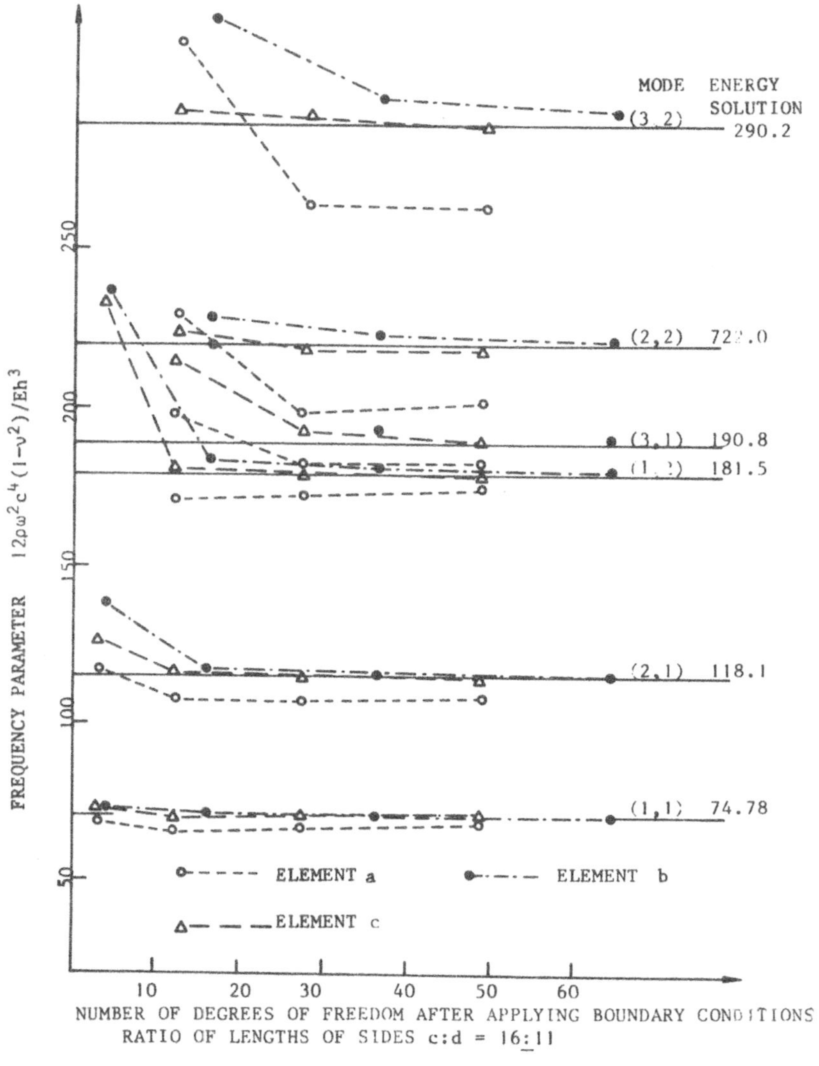


**Figure 6** **Convergence for Clamped Rectangular Plate**

References


1. Love, A.E.H., *The Mathematical Theory of Elasticity,* 4th Edition, Dover Publications, New York, 1944.

2. Dawe, D.J., A finite element approach to plate vibration problems, *Journal of Mechanical Engineering Science,* Vol. 7, No.1, 28-32, 1965.

3. Bogner, F.K., R.L. Fox and L.A. Schmit, The generation of inter-element compatible stiffness and mass matrices by the use of interpolation formulae, *Proceedings of the Conference on Matrix Methods in Structural Mechanics,* AFFDL-TR-66-80, 397-444, Wright-Paterson Air Force Base, Ohio, 1965.

4. Henshell, R.D., D. Walters and G.B. Warburton, A new family of curvilinear plate bending elements for vibration and stability, *Journal of Sound & Vibration,* 1972, Vol.20, 381-397.

5. Deak, A.L. and T.H.H. Pian, Application of the smooth surface interpolation to the finite element analysis, *AIAA Journal,* Vol.5, No.1, 187-189, 1967.

6. Timoshenko, S.P. *Vibration Problems in Engineering,* 3rd Edition, Van Nostrand, Princeton, 1955.

7. Warburton, G.B. The vibration of rectangular plates, *Proceedings of the Institution of Mechanical Engineers,* Vol.168, 371-384, 1954.

8. Thomas, D.L., J.M. Wilson and R.R. Wilson, Timoshenko beam finite elements, *Journal of Sound and Vibration,* Vol.31, No.3, 315-330, 1973.

## Appendix 1

## In-plane Plate Element

The degrees of freedom are in the order $(u_1, v_1, u_2, v_2, u_3, v_3, u_4, v_4)$, where the nodes are numbered as shown in Figure 1.

mass matrix

$$\frac{\rho abh}{36}\begin{bmatrix} 4 & & & & & & & \\ 0 & 4 & & & & & & \\ 2 & 0 & 4 & & & & & \\ 0 & 2 & 0 & 4 & & & & \\ 2 & 0 & 1 & 0 & 4 & & & \\ 0 & 2 & 0 & 1 & 0 & 4 & & \\ 1 & 0 & 2 & 0 & 2 & 0 & 4 & \\ 0 & 1 & 0 & 2 & 0 & 2 & 0 & 4 \end{bmatrix}$$

stiffness matrix

$$\frac{Eh}{180(1-\nu^2)}\begin{bmatrix} 2a_1 & & & & & & & \\ a_4 & 2a_2 & & & & & & \\ a_5 & -a_3 & 2a_1 & & & & & \\ a_3 & a_7 & -a_4 & 2a_2 & & & & \\ a_8 & a_3 & -a_1 & a_4 & 2a_1 & & & \\ -a_3 & a_6 & a_4 & -a_2 & -a_4 & 2a_2 & & \\ -a_1 & -a_4 & a_8 & -a_3 & a_5 & a_3 & 2a_1 & \\ -a_4 & -a_2 & a_3 & a_6 & -a_3 & a_7 & a_4 & 2a_2 \end{bmatrix}$$

$$a_1 = \frac{60b}{a} + \frac{30(1-\nu)a}{b}, \quad a_2 = \frac{60a}{b} + \frac{30(1-\nu)b}{a}, \quad a_3 = \frac{45(1-3\nu)}{2},$$

$$a_4 = \frac{45(1+\nu)}{2}, \quad a_5 = \frac{30b}{a} - \frac{30(1-\nu)a}{b}, \quad a_6 = \frac{30a}{b} - \frac{30(1-\nu)b}{a},$$

$$a_7 = \frac{-60a}{b} + \frac{15(1-\nu)b}{a}, \quad a_8 = \frac{-60b}{a} + \frac{15(1-\nu)a}{b}$$

where $E$ is Young's modulus

$\rho$ is density

$\nu$ is Poisson's ratio

$a$ and $b$ are element length and breadth

$h$ is thickness.

## Appendix 2

### Beam Element

The degrees of freedom are in the order

$$\{ w_1,\ \left(\frac{dw}{dx}\right)_1,\ v_1,\ \left(\frac{dv}{dx}\right)_1,\ u_1,\ \theta_1,\ w_2,\ \left(\frac{dw}{dx}\right)_2,\ v_2,\ \left(\frac{dv}{dx}\right)_2,\ u_2,\ \theta_2 \}$$

Mass matrix

$$\frac{\rho A a}{420}\begin{bmatrix}
156 & & & & & & & & & & & \\
22a & 4a^2 & & & & & & & & & & \\
0 & 0 & 156 & & & & & & & & & \\
0 & 0 & 22a & 4a^2 & & & & & & & & \\
0 & 0 & 0 & 0 & 140 & & & & & & & \\
0 & 0 & 0 & 0 & 0 & \frac{140 I_p}{A} & & & & & & \\
54 & 13a & 0 & 0 & 0 & 0 & 156 & & & & & \\
-13a & -3a^2 & 0 & 0 & 0 & 0 & 22a & 4a^2 & & & & \\
0 & 0 & 54 & 13a & 0 & 0 & 0 & 0 & 156 & & & \\
0 & 0 & -13a & -3a^2 & 0 & 0 & 0 & 0 & 22a & 4a^2 & & \\
0 & 0 & 0 & 70 & 70 & 0 & 0 & 0 & 0 & 0 & 140 & \\
0 & 0 & 0 & 0 & 0 & \frac{70 I_p}{A} & 0 & 0 & 0 & 0 & 0 & \frac{140 I_p}{A}
\end{bmatrix}$$

where $I_p$ is the polar moment of inertia

$A$ is cross-sectional area

$\rho$ is density

$a$ is element length

Stiffness matrix

$$\frac{E}{a^3}\begin{bmatrix}
12I_z \\
6aI_z & 4a^2I_z \\
0 & 0 & 12I_y \\
0 & 0 & 6aI_y & 4a^2I_y \\
0 & 0 & 0 & 0 & a^2A \\
0 & 0 & 0 & 0 & 0 & \frac{a^2GJ}{E} \\
-12I_z & -6aI_z & 0 & 0 & 0 & 0 & 12I_z \\
6aI_z & 2a^2I_z & 0 & 0 & 0 & 0 & 6aI_z & 4a^2I_z \\
0 & 0 & -12I_y & -6aI_y & 0 & 0 & 0 & 0 & 12I_y \\
0 & 0 & 6aI_y & 2a^2I_y & 0 & 0 & 0 & 0 & 6aI_y & 4a^2I_y \\
0 & 0 & 0 & 0 & -a^2A & 0 & 0 & 0 & 0 & 0 & a^2A \\
0 & 0 & 0 & 0 & 0 & \frac{-a^2GJ}{E} & 0 & 0 & 0 & 0 & 0 & \frac{a^2GJ}{E}
\end{bmatrix}$$

where $E$ is Young's modulus

$G$ is the shear mcdulus

$A$ is cross-sectional area

$a$ is element length

$J$ is polar 2nd moment of area

$I_z, I_y$ are the 2nd moments of area about the relevant axes.

CHAPTER 9

# TRANSIENT RESPONSE OF STRUCTURES

by

J. Wilson

## 1. Introduction

A structural system undergoes transient vibration when the system is subject to a change from one steady state of vibration to another. Strictly speaking, the term 'transient' should be applied to changes lasting a brief time only. However, it is often applied to a continuous changing situation for an indefinite period of time.

It has been demonstrated earlier that a structure will settle down to a steady state of vibration, when under the action of a constant amplitude harmonic force (or set of forces, each of which is characterised by the same frequency), such that all particles in the structure have the same frequency of vibration and a phase angle which depends on the nature of the damping. When the frequency of the harmonic force coincides with a natural frequency of the structure (or closely approaches it), then resonance occurs and the amplitude of vibration may become very large.

When the force is not steady-state harmonic but is repetitive, the force can be split into component steady state forces with different frequencies by means of a Fourier series. Each component is treated separately as a steady state force, giving rise to a steady state response. The total response can be obtained by summing all the response components. An example of a repetitive force would be found in the forces on the rubber mounting of a motor car engine running at constant speed. In certain cases, the vibration resulting from a transient force may be much more severe than that occurring at steady state. If we return to the motor car example and take our foot off the accelerator so that the engine slows down suddenly to idling speed, the engine may undergo large vibrations on its mountings if it comes near to stalling.

## 2. Transient Response Without Damping

Consider a multi-degree of freedom system under the action of a set of forces varying continuously with time and of an irregular nature. For the general case we may take the discrete model shown in Figure 1.

The equations of motion may be written as

$$m_1 \ddot{u}_1 + (k_1+k_2)u_1 - k_2u_2 = p_1(t) \tag{1a}$$

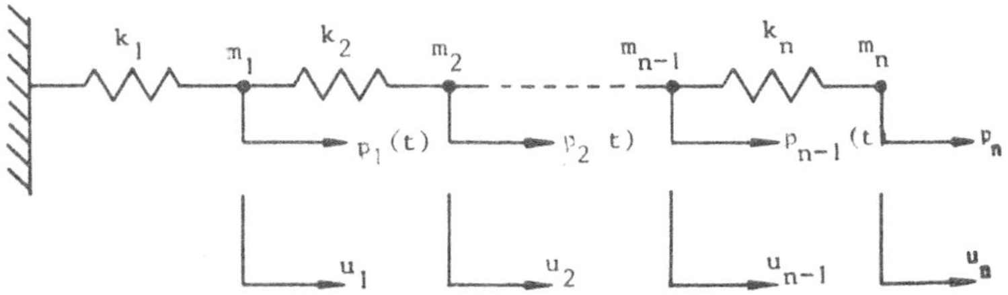


**Figure 1.**

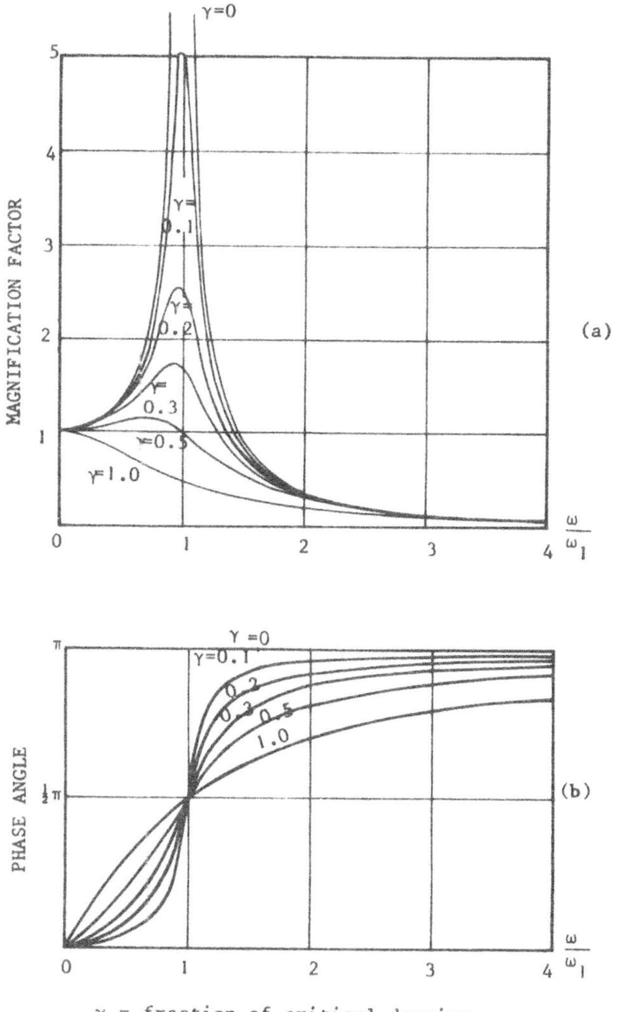


γ = fraction of critical damping

Figure 2 **Steady State Response**

$$m_j \ddot{u}_j + (k_j+k_{j+1})u_j - k_j u_{j-1} - k_{j+1}u_{j+1} = p_j(t) \quad (j = 2,n-1) \tag{1b}$$

$$m_n \ddot{u}_n + k_n u_n - k_n u_{n-1} = p_n(t) \tag{1c}$$

They may also be expressed in matrix form as

$$\underset{(n\times n)}{\tilde{M}}\,\underset{(n\times 1)}{\ddot{\tilde{U}}} + \underset{(n\times n)}{\tilde{K}}\,\underset{(n\times 1)}{\tilde{U}} = \underset{(n\times 1)}{\tilde{P}}(t)$$

where $\tilde{K}$, $\tilde{M}$ and $\tilde{U}$ are respectively the stiffness matrix, mass matrix and displacement vector. $\underset{(n\times 1)}{\tilde{P}}(t)$ is the force vector

$$\{ p_1(t) \quad p_2(t) \quad \ldots \quad p_j(t) \quad \ldots \quad p_n(t) \}$$

In general the actual loading on the structure will not be concentrated at discrete points. However if we know the actual distributed loading we can always find a statically equivalent loading system which is discrete.

As in Chapter 3, we can carry out a coordinate transformation from the original displacement vector, $\underset{(n\times 1)}{\tilde{U}}$, to a normal coordinate vector $\underset{(n\times 1)}{\tilde{q}}$ using equation (23) of Chapter 3

$$\underset{(n\times 1)}{\tilde{U}} = \underset{(n\times n)}{\tilde{Z}}\,\underset{(n\times 1)}{\tilde{q}} \tag{2}$$

where $\underset{(n\times n)}{\tilde{Z}}$ is the model matrix formed by the n normalized mode shape vectors, $\underset{(n\times 1)}{\tilde{z}_r}$.

Thus
$$\underset{(n\times n)}{\tilde{M}}\,\underset{(n\times n)}{\tilde{Z}}\,\underset{(n\times 1)}{\ddot{\tilde{q}}} + \underset{(n\times n)}{\tilde{K}}\,\underset{(n\times n)}{\tilde{Z}}\,\underset{(n\times 1)}{\tilde{q}} = \underset{(n\times 1)}{\tilde{P}}(t) \tag{3}$$

If we premultiply equation (3) by $\underset{(n\times n)}{\tilde{Z}^T}$ we find

$$(\tilde{Z}^T \tilde{M} \tilde{Z})\,\ddot{\tilde{q}} + (\tilde{Z}^T \tilde{K} \tilde{Z})\,\tilde{q} = \tilde{Z}^T \tilde{P}(t) \tag{4}$$

Again from Chapter 3, $\tilde{Z}^T \tilde{M} \tilde{Z}$ is the unit matrix (by definition of $\tilde{Z}$) and $\tilde{Z}^T \tilde{K} \tilde{Z}$ is the matrix of natural frequencies $\tilde{\Omega}$. Thus equation (1) can be uncoupled into a set of independent equations of the type

$$\ddot{q}_r + \omega_r^2 q_r = z_{1r} p_1(t) + z_{2r} p_2(t) \ldots + z_{nr} p_n(t) \tag{5}$$

by use of the transformation of coordinates given in equation (2).

The solution of equations (5) for $q_r(t)$ can be carried out by any suitable method given for a one-degree of freedom system (e.g. Duhamel integral), and the dynamic response of the whole system in terms of the original coordinates by using the transformation, $\underset{\sim}{U} = \underset{\sim}{Z}\,\underset{\sim}{q}$ again. Alternatively a numerical method of solution may be employed. These methods will be discussed later.

## 3. Damping

### 3.1 Introduction

So far we have not considered the effect of the amount or nature of damping on the response of the structure. If we keep to the strict definition of transient, i.e. very brief changes, and if we assume that the damping is small, there is little difference between the undamped transient response and the damped transient response. However if we consider the term transient in the broadest sense of a continuously changing irregular force and response then it may become necessary to introduce the effects of damping. We have seen previously how the steady state behaviour of a one-degree-of-freedom structural system may be considerably modified by the presence of damping. The amount of modification depends upon the extent and type of damping present.

Damping always exists in any structure although it is generally slight when compared to the levels of damping found in mechanical systems. It may be so slight as to render the response of the structure scarcely different from that predicted for the undamped system. If we look at Figure 2 which shows the response of a viscously damped one-degree-of-freedom system to a constant forcing function with varying amounts of damping, we find that away from the natural frequency of the system, there is very little difference in response for a large range of damping factors. The same will be true of a multi-degree of freedom system, although of course there is much more chance of encroaching on a natural frequency in this case. In fact it is only when the exciting force has a frequency close to a natural frequency that we need to take notice of damping.

### 3.2 Nature of Damping

Damping is basically a dissipation of energy which occurs in vibrating systems. How can energy be dissipated in structures ? There are three ways in which energy is dissipated.

(i) Energy dissipated within the materials of construction. We will call this *material damping*. This varies from material to material. It can be small for most structural steels and some reinforced concrete, although materials like timber laminates may possess high damping.

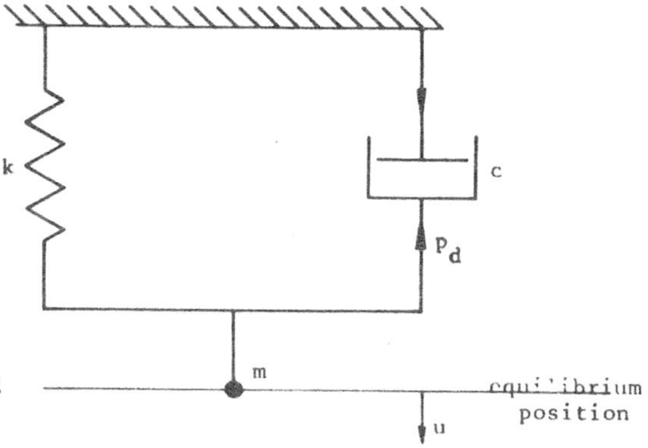


Figure 3a
Viscous damping

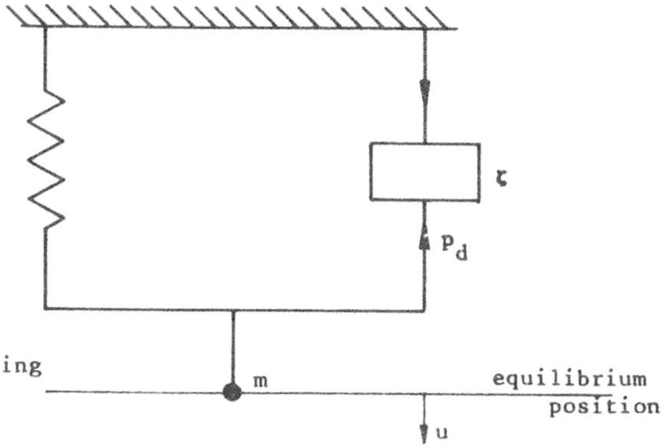


Figure 3b
Hysteretic damping

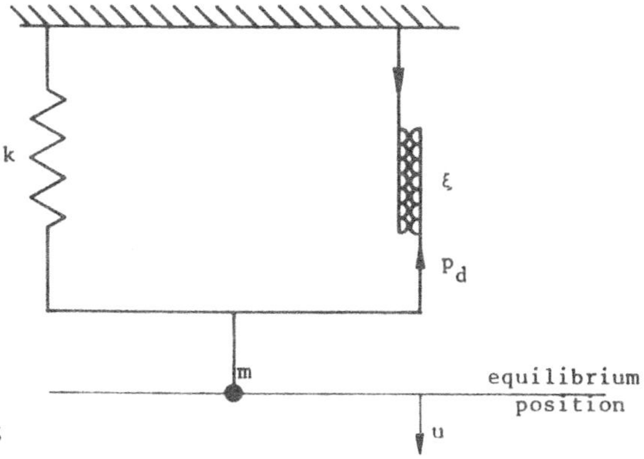


Figure 3c
Coulomb damping

(ii) Energy dissipated at structural discontinuities, e.g. bolted and riveted connections, construction joints in reinforced concrete structures. This we will term *discontinuity damping*. Whereas material damping can be measured and predicted by testing, it is much more difficult to predict the level of discontinuity damping. It may vary greatly even for structures which are nominally identical.

(iii) Energy dissipated by the structures's environment e.g. air and water resistance, and foundations. Again this is difficult to predict. Dissipation of energy in the foundations occurs even when the foundation material is linear elastic because of the propagation of stress waves through the foundation.

## 3.3 Representation of Damping

Damping from all these sources is conventionally represented in three ways, not necessarily specifically related to a particular method of energy dissipation listed above.

(a) Viscous damping
(b) Coulomb damping
(c) Hysteretic damping

It must be stressed that these are only ways of *representing* damping. They do not imply a mechanism for damping.

(a) Viscous damping, analogous to the damping produced by motion of fluids, is defined such that the damping force is proportional to a velocity. In the viscously damped system shown in Figure 3a,

$$f_d = - c \dot{u} \qquad (6)$$

where $c$ is the viscous constant
$\dot{u}$ is the velocity of mass
$f_d$ is the damping

The minus sign shows that the force on the mass acts in the opposite direction to the velocity of the body.

For a freely vibrating single degree of freedom system damped viscously as represented by Figure 3a, the equation of motion for this case becomes

(a) $$m \ddot{u} + c \dot{u} + k u = C \qquad (7)$$

We have already seen the solution for equation (7) is

$$u = Ae^{-\frac{c}{2m}t} \sin(\omega_d t + \alpha) \tag{8a}$$

where $\omega_d = \sqrt{\frac{k}{m} - (\frac{c}{2m})^2}$, the damped natural frequency.

By writing $\omega_1 = \sqrt{\frac{k}{m}}$ where $\omega_1$ is the natural frequency

$$u = Ae^{-\frac{c\omega_1^2}{2k}t} \left[ \sin \{\omega_1\sqrt{1 - (\frac{c\omega_1}{2k})^2}\; t + \alpha\} \right] \tag{8b}$$

where A and $\alpha$ are constants depending on the initial conditions. The logarithmic decrement, $\delta$, for the free vibration is defined by

$$\delta = \log_e (\frac{a_n}{a_{n+1}}) = (\frac{cT_d}{2m}) = \frac{\pi c}{m\omega_d} \simeq \frac{\pi c}{m\omega_1} = \frac{\pi c \omega_1}{k} \tag{9}$$

where $T_d$ is the damped natural period.

If we plot the form of the above curve we find that the motion is oscillatory with exponentially decreasing amplitude such that logarithm of the ratio between any two successive amplitudes is constant.

In general it can be shown that the energy dissipated per cycle when the system is steadily excited by imposed forcing function of frequency $\Omega$ such that the amplitude is a, is given by

$$\Delta U = \pi\ c\ \Omega\ a^2 \tag{10}$$

Hence energy loss increases as square of amplitude and is proportional to the exciting frequency and the dashpot constant.

(b) Coulomb or frictional damping can be regarded as existing when the damping force is a constant (depending only on the normal reaction) and opposes the motion of the body

$$f_d = -\ \xi\ (\text{sign}\ \dot{u}) = -\ \xi\ \frac{\dot{u}}{|\dot{u}|} \tag{11a}$$

where $\xi$ is a constant.

Thus the equation of free vibration for a single degree of freedom becomes

$$m\ddot{u} + \xi \frac{\dot{u}}{|\dot{u}|} + ku = 0 \tag{11b}$$

In the third case the solution for the free vibration for equation (7c) is

$$u = (C + \frac{\xi}{k}) \sin(\omega_n t + \gamma) - \frac{\xi}{k} \quad \text{with } \frac{du}{dt} \text{ positive} \tag{12a}$$

$$u = (C - \frac{\xi}{k}) \sin(\omega_n t + \gamma) + \frac{\xi}{k} \quad \text{with } \frac{du}{dt} \text{ negative} \tag{12b}$$

If we assume that our system starts off displaced from the equilibrium position by an amount C with zero initial velocity, the displacement will look like that shown in Figure 6. The envelope of successive amplitudes is a straight line and the motion does cease after a finite time. The frequency of the damped motion is not altered by the presence of frictional damping as in the previous case.

(c) Hysteretic Damping. This is sometimes called 'structural' damping in some text books. It was noticed that for some structures when excited by steady amplitude forces, the measured energy dissipated per cycle (or the work done by the applied forces) was independent of the frequency of the exciting force. (Compare the case of viscous damping when $\Delta u = \pi c \Omega a^2$).

In order to model this behaviour the term hysteretic damping (reference [1]) was invented such that

$$\Delta U = \pi h a^2 \tag{13}$$

where $h$ is the hysteretic damping constant, and $a$ is the amplitude of motion.

Now this definition of hysteretic damping happens to coincide with the definition of 'structural' damping as stated by Clough and Penzien [2] for steady state excitation.

Clough and Penzien [2] define structural damping for a single degree of freedom system as being such that the damping force is proportional to displacement and opposes the motion:

i.e.

$$f_d = - h|u| \frac{\dot{u}}{|\dot{u}|} \tag{14}$$

Under any conditions other than steady state excitation, this definition is somewhat dubious. At steady state, it gives the same result as hysteretic damping, namely,

$$\Delta U = \pi h a^2$$

The problem with this definition of hysteretic damping is that it can be shown to flout the fundamental rule of causality, (see reference [3]), i.e. hysteretic

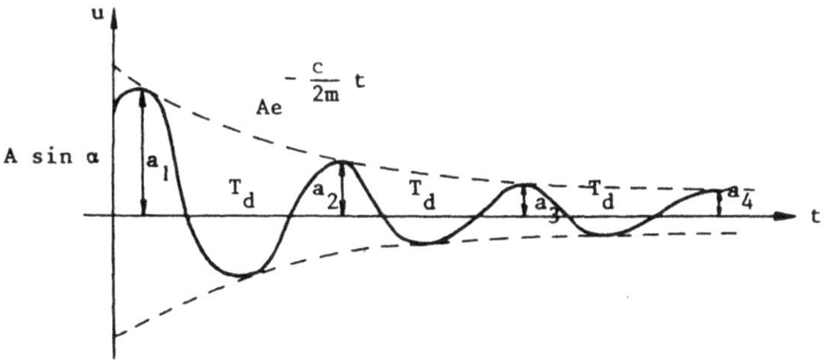


Figure 4

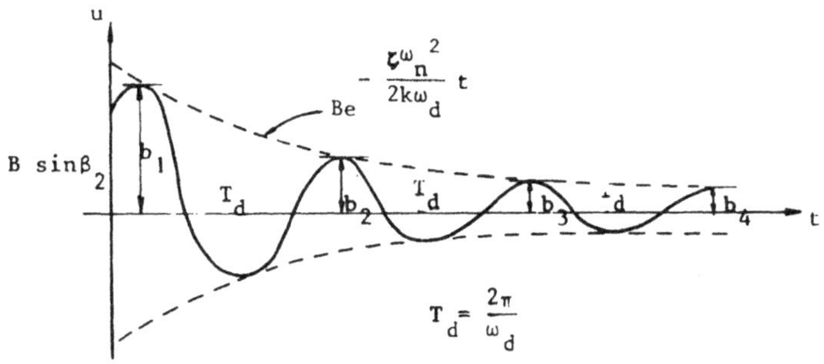


Figure 5

damping implies that the present state of system depends not only on all its past states but also on all future states. Some interesting philosophical problems then arise. However, for steady state vibration, the state of the system is steady for all time and hence present, future and past states are all known. Hence this definition can be used for this one condition.

## 4. Damped Transient Response

### 4.1 Uncoupling the Equations of Motion

We will now consider the equations of motion for a discrete n-degree of freedom system when viscous damping is present. (See Figure 6).

These may be written

$$m_1\ddot{u}_1 + (c_1+c_2)\dot{u}_1 - c_2\dot{u}_2 + (k_1+k_2)u_1 - k_2u_2 = p_1(t)$$

$$m_j\ddot{u}_j + (c_j+c_{j+1})\dot{u}_j - c_j\dot{u}_{j-1} - c_{j+1}\dot{u}_{j+1}$$

$$+ (k_j+k_{j+1})u_j - k_ju_{j-1} - k_{j+1}u_{j+1} = p_j(t) \qquad (j = 2,n-1)$$

$$m_n\ddot{u}_n + c_n\dot{u}_n - c_n\dot{u}_{n-1} + k_nu_n - k_nu_{n-1} = p_n(t) \qquad (15)$$

The equations can be written in matrix form

$$\underset{\sim}{M}\,\ddot{\underset{\sim}{U}} + \underset{\sim}{C}\,\dot{\underset{\sim}{U}} + \underset{\sim}{K}\,\underset{\sim}{U} = \underset{\sim}{P} \qquad (16)$$

where $\dot{U}$ is the velocity vector

$$\{\, \dot{u}_1 \quad \dot{u}_2 \quad \ldots \quad \dot{u}_j \quad \ldots \quad \dot{u}_n \,\}$$

and $\underset{\sim}{C}$ is the damping matrix (which in this case has a similar form to the stiffness matrix). Moreover if each damping constant has the form

$$c_j = \lambda_k\, k_j + \lambda_m\, m_j \qquad (j = 1,n) \qquad (17a)$$

then the damping matrix can be written as

$$\underset{\sim}{C} = \lambda_k\, \underset{\sim}{K} + \lambda_m\, \underset{\sim}{M} \qquad (17b)$$

and the damping is said to be proportional. Substituting (17b) in (16) we obtain

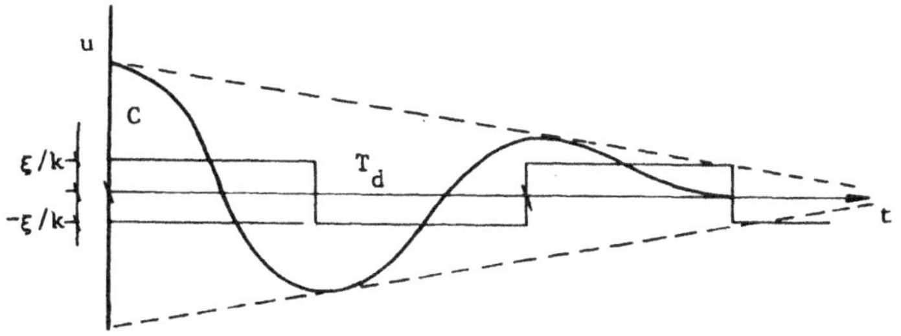


Figure 6

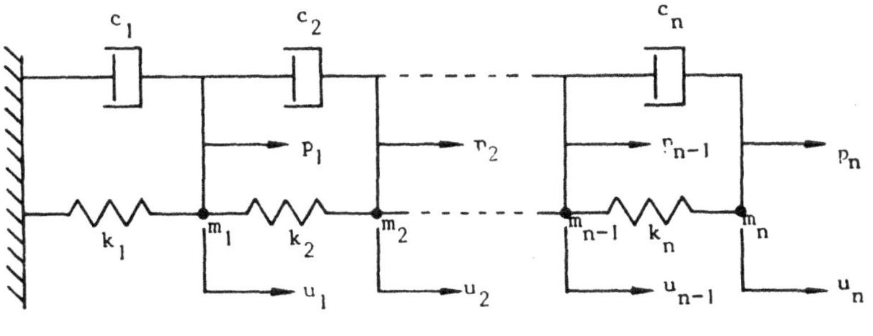


Figure 7

$$\underset{\sim}{M}\,\ddot{\underset{\sim}{U}} + (\lambda_k \underset{\sim}{K} + \lambda_m \underset{\sim}{M})\dot{\underset{\sim}{U}} + \underset{\sim}{K}\,\underset{\sim}{U} = \underset{\sim}{P} \tag{18}$$

Again a coordinate transformation can be introduced such that

$$\ddot{\underset{\sim}{U}} = \underset{\sim}{Z}\,\ddot{q} ; \quad \dot{\underset{\sim}{U}} = \underset{\sim}{Z}\,\dot{q} ; \quad \underset{\sim}{U} = \underset{\sim}{Z}\,q .$$

By pre-multiplying equation (18) by $\underset{\sim}{Z}^T$ throughout and using the transformations we have

$$(\underset{\sim}{Z}^T \underset{\sim}{M}\,\underset{\sim}{Z})\ddot{\underset{\sim}{q}} + \lambda_k(\underset{\sim}{Z}^T \underset{\sim}{K}\,\underset{\sim}{Z})\dot{\underset{\sim}{q}} + \lambda_m(\underset{\sim}{Z}^T \underset{\sim}{M}\,\underset{\sim}{Z})\dot{\underset{\sim}{q}} + (\underset{\sim}{Z}^T \underset{\sim}{K}\,\underset{\sim}{Z})\underset{\sim}{q} = \underset{\sim}{Z}^T\,\underset{\sim}{P} \tag{19}$$

leading to

$$\ddot{\underset{\sim}{q}} + (\lambda_k\,\underset{\sim}{\Omega} + \lambda_m \underset{\sim}{I})\dot{\underset{\sim}{q}} + \underset{\sim}{\Omega}\,\underset{\sim}{q} = \underset{\sim}{Z}^T\,\underset{\sim}{P} = \underset{\sim}{F}(t) \tag{20}$$

The equations are again uncoupled so that

$$\ddot{q}_r + (\lambda_k\,\omega_r^{\,2} + \lambda_m)\dot{q}_r + \omega_r^{\,2}\,q_r = f_r(t) \tag{21}$$

Again we can solve each equation for $q_r$ by a suitable method and obtain the solution in terms of the original coordinate system by writing

$$\underset{\sim}{U} = \underset{\sim}{Z}\,\underset{\sim}{q}$$

Thus if we have a structure for which the damping is viscous and the matrix of damping constants is proportional to the stiffness matrix and/or mass matrix, the equations of motion can be uncoupled in terms of the normal mode shapes.

When the damping is not proportional the equations do not uncouple in terms of the normal modes. However the equations of motion can be uncoupled by use of a complex transformation matrix

$$\underset{\sim}{U} = \underset{\sim}{Z}^{*}\,\underset{\sim}{q}^{*}$$

where $\underset{\sim}{Z}^{*}$ is a complex matrix and $\underset{\sim}{q}^{*}$ is a complex vector. (See reference [2]).

4.2 Free Damped Vibrations

For free damped vibrations we may use equation (21) with $f_r(t) = 0$ for all values of r.

$$\ddot{q}_r + (\lambda_k\,\omega_r^{\,2} + \lambda_m)\dot{q}_r + \omega_r^{\,2}\,q_r = 0 \tag{22}$$

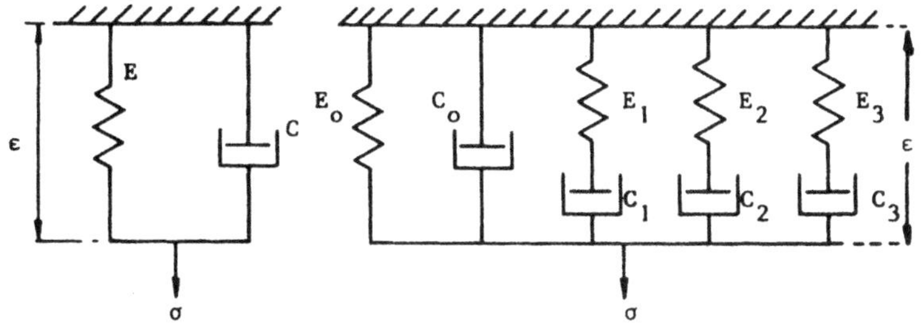


Kelvin Model

General Maxwell Model

Figure 8a

Figure 8b

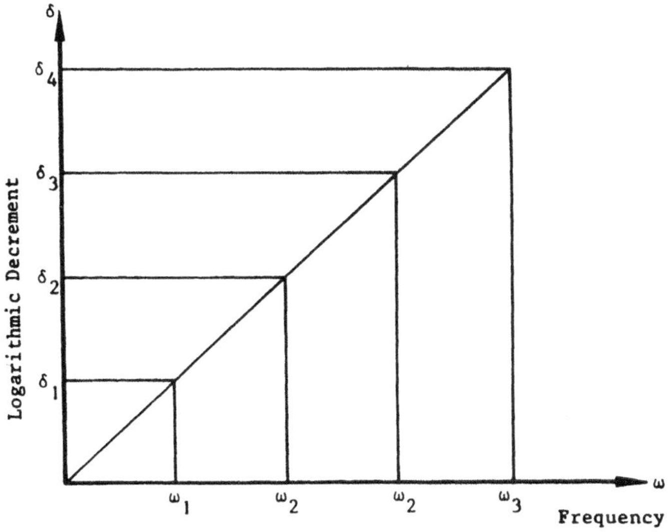


Figure 9

Let us take $\lambda_k \omega_r^2 + \lambda_m = \lambda_r$ (23)

(a constant), then equation (22) becomes

$$\ddot{q}_r + \lambda_r \dot{q}_r + \omega_r^2 q_r = 0 \tag{24}$$

The solution to this is given by equations (8). The logarithmic decrement for free vibration in the $r^{th}$ mode is given by

$$\delta_r \simeq \frac{\pi \lambda_r \omega_r}{\omega_r} \qquad \text{assuming } \delta_r \ll 1 \tag{25}$$

An interesting feature emerges from this equation. If the damping is assumed to be proportional to the stiffness only (i.e. $\lambda_m = 0$ in equation (23)), then

$$\frac{\lambda_r}{\omega_r} = \lambda_k \quad \text{and} \quad \delta_r \simeq \pi \lambda_k \omega_r \tag{26}$$

This form of damping results when the material behaviour can be represented by a Kelvin model consisting of a spring and a dashpot in parallel (see Figure 8a) and the logarithmic decrements are proportional to frequency.

If logarithmic decrements for different modes of vibration are found to lie on a straight line when plotted against the frequencies, we can say that the material of which the structure is made is Kelvin. The damping may not be *actually* attributable to the material source, but if it behaves so it can be treated in this manner.

Suppose now that a structure built of a Kelvin material has natural frequencies $\omega_1$, $\omega_2$, $\omega_3$ etc. and two or more of these natural frequencies coincide. Then the logarithmic decrements for the modes of vibration associated with these frequencies will be the same. Also if two different structures composed of the same material, happen to have a mutual natural frequency, then the logarithmic decrements will be equal. (This statement assumes that only material damping is present). Furthermore, for any structure composed of the same Kelvin material, the logarithmic decrement can always be found from one curve, (See Figure 8).

If we take the Kelvin model of Figure 7a and add sets of springs and dashpots in series as indicated in Figure 7b, we obtain a general Maxwell model. This model may be regarded as approximating, in the limit, the behaviour of a general linear viscoelastic material.

For this type of model it can again be shown that the equations of motion will uncouple. As a consequence of this, for a given material model we can plot $\delta$

against $\omega$ and obtain a curve of the type shown in Figure 9. We can again say that if several different structures have a common natural frequency, then the logarithmic decrements for free vibration in the modes associated with that frequency will be equal.

This fact may be used to predict logarithmic decrements of structures, built in reinforced concrete, or any material which behaves in a viscoelastic manner. (Note that concrete may be regarded as a hard viscoelastic material because it creeps). The assumption inherent in this argument is that material damping is the predominant source of damping. In practice what it means is that we can measure the logarithmic decrements of one structure, plot a graph and predict logarithmic decrements for other structures composed of the same material *provided we take account of natural frequencies.*

## 4.3 Example

### Logarithmic decrements of large reinforced chimneys

Figure 10 shows a plot of $\delta$ against $\omega$ for several large reinforced concrete multi-flue chimneys constructed at power stations throughout England and Wales. The data is obtained from reference [4] . You will see from the graph that a straight line has been drawn in through the points. (Note that the abscissae are plotted on a log scale). Not all of the points fall on this line, but remember that in each case the concrete mix was different. Moreover the Drax chimney has an artificial source of damping introduced by supporting the chimney floors on rubber bearings. If we neglect the results for the Drax chimney there remains only one point which is remote from the line. This line could be used as a basis for predicting logarithmic decrements of similar chimneys, although it might be risky to extrapolate from the line outside the range of information given.

## 5. Numerical Methods

It has been demonstrated that for undamped (or proportionally damped) structural systems, the equations of motion governing the structural behaviour can be uncoupled by a coordinate transformation involving the modal matrix $\underset{\sim}{Z}$. Thus we can treat each uncoupled equation as a single degree of freedom system. In many cases, the nature of the transient force is so complex that we cannot arrive at an analytical solution and must resort to numerical methods of solution.

These can in general be applied to the original coupled equations as they stand or the uncoupled set of equations. To illustrate some numerical methods, a single degree of freedom system will be considered since this is representative of the uncoupled equations arising from a multi-degree of freedom system.

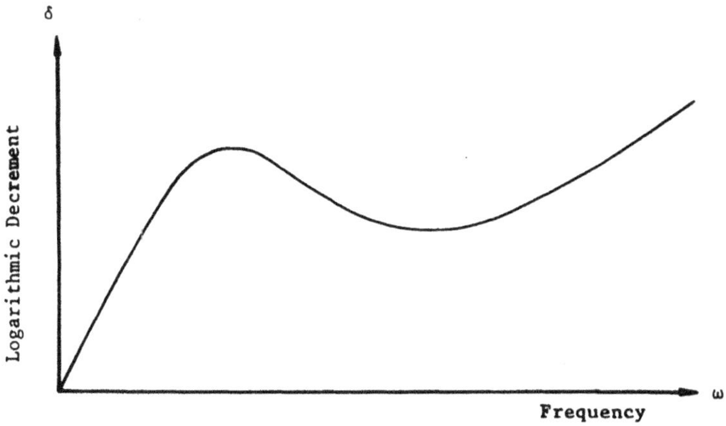


Figure 10

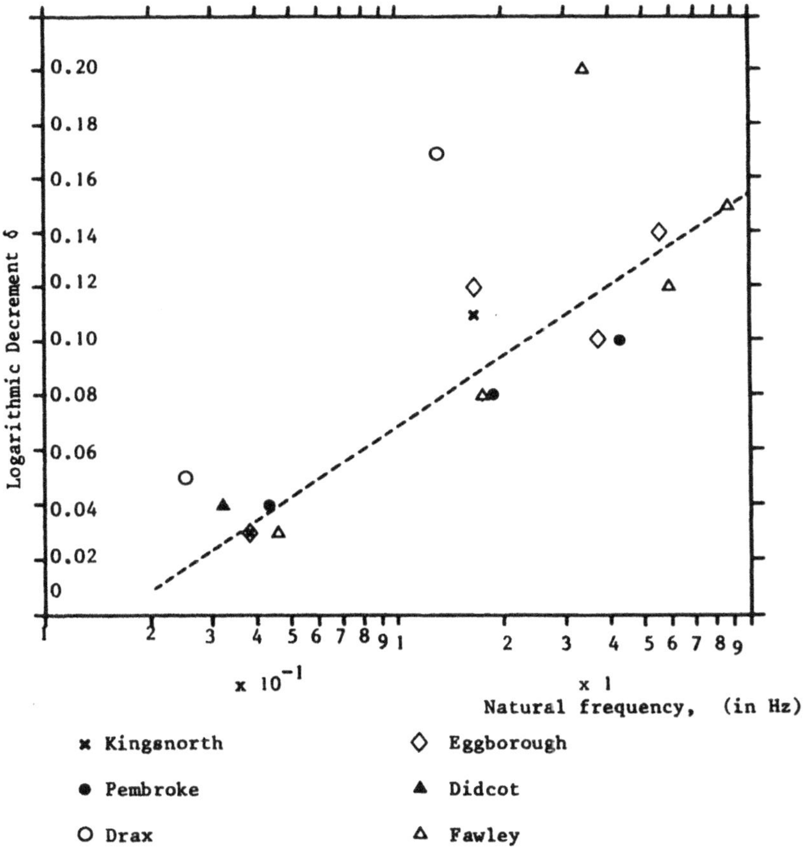


Figure 11 Natural Frequencies and Associated Logarithmic Decrements of Several Large Multi-Flue Chimneys

## 5.1 Finite Difference Method

Let us take first as an example the differential equation for a single degree of freedom undamped vibration excited by a transient (irregular) force, p(t);

$$m\,\ddot{u} + k\,u = p(t) \quad \text{where} \quad \dot{u} = 0,\ u = 0 \text{ at } t = 0$$

and write as

$$m\frac{d^2u}{dt^2} + ku = p(t) \tag{27}$$

Let us consider how p and u might vary at fixed steps in time, h, as shown in Figure 11.

Suppose that we wish to approximate the velocity of the body in any interval h, we would have (referring to Figure 12) that

$$\dot{u}_{(j+\frac{1}{2})} \simeq \frac{(u_{j+1} - u_j)}{h}$$

Similarly

$$\dot{u}_{(j-\frac{1}{2})} \simeq \frac{u_j - u_{j-1}}{h}$$

Now if we want the acceleration at time t = jh

$$\ddot{u}_j = \frac{d}{dt}(\dot{u})_j \simeq \frac{\dot{u}_{(j+\frac{1}{2})h} - \dot{u}_{(j-\frac{1}{2})h}}{h}$$

$$= \frac{u_{j+1} - 2u_j + u_{j-1}}{h} \tag{28}$$

Substituting this in equation (27) for the $j^{th}$ time step we have

$$m\left[\frac{u_{j+1} - 2u_j + u_{j-1}}{h^2}\right] + k\,u_j = p_j$$

$$\frac{m}{h^2}u_{j+1} + \left[k - \frac{2m}{h^2}\right]u_j + \frac{m}{h^2}u_{j-1} = p_j$$

Multiply by $\frac{h^2}{m}$ gives

$$u_{j+1} + \left[\frac{kh^2}{m} - 2\right]u_j + u_{j-1} = \frac{h^2}{m}\,p_j \tag{29}$$

But $\frac{k}{m} = \omega_n{}^2$ where $\omega_n$ is the natural frequency of the system. Therefore equation (29) becomes

$$u_{j+1} + (h^2\omega_n{}^2 - 2)u_j + u_{j-1} = \frac{h^2}{m}\,p_j \tag{30}$$

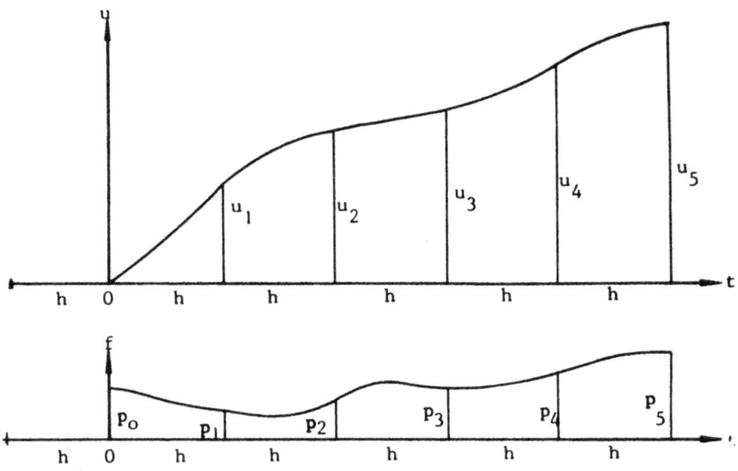


**Figure 12**

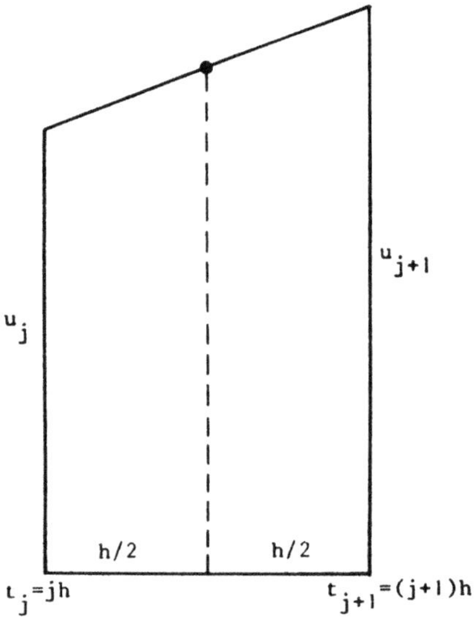


Figure 13

Example

Let us take the simple case when m = 1, $\omega_n = 1$ and h = 1. Substituting these values in equation (30) we obtain

$$u_{j+1} - u_j + u_{j-1} = p_j$$

We can rewrite this as

$$u_{j+1} = p_j + u_j - u_{j-1}$$

Let us assume that the force, p, is zero for $t < 0$ and has a constant value of 3 for $t \geq 0$. Thus $p_o = p_1 = p_2 = \text{etc.} = 3$.

Thus equation (31) for $t \leq 0$ becomes

$$u_{j+1} = 3 + u_j - u_{j-1} \tag{32}$$

We start at the first step with j = 0. At this point in time $u_j = 0$ and $u_{j-1} = 0$. Thus $u_1 = 3$. We can then go on one step in time and put j = 1. This gives

$$u_2 = 3 + u_1 + u_o = 6$$

So we can go on marching forward in time to obtain the response of our system numerically. For the first thirteen time steps, the values of u are given by applying equation (32) successively. They are listed in the following table and plotted in Figure 13.

| $u_{-1}$ | $u_o$ | $u_1$ | $u_2$ | $u_3$ | $u_4$ | $u_5$ | $u_6$ | $u_7$ | $u_8$ | $u_9$ | $u_{10}$ | $u_{11}$ | $u_{12}$ |
|---|---|---|---|---|---|---|---|---|---|---|---|---|---|
| 0 | 0 | 3 | 6 | 6 | 3 | 0 | 0 | 3 | 6 | 6 | 3 | 0 | 0 |

The periodicity is thus 6h = 6 sec. Since the natural frequency is 1 rad/sec i.e. $\frac{1}{2\pi}$ cycles/sec. the period of vibration is $2\pi$ secs = 6.28 sec. Our numerical calculation gives 6 sec. Thus the numerical analysis distorts the natural frequency. We also see that the vibration takes place about a displaced position given by u = 3 corresponding to the static displaced position, since the static displacement, $x_s$, is given by

$$x_s = \frac{p_o}{k} = \frac{p_o}{m\omega_n^2} = \frac{3}{1} = 3$$

The amplitude of the motion is seen to be slightly greater than 6 if we draw in a curve. The computed maximum displacement is thus twice the static displacement.

Now let us take a time step of ½ second. Our numerical formula becomes

$$u_{j+1} - 1.75u_j + u_{j-1} = \tfrac{1}{4}p_j \quad . \tag{33}$$

Rearranging equation (33) gives

$$u_{j+1} = 0.25p_j + 1.75u_j - u_{j-1} = 0.75 + 1.75u_j - u_{j-1} \tag{34}$$

since $p_j = 3$ for $j \geq 1$

$u_0 = 0$
$u_1 = 0.75 + 0 - 0 = 0.75$
$u_2 = 0.75 + 1.75 \times 0.75 - 0 = 2.06$
$u_3 = 0.75 + 1.75 \times 2.06 - 0.75 = 3.60$
$u_4 = 0.75 + 1.75 \times 3.60 - 2.06 = 4.98$
$u_5 = 0.75 + 1.75 \times 4.98 - 3.60 = 5.88$
$u_6 = 0.75 + 1.75 \times 5.88 - 4.98 = 6.07$
$u_7 = 0.75 + 1.75 \times 6.07 - 5.88 = 5.50$
$u_8 = 0.75 + 1.75 \times 5.50 - 6.07 = 4.31$
$u_9 = 0.75 + 1.75 \times 4.31 - 5.50 = 2.79$
$u_{10} = 0.75 + 1.75 \times 2.79 - 4.31 = 1.32$
$u_{11} = 0.75 + 1.75 \times 1.32 - 2.79 = 0.27$
$u_{12} = 0.75 + 1.75 \times 0.27 - 1.32 = -0.10$
$u_{13} = 0.75 - 1.75 \times 0.10 - 0.27 = 0.30$
$u_{14} = 0.75 + 1.75 \times 0.30 + 0.10 = 1.38$

As we take our time step smaller and smaller, so our results become more accurate. When the step is small enough we cannot distinguish between the exact solution for the problem

$$u(t) = 3(1 - \cos t) \tag{35}$$

and the approximate numerical solution.

If too large a time step is chosen, the process becomes unstable. Numerical processes do not always give solutions which agree closely with analytical. The reasons for this can be several. Sometimes if the process used is a marching one, where we know all quantities from the previous steps and need to know the current values, it may become unstable. By this we mean that any numerical error introduced in the calculation increases in magnitude until it swamps the actual solution. Sometimes an iterative process is used when the solution is expressed in the form

$$x = f(x) \quad .$$

We guess an initial value of x, $x_o$, substitute it in f(x) and work out $x_1 = f(x_o)$. We then take the new value of x, $x_1$, and work out $x_2 = f(x_1)$ until $x_{n+1} = x_n + \varepsilon$ where $\varepsilon$ is an acceptable error. In this sort of process we need convergence. Sometimes this does not happen. Fortunately in most cases by looking at the equations we can establish convergency and stability limits.

A great advantage of numerical methods is that we can easily use digital computers to work out solutions for us.

5.2 Newmark β Method

There are other numerical methods which may be used. One fairly accurate and simple method is the Newmark β method [5]. In this method we assume variations for the velocity, $\dot{u}$, and displacement, u, in the time interval h to be such that the values at beginning and end of the time step (subscripts 0 and 1 respectively) are related by equations of the form

$$\dot{u}_1 = \dot{u}_o + (1 - \gamma)h\,\ddot{u}_o + \gamma\,h\,\ddot{u}_1 \qquad (36a)$$

$$u_1 = u_o + h\,\dot{u}_o + (\tfrac{1}{2} - \beta)h^2\,\ddot{u}_o + \beta\,h^2\,\ddot{u}_1 \qquad (36b)$$

where β and γ are constants. It has been shown that unless $\gamma = \frac{1}{2}$, an error is introduced. β can have any desired value. Certain values of β and γ have physical significance, but in general this is not so. For instance $\beta = \frac{1}{2}$ and $\beta = \frac{1}{4}$ corresponds to constant accelerations in the time interval and $\beta = 1/6$ to linear acceleration. $\gamma = \frac{1}{2}$ corresponds to linear acceleration variation in the time interval.

If we apply the Newmark β method to the previous problem we have

$$m\,\ddot{u} + k\,u = p(t) \qquad \text{where} \quad p(t) = 0 \ \text{for } t < 0$$
$$p(t) = p_o \ \text{for } t > 0$$
$$u = 0,\ \dot{u} = 0 \ \text{at } t = 0$$

Then we have at the beginning of the time step

$$m\,\ddot{u}_o + k\,u_o = p_o$$

and at the end

$$m\,\ddot{u}_1 + k\,u_1 = p_1$$

Substituting for $\ddot{u}_o$ and $\ddot{u}_1$ from equations (36) we obtain

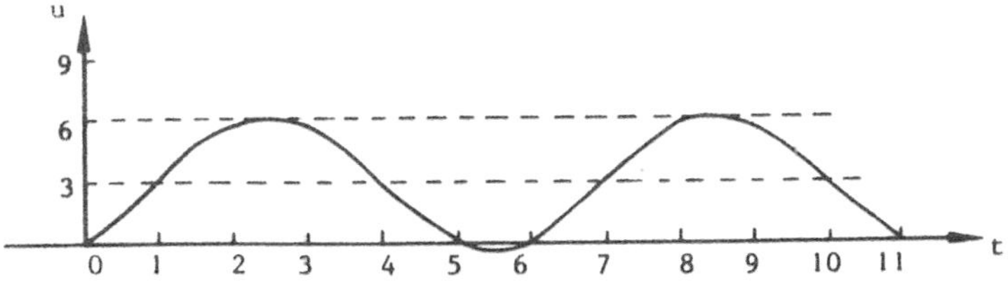


**Figure 14**

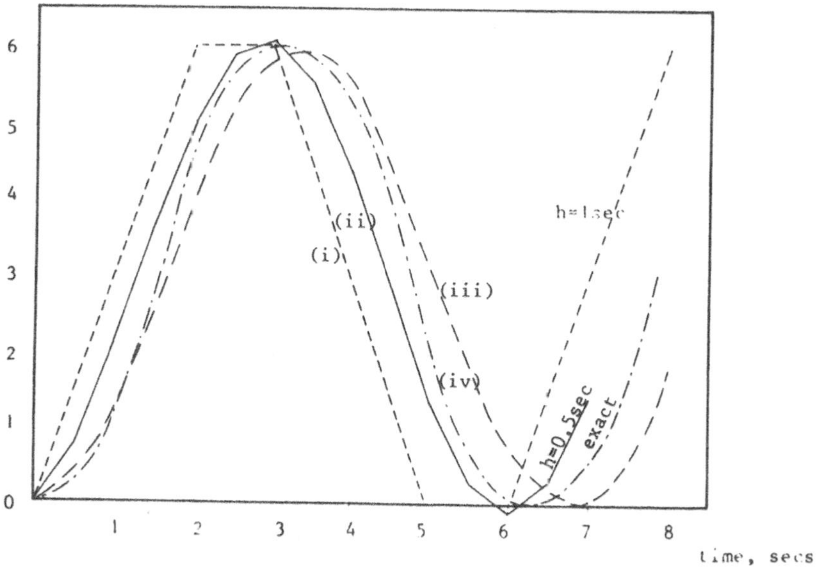


**Figure 15**

Figure 15 shows a comparison between the results obtained for the displacement, u, in the problem by four methods,

(i) Finite difference method, h = 1 sec.

(ii) Finite difference method, h = 0.5 sec.

(iii) Newmark β method, h = 1 sec.

(iv) Exact solution

$$\dot{u}_1 = \dot{u}_o + (1 - \gamma)h\left(\frac{p_o}{m} - \frac{k}{m}u_o\right) + \gamma h\left(\frac{p_1}{m} - \frac{k}{m}u_1\right) \tag{37a}$$

$$u_1 = u_o + h\dot{u}_o + (\tfrac{1}{2} - \beta)h^2\left(\frac{p_o}{m} - \frac{k}{m}u_o\right) + \beta h^2\left(\frac{p_1}{m} - \frac{k}{m}u_1\right) \tag{37b}$$

Expressing the unknown quantities at the end of the time interval in terms of the known quantities at the beginning we have (putting $\frac{k}{m} = \omega_n^2$)

$$u_1(1+\beta h^2\omega_n^2) = u_o\left[1 - (\tfrac{1}{2}-\beta)h^2\omega_n^2\right] + \dot{u}_o h + (\tfrac{1}{2}-\beta)\frac{h^2}{m}p_o + \frac{\beta h^2}{m}p_1 \tag{38a}$$

$$\dot{u}_1 + \gamma h\,\omega_n^2 u_1 = \dot{u}_o - (1-\gamma)h\omega_n^2 u_o + (1-\gamma)\frac{h}{m}p_o + \gamma\frac{h}{m}p_1 \tag{38b}$$

Thus we have

$$u_1 = \frac{1}{1+\beta h^2\omega_n^2}\left\{\left[1 - (\tfrac{1}{2}-\beta)h^2\omega_n^2\right]u_o + h\dot{u}_o + (\tfrac{1}{2}-\beta)\frac{h^2}{m}p_o + \frac{\beta h^2}{m}p_1\right\} \tag{39a}$$

$$\dot{u}_1 = \left\{-\gamma h\omega_n^2 u_1 + \dot{u}_o - (1-\gamma)h\omega_n^2 u_o + (1-\gamma)\frac{h}{m}p_o + \gamma\frac{h}{m}p_1\right\} \tag{39b}$$

As stated previously, unless $\gamma = \frac{1}{2}$ extraneous damping is introduced into the equations. Putting $\gamma = \frac{1}{2}$ in equation (39b) leads to

$$\dot{u}_1 = \left\{\dot{u}_o - \tfrac{1}{2}h\omega_n^2(u_1 + u_o) + \tfrac{1}{2}\frac{h}{m}(p_o + p_1)\right\} \tag{40}$$

Let us now carry out the calculations of the previous example with $\beta = \frac{1}{4}$, $h = 1$, $\omega_n = 1$, $m = 1$. Equations (39a) and (40) give

$$u_1 = \frac{3}{5}u_o + \frac{4}{5}\dot{u}_o + \frac{1}{5}p_o + \frac{1}{5}p_1 \tag{41a}$$

$$\dot{u}_1 = -\frac{1}{2}u_o + \dot{u}_o - \frac{1}{2}u_1 + \frac{1}{2}p_o + \frac{1}{2}p_1 \tag{41b}$$

and the values of $u$ and $\dot{u}$ for the first nine time steps ($j = 0$ to $j = 8$) are given in the following table.

| Step Number | 0 | 1 | 2 | 3 | 4 | 5 | 6 | 7 | 8 |
|---|---|---|---|---|---|---|---|---|---|
| $u$ | 0 | 1.2 | 3.84 | 5.82 | 5.51 | 3.20 | 0.73 | 0.07 | 1.76 |
| $\dot{u}$ | 0 | 2.4 | 2.88 | 1.03 | -1.63 | -2.98 | -1.95 | 0.65 | |

Figure 14 shows a comparison between the results obtained for the displacement, u, in the problem by four methods,

(i) Finite difference method, h = 1 sec.

(ii) Finite difference method, h = 0.5 sec.

(iii) Newmark β method, h = 1 sec.

(iv) Exact solution.

In addition to these methods illustrated there are many more numerical methods of solving the general differential equation of damped force vibration

$$m \ddot{u} + c \dot{u} + k u = p(t)$$

These include Runge-Kutta, Galerkin etc. All these methods will be prone to numerical instabilities etc. but by choice of a suitable time interval they can be overcome. In general the shorter the time interval the more accurate is the solution and the less likelihood there is of instability. Of course the penalty is longer time for computation.

Numerical methods have all been developed for single degree of freedom systems. They can easily (from a computer viewpoint) be modified to suit a multi-degree of freedom system. However in this case the time interval will have to be very small in order to avoid numerical difficulties with some of the higher frequency components of the vibrations. Accuracy will of course improve but the law of diminishing returns will apply in most cases.

For a structure which has proportional damping, we may uncouple the equations in terms of the normal modes and solve each equation of the type

$$m_r \ddot{q}_r + c_r \dot{q}_r + k_r q_r = p_r(t)$$

as was demonstrated previously. Finally the solutions can be recombined. The advantages of this method are that different time intervals can be used. However modern computers are so big and fast that there is really no need to uncouple the equations. It is just as easy to manipulate the matrices. For non-proportional damping this method would have to be used in any case.

References

1. Bishop, R.E.D. and D.C. Johnson, *The Mechanics of Vibrations*, Cambridge University Press, 1960, London.

2. Clough, R.W. and Penzien, J., *Dynamics of Structures*, McGraw Hill, New York, 1975.

3. Crandall, S.H., The Role of Damping in Vibration Theory, J. Sound Vibration, 11, (1), 1970, 3-18.

4. Jeary, A.P. Damping Measurements from the Dynamic Behaviour of Several Large Multi-Flue Chimneys, *Proc. Instn. Civil Engrs.*, 1974, 57, Part 2, 321-330.

5. Newmark, N.M. A Method of Computation for Structural Dynamics, *Proc. Am. Soc. Civil Engrs., J. of Mech. Eng. Div.*, 1959, EM 3, 85, 67-94.

CHAPTER 10

# MACHINE FOUNDATIONS

by

R.R. Wilson

## 1. Introduction

Many factors have to be taken into account when a foundation is designed for a machine. In particular for rotating machinery, such as pumps, blowers or turbines, a knowledge of the vibrational characteristics of a foundation is of great importance [1],[2]. In this chapter, the principles of the design of a foundation subject to dynamic loading are introduced and discussed with reference to simple mass-spring systems. The use of undamped and damped dynamic absorbers is considered and finally an analysis is described in which the finite element method is used to predict the dynamic behaviour of turbine foundations.

Three separate but closely related vibration problems can occur in the design of a machine foundation. Firstly there is the question of insulation of a machine from its surroundings. It is often important to ensure that only a small fraction of the force caused by the operation of the machine is transmitted through the foundation to the surrounding structure. For example in a hospital it is desirable that little vibration should be transmitted to the building from a compressor. The opposite situation also occurs; it may be necessary to design a foundation to protect a machine from vibrations occurring in the surroundings.

The second consideration in the design of a foundation is that the structure of the foundation must be able to withstand the amplitudes of vibration induced by the forcing from the machine. The vibration amplitudes may also be important if auxiliary equipment is mounted on the foundation or pipes supported on the structure.

Finally the influence the foundation has on the operation of the machine must be considered. The critical speeds of a machine on a flexible foundation are much lower than the critical speeds of the same machine rigidly supported.

## 2. Transmissibility of a Foundation on a Rigid Base

A single degree of freedom can be used to represent a machine and foundation on a rigid base as shown in Figure 1. The foundation is considered as a spring k and a dashpot c connecting a machine of mass m to a rigid base. Any parts of the foundation moving with the machine are included in the mass m. The operation of

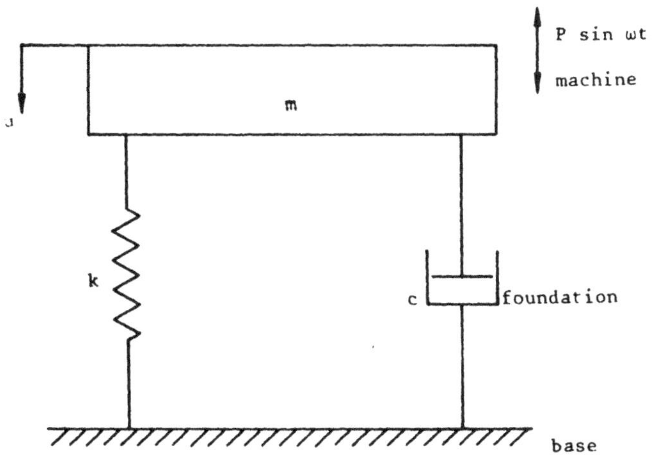


**Figure 1** Machine with Foundation on Rigid Base

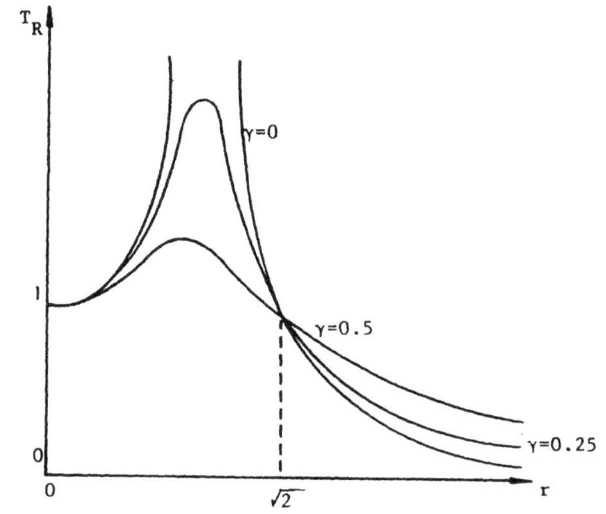


**Figure 2** Transmissibility of a Foundation on a Rigid Base.

the machine gives rise to a sinusoidally varying force P sin ωt. The equation of motion of this system is

$$m\ddot{u} + c\dot{u} + ku = P \sin \omega t \quad . \tag{1}$$

After sufficient time has elapsed for the transient component to die away, we are left with the steady state solution (as given by equation 50 in Chapter 2),

$$u = \frac{P}{\sqrt{(k-m\omega^2)^2 + \omega^2 c^2}} \sin(\omega t - \alpha) \tag{2}$$

where $$\tan \alpha = \frac{\omega c}{k-m\omega^2} \tag{3}$$

The force transferred to the supporting structure is the sum of the force in the spring (ku) and the force in the dashpot ($c\dot{u}$). Thus the amplitude of the force is given by

$$P_T = \sqrt{(ku)^2 + (c\dot{u})^2} \tag{4}$$

Substituting for u and $\dot{u}$ from equation (2)

$$P_T = \frac{P\sqrt{k^2 + c^2\omega^2}}{\sqrt{(k-m\omega^2)^2 + \omega^2 c^2}} \tag{5}$$

The *transmissibility* of a foundation is defined by

$$\text{transmissibility} = \frac{\text{transmitted force}}{\text{impressed force}} \quad . \tag{6}$$

i.e. $T_R = \frac{P_T}{P}$

$$\therefore \quad T_R = \frac{\sqrt{k^2 + c^2\omega^2}}{\sqrt{(k-m^2)^2 + \omega^2 c^2}} \tag{7}$$

The natural frequency of the system is $\omega_n = \sqrt{\frac{k}{m}}$ , and the critical damping $c_c = 2\sqrt{km}$ .

$$\text{Thus} \quad T_R = \frac{\sqrt{1 + (2\frac{c\omega}{c_c\omega_n})^2}}{\sqrt{(1 - (\frac{\omega}{\omega_n})^2)^2 + (2\frac{c\omega}{c_c\omega_n})^2}} = \frac{\sqrt{1 + 4\gamma^2 r^2}}{\sqrt{(1-r^2)^2 + 4\gamma^2 r^2}} \quad , \tag{8}$$

where $\gamma = c/c_c$ and $r = \omega/\omega_n$.

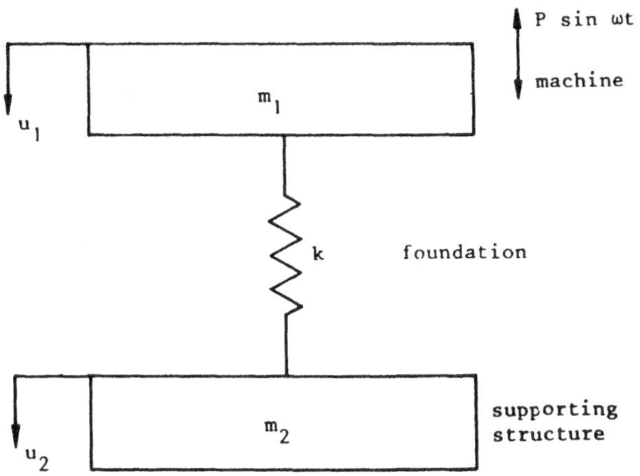


Figure 3  Machine with Foundation on Flexible Base

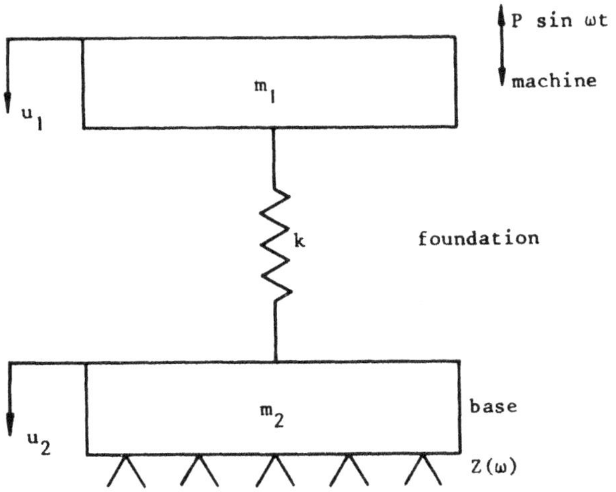


Figure 4  Mechanical Impedance of the Base

The transmissibility is a function of the ratios r and $\gamma$ as shown in Figure 2. It can be seen that for the force transmitted by the foundation to be small, r should be large, i.e. the foundation should have a natural frequency much less than the operating speed of the machine it supports.

When $\omega < \sqrt{2}\,\omega_n$ then $T_R > 1$, and the force transmitted by the foundation is greater than the force applied by the machine. In this range of operating speeds an increase in damping reduces the force transmitted. However, for operating speeds where $\omega > \sqrt{2}\,\omega_n$ then the higher the damping, the higher the force transmitted. This suggests that the damping in a machine foundation should be as low as possible to reduce the force transmitted. In practice, however, a foundation has higher resonances in addition to the fundamental resonance considered above. If the machine is operating near one of these higher resonances then it is the damping in the foundation that prevents very large forces being transmitted.

Thus, despite the result of the above analysis, it is probably not desirable to reduce the damping of a foundation in order to lower forces transmitted. Conversely, this analysis does suggest why, when faced with excessive forces being transmitted, it is often preferable to vary the stiffness of the foundation or mass of the machine rather than to attempt to increase the damping of the system.

## 3. Transmissibility of a Foundation on a Flexible Base

In the previous section it was assumed that the structure to which the foundation was connected did not move. In many situations this is not a realistic assumption. For example, with an aircraft engine mounted on the wing of the aeroplane or a turbine supported on the hull of a ship, the area surrounding the point of support will move the base of the foundation. This can be represented by the two degree of freedom system shown in Figure 3, where $m_2$ represents the mass of the supporting structure which moves with the foundation. To simplify the analysis, the damping of the system has been neglected.

The equations of motion are

$$m_1\ddot{u}_1 + k(u_1 - u_2) = P \sin \omega t \tag{9}$$

$$m_2\ddot{u}_2 - k(u_1 - u_2) = 0 \tag{10}$$

Assuming solutions of the form

$$u_1 = A_1 \sin \omega t \tag{11}$$

$$u_2 = A_2 \sin \omega t \tag{12}$$

then

$$A_1(k-m_1\omega^2) - A_2k = P \qquad (13)$$

$$-A_1k + A_2(k-m_2\omega^2) = 0 \qquad (14)$$

The natural frequencies of the system are given by the zeros of the equation

$$(k-m_1\omega^2)(k-m_2\omega^2) - k^2 = 0 \qquad (15)$$

This gives two solutions,

$$\omega_1{}^2 = 0 \quad \text{and} \quad \omega_2{}^2 = k\left[\frac{m_1 + m_2}{m_1 m_2}\right] \qquad (16)$$

The zero natural frequency arises because, since the system is not constrained, it can execute rigid body motion.

The amplitude of the steady state response is calculated by solving equations (13) and (14). Expressing $A_1$ in terms of $A_2$ from equation (14), gives

$$A_1 = A_2\,\frac{(k-m_2\omega^2)}{k} \qquad (17)$$

Substituting into equation (13),

$$A_2\left[\frac{(k-m_1\omega^2)(k-m_2\omega^2)}{k} - k\right] = P \qquad (18)$$

i.e.

$$A_2 = \frac{kP}{(k-m_1\omega^2)(k-m_2\omega^2) - k^2} \qquad (19)$$

The force transmitted to the supporting structure is $m_2\ddot{u}_2$ and hence the magnitude of this transmitted force is

$$P_T = -\omega^2\ m_2\ A_2 \qquad (20)$$

i.e.

$$P_T = \frac{-m_2\omega^2\ kP}{(k-m_1\omega^2)(k-m_2\omega^2) - k^2} \qquad (21)$$

The transmissibility of the foundation is given as

$$T_R = \frac{P_T}{P} \qquad (22)$$

Thus from equation (19),

$$T_R = \frac{-k\ m_2\ \omega^2}{(k-m_1\omega^2)(k-m_2\omega^2) - k^2} \qquad (23)$$

i.e. $$T_R = \frac{1}{\frac{m_1+m_2}{m_2} - \frac{m_1\ \omega^2}{km_2}} \tag{24}$$

$$\therefore \quad T_R = \frac{m_2}{m_1+m_2}\left[\frac{1}{1-\frac{\omega^2}{\omega_n^2}}\right] \tag{25}$$

Hence, as was shown previously for a foundation on a rigid base, we have the result that the lower the natural frequency of the system, the smaller the force transmitted.

If we now extend the analysis to the rather more realistic situation shown in Figure 4 where, instead of being completely free to move, the base of the foundation offers resistance to motion. The *mechanical impedance* $Z(\omega)$ of the base structure can be defined as the force at frequency $\omega$ required to produce a unit displacement.

i.e. $$Z(\omega) = \frac{\text{applied force}}{\text{displacement}} \tag{26}$$

The motion of the system is described by the equations

$$m_1\ddot{u}_1 + k(u_1-u_2) = P \sin \omega t \tag{27}$$

$$- k(u_1-u_2) = u_2 Z(\omega) \tag{28}$$

If the base could be represented as an unconstrained mass $m_2$, as considered previously, then $Z(\omega) = m_2\omega^2$, and equation (28) reduces to equation (10).

Equations (27) and (28) can be solved as before to give

$$A_2 = \frac{Pk}{Z(\omega)(k-m_1\omega^2) - km_1\omega^2} \tag{29}$$

The magnitude of the transmitted force is given by

$$P_T = A_2\ Z(\omega)\ , \tag{30}$$

and so the transmissibility is

$$T_R = \frac{P_T}{P} = \frac{A_2\ Z(\omega)}{P} \tag{31}$$

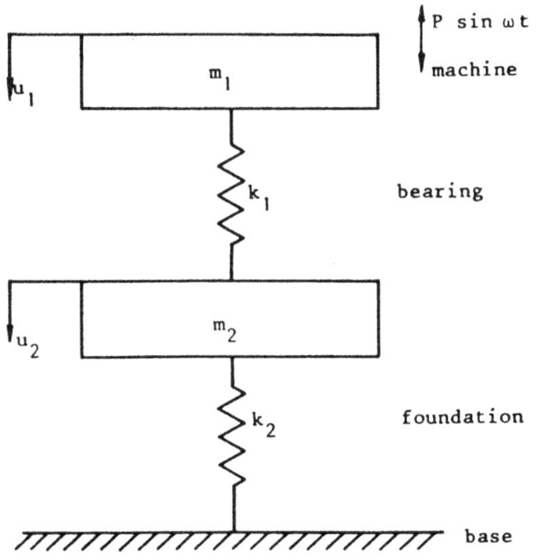


Figure 5 Machine-Bearing-Foundation System

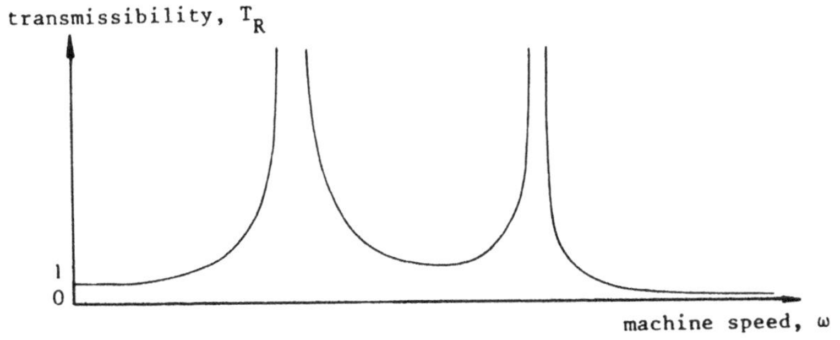


Figure 6 Variation of Transmissibility for a Machine-Bearing-Foundation System

$$\therefore \quad T_R = \frac{Z(\omega)\ k}{Z(\omega)(k-m_1\omega^2) - km_1\omega^2} \tag{32}$$

The mechanical impedance depends on the nature of the base. For example, if the foundation rests on a concrete raft on soil then the mechanical impedance can be determined in terms of a spring, mass and dashpot model of the soil, with, in general, each of the constants frequency dependent. It is also possible to find $Z(\omega)$ experimentally. The displacement produced by a vibrator applying a harmonic force can be measured as a function of the applied force.

## 4. Low Tuned and High Tuned Foundations

If a machine is flexibly mounted on its foundation through, for example, bearings, and the foundation is connected to a rigid base, then the system may be idealized as the two degree of freedom system shown in Figure 5.

The equation of motion of the system is

$$m_1\ddot{u}_1 + k_1(u_1-u_2) = P \sin \omega t \tag{33}$$

$$m_2\ddot{u}_2 - k_1(u_1-u_2) + k_2u_2 = 0 \tag{34}$$

Assuming once again solutions of the form

$$u_1 = A_1 \sin \omega t \quad \text{and} \quad u_2 = A_2 \sin \omega t \tag{35}$$

and substituting into equations (33) and (34) gives

$$A_1(k_1-m_1\omega^2) - A_2k_1 = P \tag{36}$$

$$-A_1k_1 + A_2(k_1+k_2-m_2\omega^2) = 0 \tag{37}$$

Substituting for $A_2$ from equation (36) into (37),

$$A_1\left[k_1 - m_1\omega^2 - \frac{k_1^2}{(k_1+k_2-m_2\omega^2)}\right] = P \tag{38}$$

Thus

$$A_1 = \frac{(k_1+k_2-m_2\omega^2)P}{(k_1-m_1\omega^2)(k_1+k_2-m_2\omega^2) - k_1^2} \tag{39}$$

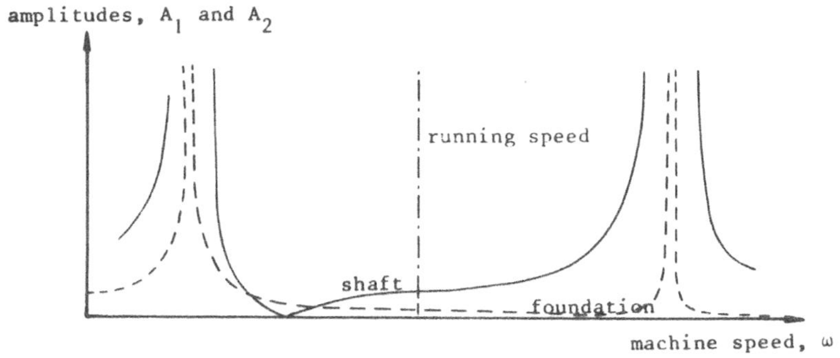


Figure 7 **Foundation and Shaft Amplitudes for a Machine on a Low Tuned Foundation**

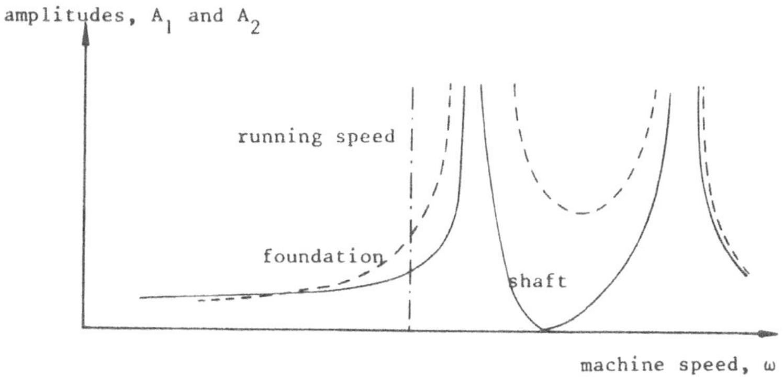


Figure 8 Foundation and Shaft Amplitudes for a Machine on a High Tuned Foundation

and

$$A_2 = \frac{k_1 P}{(k_1-m_1\omega^2)(k_1+k_2-m_2\omega^2)} \qquad (40)$$

The force transmitted to the base is $P_T = k_2A_2$ and the transmissibility $T_R$ is given by

$$T_R = \frac{P_T}{P} = \frac{k_2A_2}{P} \qquad (41)$$

$$\therefore \quad T_R = \frac{k_1k_2}{(k_1-m_1\omega^2)(k_1+k_2-m_2\omega^2)} \qquad (42)$$

The variation of the transmissibility with the frequency of the applied force is shown in Figure 6.

In designing a fixed speed machine there are two basic decisions. Firstly should the machine run above or below the first critical speed of the shaft? ($\sqrt{k_1/m_1}$ in the system considered above). Secondly should the foundation be *low tuned* or *high tuned* ? i.e. should the first natural frequency of the foundation on its own ($\sqrt{k_2/m_2}$) be below or above the running speed of the machine? [3],[4].

Table 1 summarizes the four possibilities.

| | |
|---|---|
| Low tuned foundations, low critical speed of shaft | High tuned foundations low critical speed of shaft |
| Low tuned foundations, high critical speed of shaft | High tuned foundations high critical speed of shaft |

Table 1 Design Possibilities for Fixed Speed Machine

The possibility of having a low critical speed is really only applicable to larger machines with long flexible shafts. For a particular fixed speed machine having its first critical above running speed [3], Figures 7 and 8 shows the variation of foundation and shaft amplitude with frequency for a low tuned and a high tuned foundation.

## 5. Dynamic Absorber

When a machine on a foundation is operating near its resonance, it is possible to reduce the vibration amplitudes by attaching a dynamic absorber. This can be represented by an additional mass and spring as shown in Figure 9. The equation

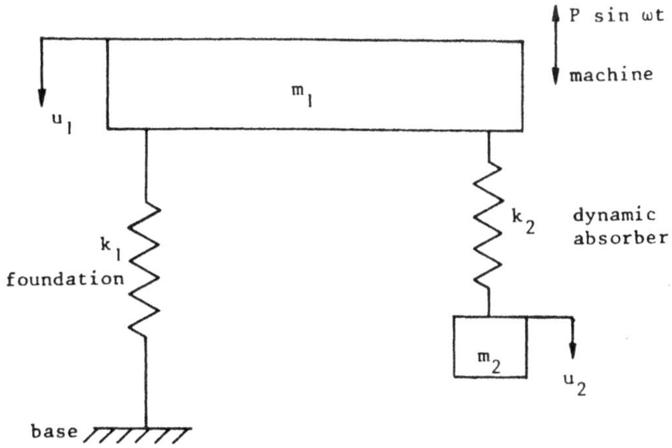


Figure 9 Undamped Dynamic Absorber

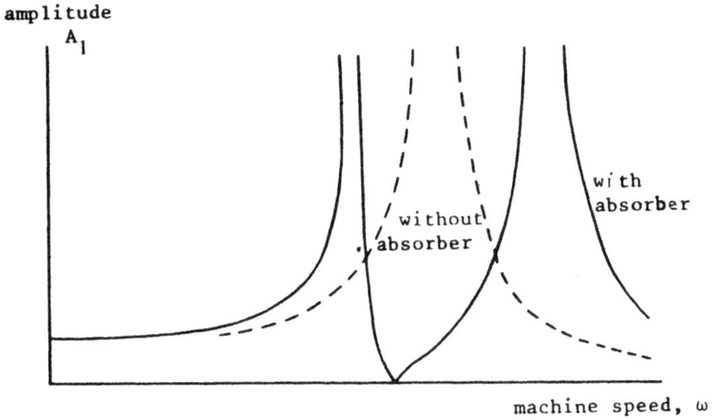


Figure 10 Effect of Undamped Dynamic Absorber on Response of Machine

of motion of this system is

$$m_1\ddot{u}_1 + k_1u_1 + k_2(u_1-u_2) = P \sin \omega t \tag{43}$$

$$m_2\ddot{u}_2 - k_2(u_1-u_2) = 0 \tag{44}$$

These equations can be solved as before to give the steady state amplitudes,

$$A_1 = \frac{(k_2-m_2\omega^2)P}{(k_1+k_2-m_1\omega^2)(k_2-m_2\omega^2) - k_2^2} \tag{45}$$

$$A_2 = \frac{k_2P}{(k_1+k_2-m_1\omega^2)(k_2-m_2\omega^2) - k_2^2} \tag{46}$$

We are interested primarily in the response of the machine at the resonant frequency of the original system, i.e. before the addition of the dynamic absorber. Substituting this frequency, $\omega = \sqrt{k_1/m_1}$, into equation (45) gives

$$A_1 = \frac{(k_2-m_2\,\frac{k_1}{m_1})P}{(k_1+k_2-k_1)(k_2-m_2\,\frac{k_1}{m_1}) - k_2^2} \tag{47}$$

$$\therefore \quad A_1 = \frac{P}{k_2}\left(1 - \frac{m_1k_2}{k_1m_2}\right) \tag{48}$$

If the dynamic absorber is designed so that

$$\frac{k_1}{m_1} = \frac{k_2}{m_2}, \tag{49}$$

then from equation (48) the amplitude of the machine at the original resonant frequency will be zero. Figure 10 shows the variation of the amplitude of vibration of the machine with and without the dynamic absorber.

## 6. Damped Dynamic Absorber

The dynamic absorber described in the previous section removes the original resonance peak in the response curve for the machine but introduces two new peaks. If it is necessary to reduce the amplitude of vibration over a range of frequencies then a damped dynamic absorber as shown in Figure 11 can be used. The equations of motion are

$$m_1\ddot{u}_1 + c(u_2-u_1) + k_1u_1 + k_2(u_2-u_1) = P \sin \omega t \tag{50}$$

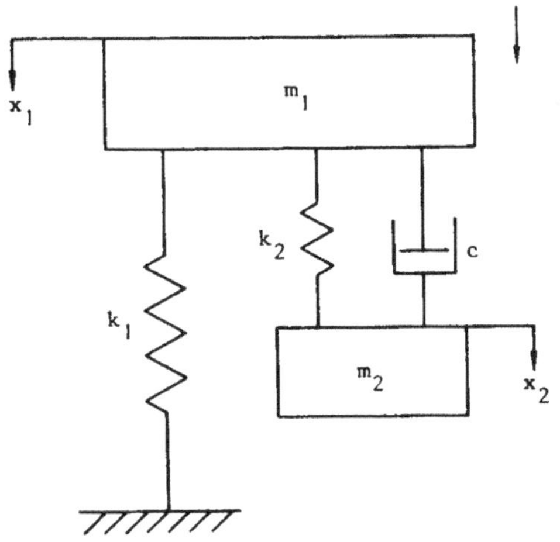


Figure 11 Damped Dynamic Absorber

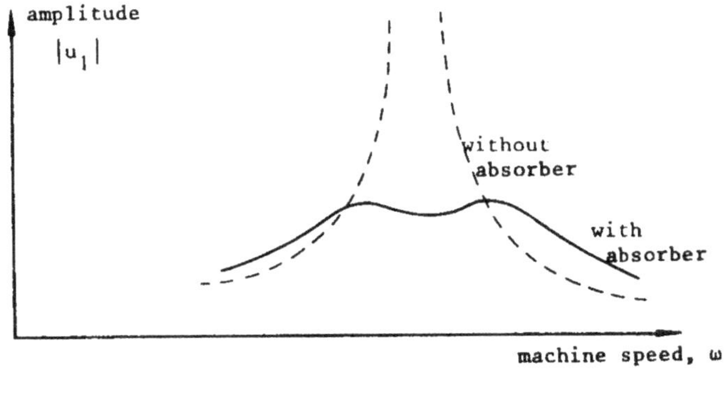


Figure 12 Effect of a Damped Dynamic Absorber on the Response of a Machine

$$m_2\ddot{u}_2 - c(\dot{u}_2-\dot{u}_1) - k_2(u_2-u_1) = 0 \tag{51}$$

These equations can be solved by the substitution

$$u_1 = A_1 \sin \omega t + B_1 \cos \omega t \tag{52}$$

$$u_2 = A_2 \sin \omega t + B_2 \cos \omega t \quad . \tag{53}$$

This leads to the following result for the amplitude of vibration of the machine [5],

$$u_1{}^2 = \frac{(k_2-m_2\omega^2)^2 + \omega^2c^2}{[(-m_1\omega^2+k_1)(-m_2\omega^2+k_2)-m_2\omega^2k_2]^2 + \omega^2c^2(-m_1\omega^2+k_1-m_2\omega^2)^2} \tag{54}$$

The values of $m_2$, $k_2$ and c can be chosen so that the response of the machine is reduced over a wide range of frequencies as shown in Figure 12.

## 7. Design Codes

The design of a machine foundation should ideally be considered as part of the process of designing the machine itself. Often this is not what happens contractually. The foundation designer is presented with a specification which the foundation should meet. It is probably in these situations that the machine foundation standards are most widely used.

The German standard DIN 4024 [6], issued in 1955, contains recommendations on the type of vibrational analysis that should be carried out when designing a machine foundation. It is suggested that all the natural frequencies of a foundation should differ from the running speed of the machine by at least 20%. In calculating the forces to which the machine is subjected, it is to be assumed that the structure behaves as an undamped single degree of freedom with a natural frequency equal to the natural frequency of the foundation which is closest to the running speed of the machine.

In 1971, a British standard (B.S. 4675:1971) [7] for the evaluation of mechanical vibration in machines was introduced. It is based on the German standard VDI 2056 [8]. Eight types of machine and foundation combinations are considered. For example, type IV is large turbines on flexible foundations. The code recommends that, as part of the acceptance procedure for a machine, vibration measurements should be made at a number of points on the supporting structure (e.g. at the foot of each bearing pedestal).

Although the standard does not specify general levels of acceptable vibration since these will vary with design, it is suggested that the customer and supplier should agree on the maximum allowable velocity levels. For each type of machine, performance classifications are defined in terms of a range of velocity levels. Table 2 shows these classifications for machine type IV.

| | | | | |
|---|---|---|---|---|
| Performance classification | A | B | C | D |
| Velocity (mm/sec) | < 2.8 | 2.8-7.1 | 7.1-18 | > 18 |

Table 2 Classification of Velocity Severity

## 8. Steel Foundations for Turbo-alternators

In the U.K., turbo-alternators have traditionally been mounted on large concrete blocks. however, with the increasing size of modern turbines this means of support presented several difficulties. Time and ease of erection, the possibility of off-site prefabrication and the need to place auxiliary equipment close to the main machine have led to the introduction of steel frame foundations.

Because of the flexible nature of these structures, it is important to be able to predict their vibrational behaviour at the design stage. A study of the use of finite elements for the dynamic analysis of steel foundations was carried out, and the results compared with experimental measurements made on foundations in operation [9].

A steel foundation is made up of box-section beams, columns and plates. It was idealized using beam and plate elements to represent each of the components, as shown in Figure 13.

As an initial step the natural frequencies and mode shapes were calculated. The mass and stiffness matrices were formed for each element and combined to give system matrices for the complete structure. The boundary conditions were then applied and the resulting eigenvalue equation solved. (The method used was to combine the system matrices, reduce the resultant motion to tridiagonal form and apply the Stürm sequence technique). A number of different idealizations of the foundation were used to establish the coarsest (and hence most efficient) mesh which gave acceptable results.

For particular foundations being analysed, measurements had been made of the displacements at a number of points on the foundation as the turbine was run up to speed. Figure 14 shows the experimental results for one position together with some of the calculated natural frequencies. It can be seen that the natural

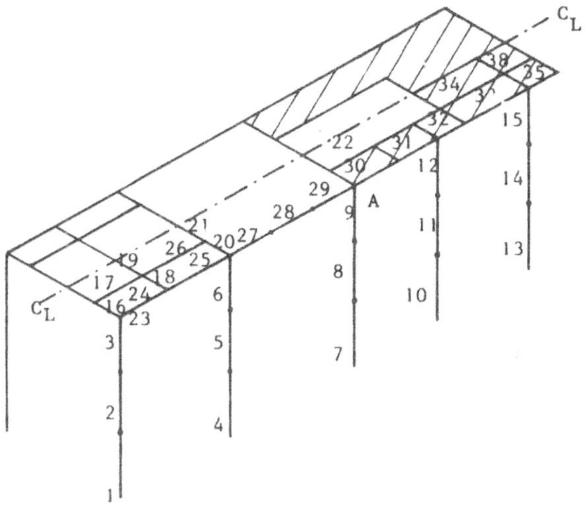


Figure 13 Idealization of Turbine Foundation

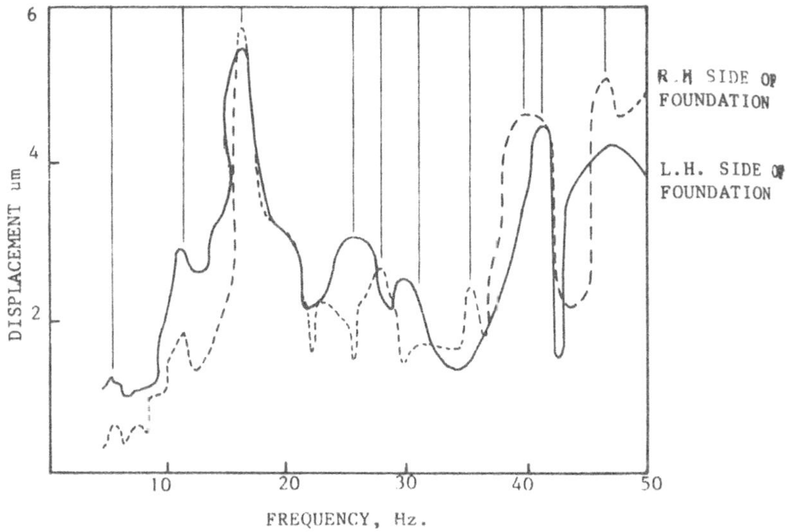


Figure 14 Measure Displacement and Calculated Natural Frequencies

frequencies correspond to peaks in the measured response.

To develop a method of estimating the amplitude of vibration to be expected, a response analysis was carried out. The damping was estimated from measurements taken when the foundations were shaken at varying frequencies using an actuator. For the response analysis, the same mass and stiffness system matrices were used as for the free vibration case. It was found that a satisfactory agreement between theory and experiment was obtained by using a damping matrix proportional to the stiffness matrix with a constant of proportionality of 0.0002.

The response when the turbine was running was estimated by assuming that the forcing was caused by eccentricity of the shaft. The magnitude of the forces were taken as being proportional to the rotating masses. The results obtained in this way gave reasonable estimates of the variation of the general levels of vibration with frequency.

The detailed analysis of the two foundations in this study resulted in the development of guidelines for use in the analysis of similar structures. By doing this, it was then possible to carry out analyses of other foundations and have some confidence that the results were reasonably realistic.

## 9. Conclusions

The discussions in Sections 2 to 6 of this chapter are meant merely to indicate the main features of the vibrational behaviour of a machine foundation. When it is necessary to study the behaviour of a foundation in practice, it is not realistic to expect that much useful information can be obtained by considering the structure as a one or two degree of freedom system. For most machines, especially the larger ones, the supporting structure is quite complex and requires many degrees of freedom to give an adequate representation. Clearly it is not always possible to perform as detailed an analysis as outlined in the previous section. However, it is desirable that all the resonances of a foundation in the range of frequencies in which the machine is operating should be known. The representation of the foundation should be such that the frequencies of modes which correspond to swaying, bending or torsion of the structure can be calculated in addition to those consisting of the simple vertical motion. If possible any analysis should examine the behaviour of the combined machine-bearing-foundation system.

References

1. Barkan, D. *Dynamics and Bases of Foundations*, McGraw-Hill, 1962.

2. Major, A. *Vibration Analysis and Design of Foundation for Machines and Turbines*, Collets, 1962.

3. Flint, I.J. Plant aspects of turbine generator foundations, *Proceedings of the Institution of Mechanical Engineers*, Vol.181, Part 1, No.22, 567-578, 1967.

4. Wilson, R.R. Flexible supports for rotating machinery, *Shock and Vibration Digest*, Vol.6, No.6, 1-5, 1974.

5. Den Hartog, J.P. *Mechanical Vibrations*, McGraw-Hill, New York, 1956.

6. D.I.N. 4024, Supporting structures for rotating machinery, particularly table-type foundations for steam turbines, German standard, 1955. (Translation I.B. 1751).

7. B.S. 4675:1971, Recommendations for a basis for comparative evaluation of vibration in machinery, British Standard Institution, London, 1971.

8. V.D.I. 2056, Evaluation criteria for mechanical vibrations in machines, German standard, 1964. (English Electric translation, May 1968).

9. Wilson, R.R. and C.A. Brebbia Dynamic behaviour of steel foundations for turbo-alternators, *Journal of Sound and Vibration*, Vol.18, No.3, 405-416. 1971.

CHAPTER 11

# VIBRATION OF AXISYMMETRIC SHELLS

by

J. Wilson

## 1. Introduction

Axisymmetric shells, also termed shells of revolution, form an important class of shell structures used in a variety of engineering applications. Examples of their use are cooling towers at power stations, industrial chimneys and containment vessels.

A shell of revolution is defined as the shell formed by complete or partial rotation of a generator about an arbitrary axis of symmetry. We will consider shells of revolution formed by the complete rotation of a generator and bounded by planes perpendicular to the axis of rotation.

In this Chapter, equations governing the elastic deformation of a shell of revolution are derived from the thin shell theory of Novozhilov [8] and the assumptions underlying Novozhilov's theory are outlined.

## 2. Novozhilov's Thin Shell Theory

### 2.1 Assumptions

Shells of revolution have been studied extensively and numerous theories have been proposed. The books of Novozhilov [8] and Kraus [7] between them give a good historical account of the development of these theories. The basis of all current shell theories are the Kirchoff-Love hypotheses. These assume:

(i) a thin shell.

If the ratio of the shell thickness, h, to the smaller principle radius of curvature, $r_o$, is small, the shell is termed thin and its geometry may be defined by the shape of its mid-surface and its thickness normal to the mid-surface. The errors incurred by this assumption are of the order of $h/r_o$. Novozhilov has suggested that this ratio should be less than 0.05 to justify the use of thin shell theory. The accuracy of structural calculations could thus be contained within an acceptable error.

(ii) small deformations.

Linear theory may be used if the deformations are small compared with the shell thickness.

(iii) lines normal to the mid-surface before deformation remain so after deformation and do not change their length.

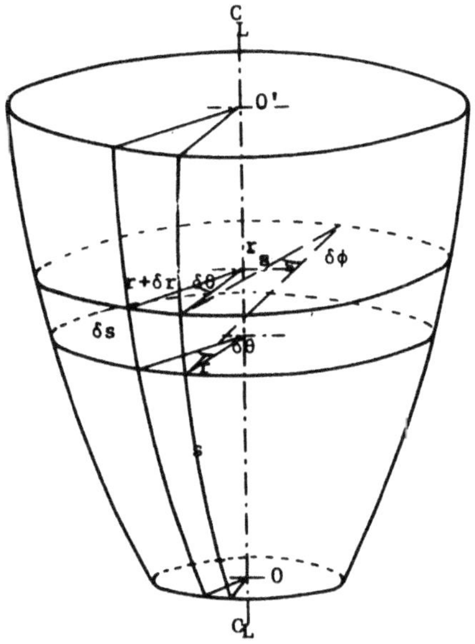


Figure 1 Mid-surface of Shell of Revolution

(iv) normal stresses acting on planes parallel to the mid-surface are negligible compared with other stresses.

Assumptions (iii) and (iv) are similar to the assumptions made in simple beam theory and they imply that transverse shear stresses are negligible. The following assumptions, additional to the Kirchoff-Love hypotheses are made:

(v) uniform boundary conditions. For many structures such boundary conditions approximately hold.

(vi) the shell materials are isotropic, homogeneous and linear elastic and each portion of shell composed of a single material is itself a complete shell of revolution.

(vii) for dynamic behaviour of shells, it is assumed that the contribution of the rotatory inertia terms to the kinetic energy is negligible.

More refined theories may be obtained by removing or delaying one or more of these assumptions in the derivation of the equations of motion for a shell of revolution. (See Kraus [7], Chapter 3).

For practical structures, a simple theory based on the Kirchoff-Love hypotheses will be sufficiently accurate.

## 2.2 Equilibrium Equations

To obtain the equilibrium equations for a shell of revolution we consider a portion which has been cut from the shell. The shell portion is bounded by two generators and two arcs which lie in parallel planes perpendicular to the shell axis. The applied external forces consist of body forces acting on the portion and surface forces acting on both inner and outer faces of the portion. Internal stresses act on the surfaces exposed by the cuts.

The Kirchoff-Love assumptions enable the thin shell to be replaced by its mid-surface. Thus the shell portion is replaced by a portion of the shell mid-surface defined by the coordinates r, s and $\theta$, and having meridional sides of length $\delta s$ and parallel arcs of length $r\,\delta\theta$ and $(r+\delta r)\,\delta\theta$. (See Figure 1).

r is the perpendicular radial distance from the axis of symmetry to the shell mid-surface. s is the distance measured along the meridian or generator. $\theta$ is the longitudinal angular coordinate of the meridian.

The applied external forces are replaced by equivalent uniformly distributed stresses $q_n$, $q_s$ and $q_\theta$ acting on the surface of the mid-surface portion in the normal,

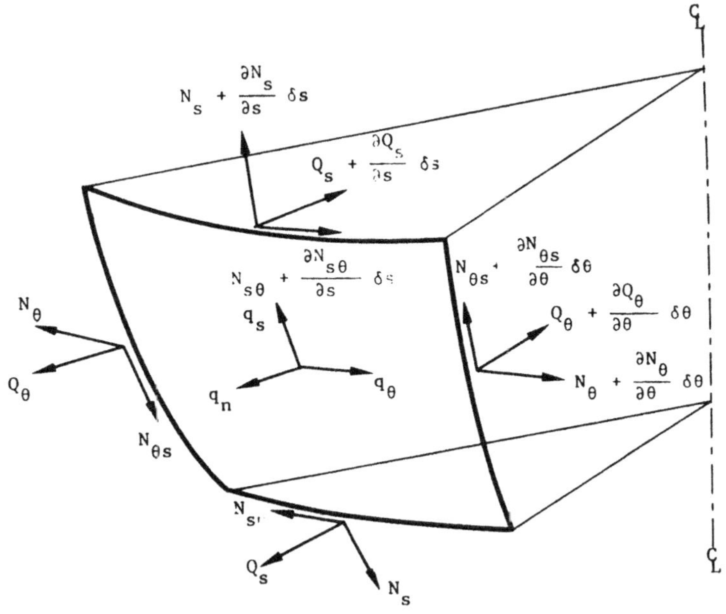


Figure 2a Direct and Shear Stress Resultants

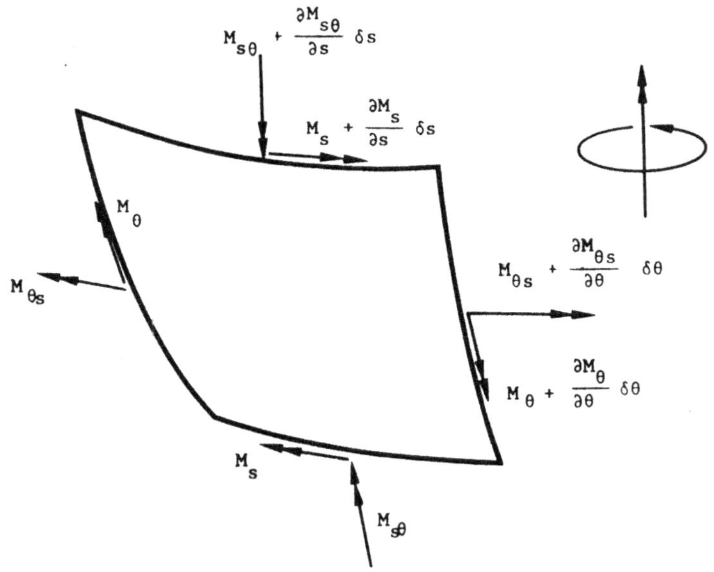


Figure 2b Bending and Torsional Stress Resultants

meridional and longitudinal directions respectively. (Moments due to external forces are neglected). The internal stresses are replaced by *stress resultants* acting along the boundaries of the mid-surface portion. Stress resultants are expressed in terms of forces or moments per unit length of boundary.

Figures 2a and 2b show the stresses and stress resultants acting on the portion of the shell mid-surface. $N_s$ and $N_\theta$ are the direct stress resultants, $N_{s\theta}$ and $N_{\theta s}$ are the in-plane shear stress resultants, $Q_s$ and $Q_\theta$ are the transverse shear stress resultants, $M_s$ and $M_\theta$ are the bending stress resultants and $M_{s\theta}$ and $M_{\theta s}$ are the torsional stress resultants. (N.B. The total forces acting on the mid-surface portion are obtained by multiplying the stress or stress resultants by the area or arc lengths of mid-surface portion respectively.)

The equilibrium equations for a portion of a shell of revolution may be derived from Novozhilov's general theory of shells. They are

$$r_s \frac{\partial}{\partial s}(rN_s) + r_s \frac{\partial N_{\theta s}}{\partial \theta} - N_\theta r_s \sin\phi - rQ_s + r\, r_s q_s = 0 \qquad (1a)$$

$$r_s \frac{\partial}{\partial s}(rN_{s\theta}) + r_s \frac{\partial N_\theta}{\partial \theta} + N_{\theta s} r_s \sin\phi - r_s Q_\theta \cos\phi + r\, r_s q_\theta = 0 \qquad (1b)$$

$$r_s \frac{\partial}{\partial s}(rQ_s) + r_s \frac{\partial Q_\theta}{\partial \theta} + rN_s + r_s N_\theta \cos\phi + r\, r_s\, q_n = 0 \qquad (1c)$$

$$r_s \frac{\partial}{\partial s}(rM_s) + r_s \frac{\partial M_{\theta s}}{\partial \theta} - M_\theta r_s \sin\phi + r\, r_s Q_s = 0 \qquad (1d)$$

$$r_s \frac{\partial}{\partial s}(rM_{s\theta}) + r_s \frac{\partial M_\theta}{\partial \theta} + M_{\theta s} r_s \sin\phi - r\, r_s\, Q_\theta = 0 \qquad (1e)$$

$$N_{s\theta} - N_{\theta s} + \frac{M_{\theta s}\cos\phi}{r} - \frac{M_{s\theta}}{r_s} = 0 \qquad (1f)$$

where $r_s$ is the radius of curvature of the meridian at distance s. The curvature is defined as positive if the axis of the shell and the centre of curvature lie on the same side of the meridian (see Figure 1). $\phi$ is the angle between the tangent to the meridian and the axis of symmetry.

## 2.3 Strain-displacement Relations

The strain-displacement equations of Novozhilov may be written for a shell of revolution in terms of the r, θ, s coordinate system illustrated in Figure 1 as

$$\varepsilon_s = \frac{\partial u}{\partial s} + \frac{w}{r_s}$$

$$\varepsilon_\theta = \frac{\sin\phi}{r} u + \frac{1}{r}\frac{\partial v}{\partial \theta} + \frac{\cos\phi}{r} w$$

$$\varepsilon_{s\theta} = \frac{1}{2}\left[\frac{1}{r}\frac{\partial u}{\partial \theta} + \frac{\partial v}{\partial s} - \frac{v \sin\phi}{r}\right]$$

$$\kappa_s = \frac{1}{r_s}\left[\frac{\partial u}{\partial s} - \frac{1}{r_s}\frac{dr_s}{ds} u\right] - \frac{\partial^2 w}{\partial s^2} \qquad (2)$$

$$\kappa_\theta = \frac{\sin\phi}{rr_s} u + \frac{\cos\phi}{r^2}\frac{\partial v}{\partial \theta} - \frac{1}{r^2}\frac{\partial^2 w}{\partial \theta^2} - \frac{\sin\phi}{r}\frac{\partial w}{\partial s}$$

$$\kappa_{s\theta} = \frac{1}{rr_s}\frac{\partial u}{\partial \theta} + \frac{\cos\phi}{r}\left[\frac{\partial v}{\partial s} - \frac{\sin\phi}{r} v\right] + \frac{\sin\phi}{r^2}\frac{\partial w}{\partial \theta} - \frac{1}{r}\frac{\partial^2 w}{\partial s \partial \theta}$$

where $s$ is the meridional coordinate

$\theta$ is the longitudinal coordinate

$\varepsilon_s, \varepsilon_\theta$ are the direct strains in the s and $\theta$ directions

$\varepsilon_{s\theta}$ is the 'mathematical' shear strain (= ½ 'engineering' shear strain)

$\kappa_s, \kappa_\theta$ are the changes of curvature of the shell mid-surface in the s and directions

$\kappa_{s\theta}$ is the change in twist of the surface

u, v and w are the shell mid-surface displacements in the meridional, longitudinal and normal directions respectively (see Figure 3).

$\phi$ is the angle between the tangent to the meridian and the axis of the shell.

The equations (2) are such that, if u, v and w represent rigid body displacements given by Dawe [2], the strains and curvature changes are zero.

## 2.4 Stress Strain Relationships

For a thin shell of revolution, the stress-strain relationships may be written for an isotropic material as

$$N_s = C\,(\varepsilon_s + \nu\varepsilon_\theta)$$
$$N_\theta = C\,(\varepsilon_\theta + \nu\varepsilon_s)$$

$$N_{s\theta} = N_{\theta s} = \frac{Eh}{(1+\nu)}\varepsilon_{s\theta} \qquad (3)$$

$$M_s = D\,(\kappa_s + \nu\kappa_\theta)$$

$$M_\theta = D\,(\kappa_\theta + \nu\kappa_s)$$

$$M_{s\theta} = M_{\theta s} = \frac{Eh^3}{12(1+\nu)}\kappa_{s\theta}$$

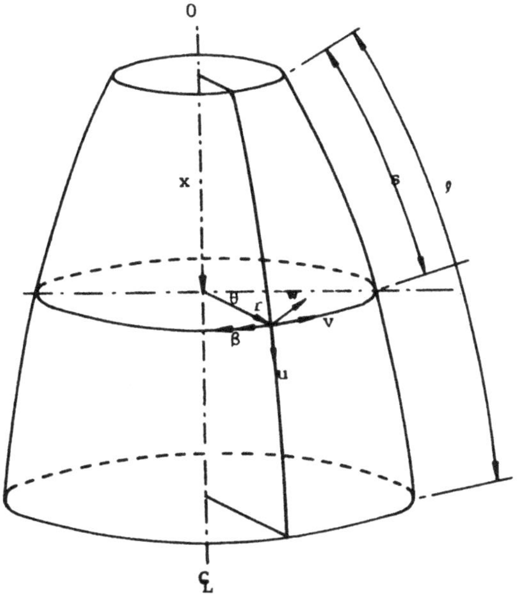


NOTE: The rotation β is clockwise about the direction of the double arrow.

Figure 3   Displacements of an Element of a Shell of Revolution

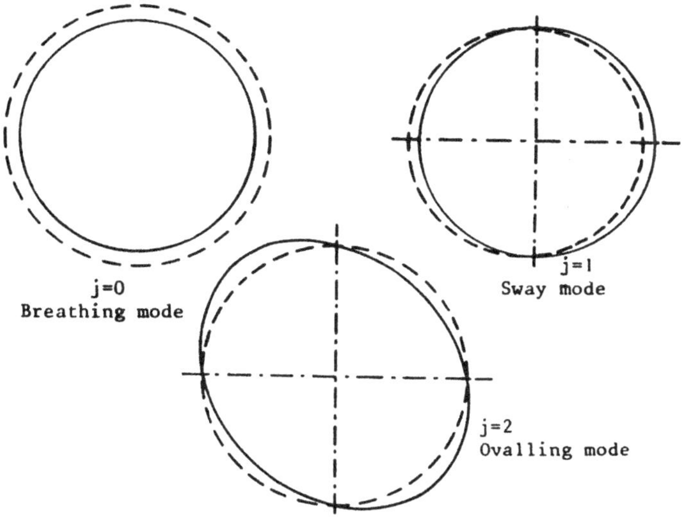


Figure 4   Mode Shapes for Shells of Revolution

where $C = \frac{Eh}{1-\nu^2}$

$D = \frac{Eh^3}{12(1-\nu^2)}$

E is the Young's modulus

$\nu$ is the Poisson's ratio

and h is the shell thickness.

The above expression is given by Novozhilov [8] who derives it from a more complex expression by assuming that the shell is thin. This approximation does lead to inconsistencies in the theory of shells, because it assumes that

$$N_{s\theta} = N_{\theta s} \tag{4a}$$

$$M_{s\theta} = M_{\theta s} \tag{4b}$$

Equations (4) are thus incompatible with the identity (1f). Flugge [4] shows that this identity is not required in solving the shell equations.

2.5 Form of General Solution

For shells of revolution having simple geometries and loading distributions, analytical solutions to equations (1), (2) and (3) may be found (Flugge, [4]; Novozhilov, [8]; Kraus, [7]). Indeed even for shells having complex geometries and loading patterns, the solution for any linear structural property, $\chi$, (i.e. displacements, strains, stresses etc.) may be expressed as a Fourier series as in the equation,

$$\chi(s,t) = \sum_{j=0}^{\infty} a_j(s,t) \cos j\theta + \sum_{j=1}^{\infty} b_j(s,t) \sin j\theta \tag{5}$$

where t is time, and a and b are functions (not necessarily analytic) of s and t, j is an integer which has significance in vibration analysis. It is termed the circumferential mode number and is used to define mode shapes in which a shell of revolution vibrates. These modes of vibration are often termed 'classical'. Figure 4 shows the mode shape for each component of displacement at every cross section of a vibrating shell of revolution for j = 0, 1 and 2. The value, j = 0, corresponds to a breathing mode (or torsional mode) of vibration; j = 1 to a sway type of mode of vibration; j = 1 to a sway type of mode and j = 2 or more to an ovalling mode of vibration.

The classical mode shapes of a thin shell of revolution have the following form:

$$u = f_1(s) \cos j\theta$$
$$v = f_2(s) \sin j\theta \qquad (6a)$$
$$w = f_3(s) \cos j\theta$$

where $f_1$, $f_2$, $f_3$ are functions of s.

When the quantity, $\chi$, is associated with steady state vibration, $a_j$ and $b_j$ become separable in s and t and equation (5) may be written

$$\chi(s,t) = \chi(s)\, e^{i\Omega t} = \{ \sum_{j=0}^{\infty} a_j(s) \cos j\theta + \sum_{j=1}^{\infty} b_j(s) \sin j\theta \}\, e^{i\Omega t} \qquad (6)$$

When a closed form solution for the quantities $a_j$ and $b_j$ of equations (5) and (6) cannot be obtained, a numerical method of solution must be used. One very important class of numerical methods which may be used to obtain a solution to the shell equations (1), (2) and (3) is the finite element method. In the finite element displacement method, which is described next, it is not necessary to satisfy the equilibrium equations.

## 3. Finite Element Displacement Method Applied to Axisymmetric Shells

### 3.1 Axisymmetric Shell Finite Element

Figure 5 shows the axisymmetric shell finite element with generally curved meridian and varying shell thickness. It is bounded by two parallel planes perpendicular to the axis of rotation which intersect the element at nodal circles.

### 3.2 Displacement Functions

The displacement functions chosen to describe the displacement of an axisymmetric shell element are given in equation (7) for $j^{th}$ harmonic number.

$$\begin{aligned} u_j &= (\alpha_{1j} + \alpha_{2j}s + \alpha_{3j}s^2 + \alpha_{4j}s^3) \cos j\theta \\ &\quad + (\beta_{1j} + \beta_{2j}s + \beta_{3j}s^2 + \beta_{4j}s^3) \sin j\theta \\ v_j &= (\alpha_{5j} + \alpha_{6j}s + \alpha_{7j}s^2 + \alpha_{8j}s^3) \sin j\theta \\ &\quad + (\beta_{5j} + \beta_{6j}s + \beta_{7j}s^2 + \beta_{8j}s^3) \cos j\theta \\ w_j &= (\alpha_{9j} + \alpha_{10j}s + \alpha_{11j}s^2 + \alpha_{12j}s^3) \cos j\theta \\ &\quad + (\beta_{9j} + \beta_{10j}s + \beta_{11j}s^2 + \beta_{12j}s^3) \sin j\theta \end{aligned} \qquad (7)$$

These displacement functions are of the same form as equation (6).

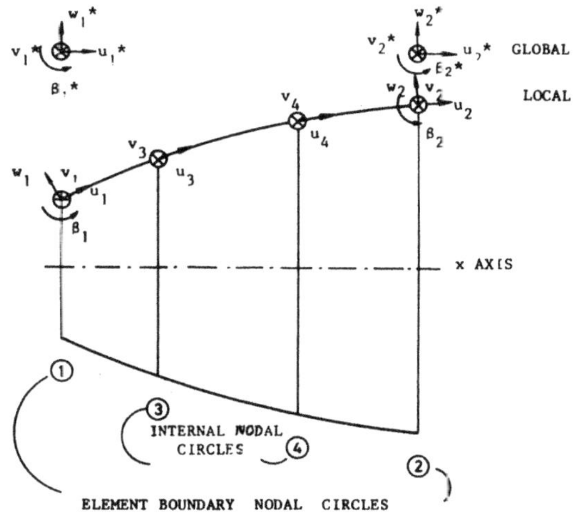


Figure 5 Local and Global Degrees of Freedom at Nodal Circles of a Shell of Revolution Element

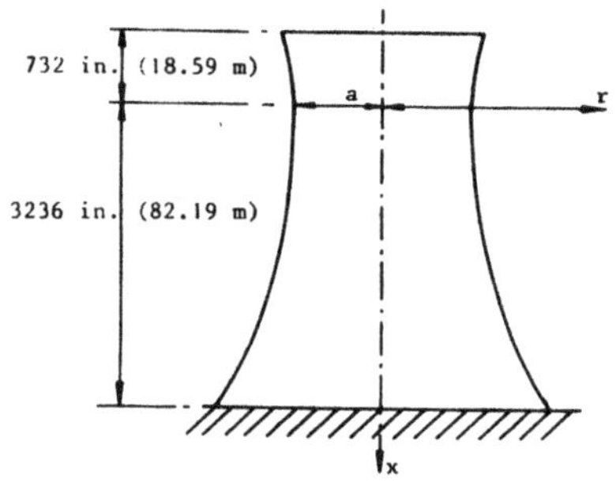


Note The meridian is defined by the hyperbola

$$\frac{r^2}{a^2} - \frac{x^2}{b^2} = 1$$

where a = 1008 in. (25.60 m)
b = 2516 in. (63.91 m)

Meridional thickness = 5 in. (0.127 m)
Young's modulus = $3 \times 10^6$ lbf in$^{-2}$ ($2.069 \times 10^{10}$ Nm$^{-2}$)
Density = 150.111 lbft$^{-3}$ (2405 kgm$^{-3}$)
Poissons ratio = 0.15

Figure 6 Dimensions and Material Properties of Hypothetical Cooling Tower.

The $\alpha_{ij}$ and $\beta_{ij}$ terms in equations (7) are separable and their form is such that any operation on one set of terms may be derived from the same operation on the other set by replacing j by -j (Percy et al. [9]). In the subsequent treatment, the $\beta_{ij}$ terms are ignored except where otherwise stated.

## 3.3 The Degrees of Freedom of the Axisymmetric Element

The displacement functions can be represented in terms of certain degrees of freedom existing at the nodal circle boundaries of the element. The only degrees of freedom which must be continuous across the element interface are u, v, w and the meridional rotation, b, where b is defined by equation (8)

$$b = \frac{\partial w}{\partial s} - \frac{u}{r_s} \tag{8}$$

Then senses of these degrees of freedom are shown in Figure 4 where the subscripts 1 and 2 refer to the first and second nodal circles respectively. If these degrees of freedom are selected as the nodal variables then the total number of degrees of freedom per element is therefore eight.

Because in equation (7) there are 12 $\alpha$ coefficients for each harmonic number, four extra degrees of freedom must be introduced into the element. This is done by the introduction of two internal nodes, and corresponding extra degrees of freedom, internal displacements $u_I$ and $v_I$ (Figure 5).

## 3.4 Derivation of Stiffness and Mass Matrices for the Axisymmetric Shell Element

The stiffness and mass matrices of the axisymmetric shell element are derived as shown below for the $j^{th}$ harmonic number. The subscript e denotes an element.

By substituting equations (7) in (2) we obtain in matrix form

$$\underset{\sim}{\varepsilon}_e^j = \underset{\sim}{\theta}^j \underset{\sim}{B}_e^j \underset{\sim}{\alpha}_e^j \tag{9}$$

where $\underset{\sim}{\varepsilon}_e^j$ is the strain vector $\{\varepsilon_s \; \varepsilon_\theta \; \varepsilon_{s\theta} \; K_s \; K_\theta \; K_{s\theta}\}_e^j$ .

$\underset{\sim}{\alpha}_e^j$ is the vector $\{\alpha_1 \; \alpha_2 \; \alpha_3 \; \ldots\ldots\; \alpha_9 \; \alpha_{10} \; \alpha_{11} \; \alpha_{12}\}_e^j$

Matrices $\underset{\sim}{\theta}^j$ and $\underset{\sim}{B}_e^j$ are given in the Appendix.

The nodal displacement vector, $\underset{\sim}{u}_e^j$, can now be expressed in terms of the displacement coefficients $\underset{\sim}{a}_e^j$ .

$$\underset{\sim}{U}_e^j = \underset{\sim}{L}_e \underset{\sim}{\alpha}_e^j \tag{10}$$

where $\underset{\sim}{u}_e^j$ is the vector $\{u_1\ v_1\ w_1\ b_1\ u_{I1}\ v_{I1}\ u_2\ v_2\ w_2\ b_2\ u_{I2}\ v_{I2}\}_e^j$ and $\underset{\sim}{L}_e$ is obtained by putting s = 0, ℓ/3, 2ℓ/3 or ℓ in equations (7). ℓ is the length of the element meridian.

The matrix $\underset{\sim}{L}_e$ is given in the Appendix.

Hence we have
$$\underset{\sim}{\alpha}_e^j = \underset{\sim}{L}_e^{-1}\ \underset{\sim}{u}_e^j \tag{11}$$

or
$$\underset{\sim}{\varepsilon}_e^j = \underset{\sim}{\theta}^j\ \underset{\sim}{B}_e^j\ \underset{\sim}{L}_e^{-1}\underset{\sim}{u}_e^j \tag{12}$$

For a thin shell, the strain energy stored in each element can be written as

$$U_e^j = \frac{1}{2}\int_0^{\ell}\int_0^{2\pi}(N_s\varepsilon_s + N_\theta\varepsilon_\theta + N_{s\theta}\varepsilon_{s\theta} + N_{\theta s}\varepsilon_{\theta s} + M_sK_s + M_\theta K_\theta + M_{s\theta}K_{s\theta} + M_{\theta s}K_{\theta s})_e^j\ r\ d\theta\ ds \tag{13}$$

By substituting from equations (3) in equation (13) and expressing this in matrix form,we have
$$U_e^j = \frac{1}{2}\int_0^{\ell}\int_0^{2\pi}[\underset{\sim}{\varepsilon}_e^j]^T\ \underset{\sim}{E}_e\ \underset{\sim}{\varepsilon}_e^j\ r\ d\theta\ ds \tag{14}$$

The matrix $\underset{\sim}{E}$ is given in the Appendix.

On substituting equation (12) in (14), equation (15) is obtained

$$U_e^j = \frac{1}{2}[\underset{\sim}{u}_e^j]^T\ [L_e^{-1}]^T \int_0^{\ell} r[\underset{\sim}{B}_e^j]^T \int_0^{2\pi}[\underset{\sim}{\theta}^j]^T\ \underset{\sim}{E}_e\ \underset{\sim}{\theta}_j\ d\theta\ \underset{\sim}{B}_e^j\ ds\ \underset{\sim}{L}_e^{-1}\ \underset{\sim}{u}_e \tag{15}$$

Equation (15) can be simplified because all the integrals involving the circumferential coordinate, θ, are either of the form

$$\int_0^{2\pi}\cos^2 j\theta\ d\theta \quad \text{or} \quad \int_0^{2\pi}\sin^2 j\theta\ d\theta \tag{16}$$

These expressions have the value nπ where n takes the value 2 for j = 0 and 1 for j ⩾ 1. Thus equation (15) may be written

$$U_e^j = \frac{1}{2}\ [\underset{\sim}{u}_e^j]^T\ \underset{\sim}{k}_e^j\ \underset{\sim}{u}_e^j \tag{17}$$

where $\underset{\sim}{k}_e^j$, the element stiffness matrix for the $j^{th}$ circumferential mode of vibration,is defined by

$$\underset{\sim}{k}_e^j = n\pi\ [\underset{\sim}{L}_e^{-1}]^T\ \{\int_0^{\ell} r[\underset{\sim}{B}_e^j]^T\ \underset{\sim}{E}_e\ \underset{\sim}{B}_e^j\ ds\}\ \underset{\sim}{L}_e^{-1} \tag{18}$$

The kinetic energy of the element may be expressed as

$$T_e^j = \frac{1}{2} \int_0^{\ell} \int_0^{2\pi} \rho h \, (\dot{u}^2+\dot{v}^2+\dot{w}^2)_e^j \, d\theta \, ds \tag{19}$$

where ρ is the density of the material forming the shell element and rotary inertia terms are neglected. By substituting equation (7) and (11) in (19) and making use of the integrals of equation (16), equation (19) becomes

$$T_e^j = \frac{1}{2} \, [\dot{u}_e^j]^T \, \underset{\sim}{m}_e^j \, \underset{\sim}{\dot{u}}_e^j \tag{20}$$

where $\underset{\sim}{m}_e^j$ is the element mass matrix which is defined by

$$\underset{\sim}{m}_e^j = n\pi \, [\underset{\sim}{L}_e^{-1}]^T \int_0^{\ell} \rho h \, \underset{\sim}{X}_e \, ds \, \underset{\sim}{L}_e^{-1} \tag{21}$$

The matrix $\underset{\sim}{X}_e$ is given in the Appendix. It is seen that $\underset{\sim}{X}_e$ is independent of j.

Note that the $\underset{\sim}{m}_e^j$ is independent of the circumferential mode number except for the value of the constant n. The inversion of the matrix $\underset{\sim}{L}_e$ and the integration of the equations (18) and (21) is carried out numerically by computer.

### 3.5 Reduction of Mass and Stiffness Matrices

In Section 3.2 it was stated that only four degrees of freedom at each boundary node must remain continuous to preserve the integrity of the shell, and therefore the element stiffness and mass matrices need only have dimensions 8 x 8 instead of 12 x 12. The use of smaller matrices would enable larger numbers of elements to be used and hence more complex problems to be solved.

The size of the structural stiffness and mass matrices may be reduced by a technique due to Guyan [5] & subsequently Henshell et al.[6] applied this method to element matrices. The methods require that the element mass and stiffness matrices be partitioned so that the strain and kinetic energies of the element for the $j^{th}$ harmonic may be written in terms of vectors involving only the external boundary degrees of freedom $\underset{\sim}{u}_{B_e}^j$ .

$$U_e^j = \frac{1}{2} \left[ \underset{\sim}{u}_B^T \,\vdots\, \underset{\sim}{u}_I^T \right]_e^j \begin{bmatrix} \underset{\sim}{k}_{BB} & \underset{\sim}{k}_{BI} \\ \underset{\sim}{k}_{IB} & \underset{\sim}{k}_{II} \end{bmatrix}_e^j \begin{bmatrix} \underset{\sim}{u}_B \\ \underset{\sim}{u}_I \end{bmatrix}_e^j \tag{22}$$

$$T_e^j = \frac{1}{2} \left[ \underset{\sim}{\dot{u}}_B^T \,\vdots\, \underset{\sim}{\dot{u}}_I^T \right]_e^j \begin{bmatrix} \underset{\sim}{m}_{BB} & \underset{\sim}{m}_{BI} \\ \underset{\sim}{m}_{IB} & \underset{\sim}{m}_{II} \end{bmatrix}_e^j \begin{bmatrix} \underset{\sim}{\dot{u}}_B \\ \underset{\sim}{\dot{u}}_I \end{bmatrix}_e^j \tag{23}$$

where $\underset{\sim}{u}^j_{B_e}$ is the vector $\{u_1\ v_1\ w_1\ b_1\ u_2\ v_2\ w_2\ b_2\}^j_e$

$\underset{\sim}{u}^j_{I_e}$ is the vector $\{u_{I1}\ v_{I1}\ u_{I2}\ v_{I2}\}^j_e$

The reduced element stiffness and mass matrices may then be written as

$$\left[\underset{\sim}{k}_r\right]^j_e = \left[\underset{\sim}{k}_{BB} - \underset{\sim}{k}_{BI}\ \underset{\sim}{k}^{-1}_{II}\ \underset{\sim}{k}_{IB}\right]^j_e \tag{24}$$

$$\left[\underset{\sim}{m}_r\right]^j_e = \left[\underset{\sim}{m}_{BB} - \underset{\sim}{k}_{BI}\ \underset{\sim}{k}^{-1}_{II}\ \underset{\sim}{m}_{IB} - \underset{\sim}{m}_{BI}\ \underset{\sim}{k}^{-1}_{II}\ \underset{\sim}{k}_{IB} + \underset{\sim}{k}_{BI}\ \underset{\sim}{k}^{-1}_{II}\ \underset{\sim}{m}_{II}\ \underset{\sim}{k}^{-1}_{II}\ \underset{\sim}{k}_{IB}\right]^j_e \tag{25}$$

Henshell et al. point out that there will be errors in the mass matrix but these are assumed negligible.

## 3.6 Transformation from Local to Global Coordinates

Transformation from local nodal displacements to global nodal displacements, $\underset{\sim}{U}^j_e$, is carried out as in equations (26) and (27) to give the global element stiffness and mass matrices $\underset{\sim}{K}^j_e$ and $\underset{\sim}{M}^j_e$

$$\underset{\sim}{K}^j_e = \underset{\sim}{R}^T_e \left[\underset{\sim}{k}_r\right]^j_e \underset{\sim}{R}_e \tag{26}$$

$$\underset{\sim}{M}^j_e = \underset{\sim}{R}^T_e \left[\underset{\sim}{m}_r\right]^j_e \underset{\sim}{R}_e \tag{27}$$

where $\underset{\sim}{U}^j_e = \{u^*_1\ v^*_1\ w^*_1\ b^*_1\ u^*_2\ v^*_2\ w^*_2\ b^*_2\}^j_e$ and the transformation matrix $\underset{\sim}{R}_e$ is given in the Appendix. The asterisks denote global quantities in the directions x, $\theta$ and r as shown in Figure 5.

## 4. Vibration Applications

The adequacy of the curved axisymmetric shell element described in this section was assessed by comparing the accuracy of computed natural frequencies of several shells of revolution and the convergence of the computed values to exact or established values of frequencies.

The vibration equation for each value of j may be written

$$[\underset{\sim}{K}_j - \omega^2\ \underset{\sim}{M}_j]\ \underset{\sim}{U}^j_o = \underset{\sim}{0} \tag{28}$$

where $\underset{\sim}{K}_j$ is the global structural stiffness matrix formed by assembling the element matrices.

$\underset{\sim}{M}_j$ is the global structural mass matrix formed by assembling the element mass matrices.

$\underset{\sim}{U}^j_o$ is a vector of the global nodal displacement amplitudes for the whole structure.

## 5. Example

To show how effective a finite element analysis can be, the natural frequencies of a notional cooling tower have been computed. The dimensions of the tower and material properties of the concrete are shown in Figure 6.

The tower was first analysed by Carter, Robinson and Schnobrich [1]. Their results are given in Table 1 which shows the first four natural frequencies for values of circumferential mode number j ranging from 0 to 4. The results B were obtained using the finite element analysis described in the previous section. The number after B shows the numbers of finite element used in the analysis. Results C were obtained by Debnath [3] using another shell finite element.

The results of Table 2 demonstrate the convergence of the natural frequencies computed by the finite element method to the established solution. An error of less than 1% is obtained using only eight elements to represent the whole shell for each of the frequencies shown.

## References

1. Carter, R.L., Robinson, A.R. and Schnobrich, W.C. Free Vibrations of Hyperboloidal Shells of Revolution, *Proc.Am. Soc. Civil Engrs. J., Eng. Mech. Div.* 1965 95, (EM5) 1033-1052.

2. Dawe, D.J. Rigid-Body Motions and Strain-Displacement Equations of Curved Shell Finite Elements, *Int. J. Mech. Sci.* 1972, 14, 569-578.

3. Deb Nath, J.M., Free Vibration, Stability and Non-classical Modes of Cooling Tower Shells, *J. Sound Vibration*, 33, (1) 1974, 79-101.

4. Flugge, W., *Stresses in Shells*, (4th Printing), Springer-Verlag 1967, Berlin.

5. Guyan, R.J. Reduction of Stiffness and Mass Matrices, *Am. Inst. Aeronaut. Astronaut. J.* 1965, 3, No.2, p.380.

6. Henshell, R.D., Bond, T.J. and Makoju, J.O. Ring Finite Elements for Axi-symmetric and Non-axisymmetric Thin Shell Analysis, Paper No. 4/45, in *Variational Methods in Engineering*, Editors, C.A. Brebbia and H. Tottenham, Southampton Univ. Press. 1973.

7. Kraus, H. *Thin Elastic Shells*, John Wiley, 1967, New York.

8. Novozhilov, V.V., *The Theory of Thin Shells*, P. Noorhoff, 1959, Groningen.

9. Percy, J.H., Pian, T.H.H., Klein, S. and Navaratna, D.R. Application of Matrix Displacement Method to Linear Elastic Analysis of Shells of Revolution, *Am. Inst. Aeronaut. Astronaut. J.* 1965, 3, (11), 2138-2145.

Table 1 The Natural Frequencies of a Hypothetical Cooling Tower

| Axial Mode Number | Method of Solution | Circumferential Mode Number | | | | |
|---|---|---|---|---|---|---|
| | | 0 | 1 | 2 | 3 | 4 |
| 1 | A | - | 3.290 | 1.765 | 1.375 | 1.181 |
| | B 1 | 7.005 | 3.328 | 1.871 | 1.576 | 1.667 |
| | 2 | 6.476 | 3.305 | 1.774 | 1.402 | 1.254 |
| | 4 | 6.298 | 3.296 | 1.768 | 1.377 | 1.184 |
| | 8 | 6.251 | 3.292 | 1.767 | 1.377 | 1.181 |
| | 16 | 6.238 | 3.291 | 1.767 | 1.376 | 1.181 |
| | C | 6.233 | 3.290 | 1.766 | 1.376 | 1.181 |
| 2 | A | 7.752 | 6.793 | 3.695 | 1.991 | 1.448 |
| | B 1 | 8.550 | 8.081 | 5.112 | 3.567 | 3.009 |
| | 2 | 7.964 | 6.965 | 3.765 | 2.051 | 1.538 |
| | 4 | 7.809 | 6.832 | 3.704 | 2.001 | 1.455 |
| | 8 | 7.769 | 6.803 | 3.693 | 1.995 | 1.450 |
| | 16 | 7.758 | 6.797 | 3.691 | 1.994 | 1.449 |
| | C | 7.752 | 6.793 | 3.689 | 1.994 | 1.448 |
| 3 | A | 11.42 | 10.52 | 6.959 | 4.327 | 2.778 |
| | B 1 | 16.64 | 17.90 | 28.55 | 40.55 | 52.57 |
| | 2 | 14.13 | 11.83 | 7.642 | 4.826 | 3.182 |
| | 4 | 12.35 | 10.73 | 7.028 | 4.386 | 2.809 |
| | 8 | 11.59 | 10.55 | 6.968 | 4.346 | 2.783 |
| | 16 | 11.43 | 10.53 | 6.960 | 4.340 | 2.780 |
| | C | 11.43 | 10.52 | 6.956 | 4.337 | 2.778 |
| 4 | A | 11.91 | - | - | - | - |
| | B 1 | 17.82 | 29.37 | 45.14 | 60.19 | 75.33 |
| | 2 | 16.89 | 15.04 | 12.82 | 9.920 | 7.642 |
| | 4 | 13.76 | 12.44 | 9.676 | 6.855 | 4.768 |
| | 8 | 12.26 | 11.55 | 9.467 | 6.736 | 4.692 |
| | 16 | 11.95 | 11.38 | 9.447 | 6.722 | 4.680 |

A Established Solution (Carter et al. [1]).

B Finite Element Solution. The following number indicates the number of elements used in the idealization.

C Finite Element Solution (Deb Nath [3]).

# APPENDIX

## Matrices used in the Text

Matrix $\underset{\sim}{\theta}^j$

$$
\begin{bmatrix}
\cos j\theta & & & & & \\
 & \cos j\theta & & & \underset{\sim}{0} & \\
 & & \sin j\theta & & & \\
 & & & \cos j\theta & & \\
 & \underset{\sim}{0} & & & \cos j\theta & \\
 & & & & & \sin j\theta
\end{bmatrix}
$$

$\theta$ is the tangential coordinate.

Matrix $\underset{\sim}{E}$

$$
\begin{bmatrix}
\frac{Eh}{1-\nu^2} & \frac{\nu Eh}{1-\nu^2} & & & & \\
\frac{\nu Eh}{1-\nu^2} & \frac{Eh}{1-\nu^2} & & & \underset{\sim}{0} & \\
 & & \frac{2\ Eh}{1+\nu^2} & & & \\
 & & & \frac{Eh^3}{12(1-\nu^2)} & \frac{\nu Eh^3}{12(1-\nu^2)} & \\
 & \underset{\sim}{0} & & \frac{\nu Eh^3}{12(1-\nu^2)} & \frac{Eh^3}{12(1-\nu^2)} & \\
 & & & & & \frac{Eh^3}{6(1+\nu)}
\end{bmatrix}
$$

Matrix $L_e$

$$
\begin{bmatrix}
1 & 0 & 0 & 0 & 0 & 0 & 0 & 0 & 0 & 0 & 0 & 0 \\
0 & 0 & 0 & 0 & 1 & 0 & 0 & 0 & 0 & 0 & 0 & 0 \\
0 & 0 & 0 & 0 & 0 & 0 & 0 & 0 & 1 & 0 & 0 & 0 \\
-\frac{1}{r_{s1}} & 0 & 0 & 0 & 0 & 0 & 0 & 0 & 0 & 1 & 0 & 0 \\
1 & \ell & \ell^2 & \ell^3 & 0 & 0 & 0 & 0 & 0 & 0 & 0 & 0 \\
0 & 0 & 0 & 0 & 1 & \ell & \ell^2 & \ell^3 & 0 & 0 & 0 & 0 \\
0 & 0 & 0 & 0 & 0 & 0 & 0 & 0 & 1 & \ell & \ell^2 & \ell^3 \\
-\frac{1}{r_{s2}} & -\frac{\ell}{r_{s2}} & -\frac{\ell^2}{r_{s2}} & -\frac{\ell^3}{r_{s2}} & 0 & 0 & 0 & 0 & 0 & 1 & 2\ell & \ell^2 \\
1 & \frac{\ell}{3} & \frac{\ell^2}{9} & \frac{\ell^3}{27} & 0 & 0 & 0 & 0 & 0 & 0 & 0 & 0 \\
0 & 0 & 0 & 0 & 1 & \frac{\ell}{3} & \frac{\ell^2}{9} & \frac{\ell^3}{27} & 0 & 0 & 0 & 0 \\
1 & \frac{2\ell}{3} & \frac{4\ell^2}{9} & \frac{8\ell^3}{27} & 0 & 0 & 0 & 0 & 0 & 0 & 0 & 0 \\
0 & 0 & 0 & 0 & 1 & \frac{2\ell}{3} & \frac{4\ell^2}{9} & \frac{8\ell^3}{27} & 0 & 0 & 0 & 0
\end{bmatrix}
$$

$\ell$ is the length of the curved meridian for the element.

Matrix $\underset{\sim}{X}_e$

$$
\underset{\sim}{X}_e = \begin{bmatrix}
\underset{\sim}{x}_e & & \underset{\sim}{0} \\
& \underset{\sim}{x}_e & \\
\underset{\sim}{0} & & \underset{\sim}{x}_e
\end{bmatrix}
$$

where

$$
\underset{\sim}{x}_e = \begin{bmatrix}
1 & s & s^2 & s^3 \\
s & s^2 & s^3 & s^4 \\
s^2 & s^3 & s^4 & s^5 \\
s^3 & s^4 & s^5 & s^6
\end{bmatrix}
$$

Matrix $\underset{\sim}{R}_e$

$$
\begin{bmatrix}
\cos\phi_1 & & -\sin\phi_1 & & & & & \\
 & 1 & & & & & \underset{\sim}{0} & \\
\sin\phi_1 & & \cos\phi_1 & & & & & \\
 & & & 1 & & & & \\
 & & & & \cos\phi_2 & & -\sin\phi_2 & \\
 & & & & & 1 & & \\
 & & & & \sin\phi_2 & & \cos\phi_2 & \\
 & \underset{\sim}{0} & & & & & & 1
\end{bmatrix}
$$

Matrix $\underset{\sim}{D}_j^{(c)}$

$$
\begin{array}{cccccccccccc}
0 & 1 & 2s & 3s^2 & 0 & 0 & 0 & 0 & \frac{1}{r_s} & \frac{s}{r_s} & \frac{s^2}{r_s} & \frac{s^3}{r_s} \\[2ex]
\frac{\sin\phi}{r} & \frac{s\sin\phi}{r} & \frac{s^2\sin\phi}{r} & \frac{s^3\sin\phi}{r} & \frac{j}{r} & \frac{js}{r} & \frac{js^2}{r} & \frac{js^3}{r} & \frac{\cos\phi}{r} & \frac{s\cos\phi}{r} & \frac{s^2\cos\phi}{r} & \frac{s^3\cos\phi}{r} \\[2ex]
-\frac{j}{2r} & \frac{-js}{2r} & \frac{-js^2}{2r} & \frac{-js^3}{2r} & -\frac{\sin\phi}{2r} & \frac{1}{2}\left(1-\frac{s\sin\phi}{r}\right) & \frac{1}{2}\left(2s-\frac{s^2\sin\phi}{r}\right) & \frac{1}{2}\left(3s^2-\frac{s^3\sin\phi}{r}\right) & 0 & 0 & 0 & 0 \\[2ex]
-\frac{1}{r_s^2}\frac{dr_s}{ds} & \frac{1}{r_s}-\frac{s}{r_s^2}\frac{dr_s}{ds} & \frac{2s}{r_s}-\frac{s^2}{r_s^2}\frac{dr_s}{ds} & \frac{3s^2}{r_s}-\frac{s^3}{r_s^2}\frac{dr_s}{ds} & 0 & 0 & 0 & 0 & 0 & 0 & -2 & -6s \\[2ex]
\frac{\sin\phi}{r\,r_s} & \frac{s\sin\phi}{r\,r_s} & \frac{s^2\sin\phi}{r\,r_s} & \frac{s^3\sin\phi}{r\,r_s} & \frac{j\cos\phi}{r^2} & \frac{js\cos\phi}{r^2} & \frac{js^2\cos\phi}{r^2} & \frac{js^3\cos\phi}{r^2} & \frac{j^2}{r^2} & \frac{j^2 s}{r^2}-\frac{\sin\phi}{r} & \frac{j^2 s^2}{r^2}-\frac{2s\sin\phi}{r} & \frac{j^2 s^3}{r^2}-\frac{3s^2\sin\phi}{r} \\[2ex]
\frac{-j}{r\,r_s} & \frac{-js}{r\,r_s} & \frac{-js^2}{r\,r_s} & \frac{-js^3}{r\,r_s} & -\frac{\cos\phi\sin\phi}{r^2} & \frac{\cos\phi}{r}\left(1-\frac{s\sin\phi}{r}\right) & \frac{\cos\phi}{r}\left(2s-\frac{s^2\sin\phi}{r}\right) & \frac{\cos\phi}{r}\left(3s^2-\frac{2s^3\sin\phi}{r}\right) & \frac{-j\sin\phi}{r^2} & \frac{j}{r}\left(1-\frac{s\sin\phi}{r}\right) & \frac{j}{r}\left(2s-\frac{s^2\sin\phi}{r}\right) & \frac{j}{r}\left(3s^2-\frac{s^3\sin\phi}{r}\right)
\end{array}
$$

CHAPTER 12

# SOME RECENT ADVANCES IN STRUCTURAL VIBRATION

by

G.B. Warburton

## 1. Introductory Remarks

In recent years the finite element method has been developed to the state where approximate mathematical models can be formulated for complex practical structures. These models have to represent the actual geometry and take into account the types of deformation which the structure will undergo in practice. Large versatile and efficient computer programs, which are based on the finite element method, exist for stress analysis; a general program will include the capability of determining response to dynamic loads. Research work on the finite element method continues, but new elements, which are economical, accurate and conform with existing elements for the other types of deformation and thus are worthy of inclusion in general programs, are of less frequent occurrence. Developments continue in elastic-plastic problems, which occur for large deformations, for example in earthquake engineering.

A modelling problem occurs when finite elements are used to idealise either the soil underlying a structure or the fluid surrounding a structure in dynamic interaction problems. The former occurs whenever vibrations are transmitted to a structure through the ground, for example by traffic, explosions, machines in the vicinity and earthquakes. It will occur also for other problems, such as the wind-induced response of a structure, unless it is assumed that the foundation is immovable. The latter problem exists for dams and offshore structures. The use of a very large number of three-dimensional elements to simulate the ground or surrounding fluid leads to excessive computation. Although considerable work has been done with two-dimensional models, either by using a plane strain idealisation (e.g. dam problems) or using an axisymmetric representation (suitable for geometrically symmetric structures), there remain many problems which have to be treated three-dimensionally in order to determine response. Artificial boundaries have to be provided for the finite element mesh representing the soil; these should not allow any reflected waves to return to the structure whose behaviour is of prime importance. (A similar problem concerning artificial boundaries arises when designing an experimental idealisation of interaction problems.) Progress has been made with the generation of satisfactory artificial boundaries, which allow a conventional finite element approach, but other methods have been developed: analytical solutions for the soil or fluid, which are combined with a finite element analysis of the structure (analytical solutions are limited to relatively simple geometries); representation of the outer domain by infinite elements; and combinations of finite element and boundary solution procedures. It is not proposed to attempt to summarise the extensive recent and current work in these areas, but instead recent advances in methods of determining response, which is the basic problem of structural dynamics and is a major subject of study in this book, will be outlined.

Although the principles of determining the response are well established, considerable research effort is being expended in order to develop satisfactory approximations, which are accurate and efficient. A realistic finite element model of any large structure leads to matrix equations of large order; allowance for interaction effects makes the order considerably larger. Displacements are often of such a magnitude that non-linear behaviour of some parts of the structure must be considered. Excitations may be simulated deterministically or stochastically. These three features have influenced developments in structural dynamics.

Recent advances for the normal mode method, namely the treatment of the damping matrix, truncation of the series solution and use of response spectrum techniques, have been described in the chapter on the determination of response, so emphasis here will be on the direct integration methods, which are the subject of considerable current research. However, before doing this some comparative comments on the two methods may be useful. In direct integration methods the full set of equations must be used; matrix algebra, at which modern computers are very efficient, forms the basis of the numerical procedure, but the sequence of calculations must be repeated many times, i.e. small time steps must be used, to obtain response accurately. There is no necessity to determine any natural frequencies or mode shapes. In the normal mode method the natural frequencies $\omega_r$ and associated normalized modal vectors $\underset{\sim}{z}_r$ must be determined for all modes which make a significant contribution to response. Thus for a large system only a fraction of the frequencies and modal vectors are required in principle; in practice, it is likely that the number of modes which are retained in the summations exceeds the optimum minimum in order to ensure that no significant modes are omitted. It must be stressed that while a change in the time-history of the excitation or in the response quantity of interest has no effect on the natural frequencies and modal vectors, it can affect substantially the number of terms that must be retained in the series for response. If a response spectrum method is used and the relevant spectra are available, the determination of maximum response is a relatively simple task. Spectra have been obtained for most simple excitation - time histories and for many well-documented earthquake records. If the spectra have to be generated, the advantages of this procedure over a conventional time-history approach are less, but still significant.

## 2. Direct Integration Methods

In these methods assumptions are made about the variation of the displacements or accelerations during small time intervals; e.g. it may be assumed that during a small time interval the displacement is a cubic function of time or the acceleration varies linearly or is constant. With such assumptions the set of n second order differential equations is replaced, in general, by n simultaneous equations. The solution of the latter gives the displacements at the end of the short time step for known conditions at the beginning. Successive applications of this procedure give the response.

Many methods exist; the established methods, which have been incorporated in computer codes (finite element or finite difference) for determining the response of large structures, include the Newmark β, Houbolt, Wilson θ and central difference methods. Methods which exhibit improved characteristics for test problems have been proposed by several authors, including Park [1] and Hilber et al [2,3]. Recent comparisons of some of the methods in current use are given in References [2-4.]

If $\underset{\sim}{U}_{s+1}$, $\underset{\sim}{U}_{s}$ and $\underset{\sim}{U}_{s-1}$ are the values of $\underset{\sim}{U}$ at times $t_{s+1}$, $t_s$ and $t_{s-1}$ respectively, where $t_{s+1} - t_s = t_s - t_{s-1} = \Delta t$, a small time interval, and velocity, acceleration and force vectors at time $t_s$ are $\dot{\underset{\sim}{U}}_s$, $\ddot{\underset{\sim}{U}}_s$ and $\underset{\sim}{P}_s$ respectively, the matrix equation for time $t_s$ is

$$\underset{\sim}{M}\, \ddot{\underset{\sim}{U}}_s + \underset{\sim}{C}\, \dot{\underset{\sim}{U}}_s + K\, \underset{\sim}{U}_s = \underset{\sim}{P}_s \tag{1}$$

If the solution at time $t_{s+1}$ is determined by considering the equilibrium condition at time $t_s$, i.e. equation (1), we have an explicit integration method. The central difference method is one of the best known methods in this class. Substituting the difference formulae

$$\dot{\underset{\sim}{U}}_s = [\underset{\sim}{U}_{s+1} - \underset{\sim}{U}_{s-1}] \,/\, (2\Delta t) \tag{2}$$

and

$$\ddot{\underset{\sim}{U}}_s = [\underset{\sim}{U}_{s+1} - 2\underset{\sim}{U}_s + \underset{\sim}{U}_{s-1}] \,/\, (\Delta t)^2 \tag{3}$$

in equation (1) and rearranging terms

$$[\underset{\sim}{M} + \tfrac{1}{2}\Delta t\, \underset{\sim}{C}]\, \underset{\sim}{U}_{s+1} = (\Delta t)^2\, \underset{\sim}{P}_s + [2\underset{\sim}{M} - (\Delta t)^2\, \underset{\sim}{K}]\, \underset{\sim}{U}_s + [\tfrac{1}{2}\Delta t\, \underset{\sim}{C} - \underset{\sim}{M}]\, \underset{\sim}{U}_{s-1} \tag{4}$$

In many applications damping is neglected. In order to preserve the explicit form of equation (4), i.e. give explicit expressions for $U_1$, $U_2$, ... at time $t_{s+1}$

rather than a set of simultaneous equations, $\underset{\sim}{M}$ must be diagonal. This is a characteristic of the finite difference method, but not of the consistent mass matrix in the finite element method. However, lumped mass approximations can be used with the latter methods; also improved diagonal mass matrices have been developed for several popular elements in order to preserve the computational advantages of explicit expressions for $U_j$. These advantages are particularly important when the matrices are of large order. Damping may be included and the explicit form preserved (provided $\underset{\sim}{M}$ is diagonal), if the backward, rather than central, difference approximation is used for $\dot{U}_s$; i.e. equation (2) is replaced by

$$\dot{\underset{\sim}{U}}_s = (\underset{\sim}{U}_s - \underset{\sim}{U}_{s-1}) / \Delta t$$

The modified form of equation (4) can be written

$$\underset{\sim}{U}_{s+1} = \underset{\sim}{M}^{-1} [(\Delta t)^2 \{\underset{\sim}{P}_s - \underset{\sim}{K}\underset{\sim}{U}_s\} - \Delta t \, \underset{\sim}{C} \, (\underset{\sim}{U}_s - \underset{\sim}{U}_{s-1})] + 2\underset{\sim}{U}_s - \underset{\sim}{U}_{s-1} \qquad (5)$$

The disadvantage of the central difference method (and all other explicit methods [5]) is its conditional stability, i.e., if too large a time step is used, the computations become numerically unstable. The stability criterion is $\Delta t/T_n \leqslant 1/\pi$, where $T_n$ is the period of the $n^{th}$ or highest mode of the system.

If the solution at time $t_{s+1}$ requires consideration of the equilibrium condition at time $t_{s+1}$, the method is implicit and requires the solution of a set of simultaneous equations at each time step. With certain provisos these methods are unconditionally stable, at least for linear systems. In the Newmark method the assumptions are [6]

$$\dot{\underset{\sim}{U}}_{s+1} = \dot{\underset{\sim}{U}}_s + [1-\gamma) \ddot{\underset{\sim}{U}}_s + \gamma\ddot{\underset{\sim}{U}}_{s+1}] \Delta t \qquad (6)$$

and $$\underset{\sim}{U}_{s+1} = \underset{\sim}{U}_s + \Delta t \, \dot{\underset{\sim}{U}}_s + (\tfrac{1}{2}-\beta)(\Delta t)^2 \ddot{\underset{\sim}{U}}_s + \beta(\Delta t)^2 \ddot{\underset{\sim}{U}}_{s+1}$$

Effectively this is a family of methods, as equations (6) contain two adjustable parameters $\beta$ and $\gamma$. If $\gamma < \tfrac{1}{2}$, negative damping is introduced by the algorithm, leading eventually to an unbounded response; if $\gamma > \tfrac{1}{2}$, positive damping is introduced. Thus in most applications $\gamma = \tfrac{1}{2}$. The general expression for unconditional stability is [7]

$$\beta \geqslant \tfrac{1}{4} (\gamma + \tfrac{1}{2})^2 \qquad (7)$$

The parameter $\beta$ is often taken to be $\tfrac{1}{4}$; then the physical interpretation of assumptions (6) is that the acceleration vector is constant during the time interval

$t_s$ to $t_{s+1}$ and is equal to the mean of $\ddot{\underset{\sim}{U}}_s$ and $\ddot{\underset{\sim}{U}}_{s+1}$ - hence the name 'average acceleration method'. It is noted that the case $\beta = 1/6$ and $\gamma = \frac{1}{2}$ (the 'linear acceleration method', as $\ddot{\underset{\sim}{U}}$ varies linearly from $\ddot{\underset{\sim}{U}}_s$ to $\ddot{\underset{\sim}{U}}_{s+1}$ over the time interval) is an implicit method, which is only conditionally stable, as condition (7) is not satisfied. ($\Delta t/T_n \leq 0.551$ is the stability criterion).

Putting $\gamma = \frac{1}{2}$, but leaving $\beta$ as an adjustable parameter, using assumptions (6) in equation (1) and two similar equations for times $t_{s-1}$ and $t_{s+1}$, the following set of simultaneous equations is obtained after eliminating terms in $\dot{\underset{\sim}{U}}$ and $\ddot{\underset{\sim}{U}}$.

$$\begin{aligned}[\underset{\sim}{M} + \tfrac{1}{2}\Delta t\,\underset{\sim}{C} + \beta(\Delta t)^2\underset{\sim}{K}]\,\underset{\sim}{U}_{s+1} &= (\Delta t)^2[\beta\,\underset{\sim}{P}_{s+1} + (1-2\beta)\,\underset{\sim}{P}_s + \beta\,\underset{\sim}{P}_{s-1}] \\ &+ [2\underset{\sim}{M} - (\Delta t)^2(1-2\beta)\,\underset{\sim}{K}]\,\underset{\sim}{U}_s - [\underset{\sim}{M} - \tfrac{1}{2}\Delta t\,\underset{\sim}{C} + \beta(\Delta t)^2\,\underset{\sim}{K}]\,\underset{\sim}{U}_{s-1}\end{aligned} \qquad (8)$$

## 3. Accuracy

With a stable implicit method, e.g. Newmark with condition (7) satisfied or Wilson with $\theta > 1.37$, considerations of accuracy govern the choice of time step $\Delta t$. Considering the response to comprise contributions from normal modes, the approximations in direct integration methods usually cause artificial attenuation of the response and some error in the period of the mode predicted by the numerical solution. These two effects increase as $\Delta t/T_r$ increases, where $T_r$ is the period of the mode whose contribution is being considered. Errors from the above methods are compared in References [2-4] and [8]. For the Newmark method with $\beta = \frac{1}{4}$ and $\gamma = \frac{1}{2}$ the error in the predicted modal period $T_r$ is less than 10 per cent, if $\Delta t/T_r \leq 0.19$. The other methods give a lower value of $\Delta t$ for the same error.

Unlike other methods, the Newmark average acceleration method does not cause any artificial attenuation of the modal contributions to response. However, this is not necessarily the best algorithm, as the period errors cause the contributions to be combined with incorrect relative phase angles, and thus the maximum response may be in error. In practice, computational economies or lack of knowledge may cause the higher significant modes to be modelled inaccurately. This leads to the recommendation [2, 3, 8] that an algorithm should possess some artificial attenuation or numerical dissipation in order that the spurious response from higher modes is damped out. Controllable dissipation is included, together with unconditional stability and second order accuracy, in the desirable attributes of any direct integration method by Hilber and Hughes [3]. They propose a family of algorithms, the $\alpha$ method, in which the parameter $\alpha$ controls the numerical dissipation. Assumption (6) is used, but in the equilibrium condition for $t_{s+1}$ [see equation (1)] $\underset{\sim}{K}\,\underset{\sim}{U}_{s+1}$ is replaced by $(1+\alpha)\,\underset{\sim}{K}\,\underset{\sim}{U}_{s+1} - \alpha\underset{\sim}{K}\,\underset{\sim}{U}_s$. Thus, when $\alpha = 0$, it is identical to the Newmark average acceleration method. The relations between the parameters are $\beta = (1-\alpha)^2/4$ and $\gamma = \frac{1}{2}-\alpha$ . The range of practical interest is $-\frac{1}{3} \leq \alpha \leq 0$, for which the required numerical dissipation is obtained. However, as expected there is some degradation in accuracy as $\alpha$ departs from zero. Accepting their criteria, the $\alpha$ methods are demonstrated to be superior to the Wilson $\theta$, Houbolt and Park methods. However, their comprehensive study is based on matrix equation (1) without the damping term $\underset{\sim}{C}\,\dot{\underset{\sim}{U}}$.

In one test, which has been applied to many integration methods [2, 3, 8], a 1 DOF system, which can represent one modal contribution of a multi-degree-of-freedom system, is given an initial displacement and the algorithmic damping or decrease in amplitude at the end of one period is plotted against $\Delta t/T_r$. For undamped systems the amplitude decay increases as $\Delta t/T_r$ and $|\alpha|$ increase. For damped systems, using the Newmark method ($\alpha$=0), the amplitude error is positive and increases as $\Delta t/T_r$ and the damping ratio increase. Thus, applying the $\alpha$ method to damped systems, increasing $|\alpha|$ and the damping ratio causes errors of opposite sign in the amplitude and leads to a pattern of results of moderate complexity. For practical problems, where $\Delta t/T_r$, and possibly also the modal damping ratio, will vary with mode number, the determination of optimum integration conditions appears to require further study.

For a stable implicit method an error criterion will be an upper bound on $\Delta t/T_i$, where $\Delta T_i$ is the period of the highest mode which makes a significant contribution to the response. For an explicit method the stability criterion is an upper bound on $\Delta t/T_n$. In seismic problems only a small fraction of the modes makes a significant contribution, i.e. $i \ll n$, and hence the maximum value of $\Delta t$ to give acceptable accuracy with an implicit method will usually be much larger than the maximum value of $\Delta t$ for stability with an explicit method. Key [4] compares methods for application to different types of problem and recommends matching of methods, i.e. the errors due to the spatial and temporal approximations should be of opposite signs. In particular, lumped (diagonal) mass matrices, which predict low values of natural frequency, should be used in conjunction with the central difference method, which distorts frequencies upward; conversely consistent mass matrices, which predict high values of natural frequency, should be used in conjunction with a stable implicit integration procedure, which depresses the frequencies. For linear seismic problems he recommends the Hilber-Hughes-Taylor $\alpha$ method with a consistent mass discretisation. For impact problems he recommends the central difference method with a lumped mass discretisation.

## 4. Non-linear Problems

In non-linear problems the restoring force vector $\underset{\sim}{K}\,\underset{\sim}{U}$ in equation (1) is replaced by $\underset{\sim}{F}(\underset{\sim}{U},t)$, i.e.

$$\underset{\sim}{M}\,\ddot{\underset{\sim}{U}} + \underset{\sim}{C}\,\dot{\underset{\sim}{U}} + \underset{\sim}{F}\,(\underset{\sim}{U},t) = \underset{\sim}{P}(t) \tag{9}$$

Applying the central difference method to equation (9), equation (5) is replaced by

$$\underset{\sim}{U}_{s+1} = \underset{\sim}{M}^{-1}\,[(\Delta t)^2(\underset{\sim}{P}_s-\underset{\sim}{F}_s) - \Delta t\,\underset{\sim}{C}\,(\underset{\sim}{U}_s-\underset{\sim}{U}_{s-1})] + 2\underset{\sim}{U}_s-\underset{\sim}{U}_{s-1} \tag{10}$$

where $\underset{\sim}{F}_s$ is the value of $\underset{\sim}{F}$ at time $t_s$. Thus all terms on the right-hand side of equation (10) have been determined from previous iterations and $\underset{\sim}{U}_{s+1}$ is given explicitly by equation (10).

If the Newmark method is applied to equation (9), $\underset{\sim}{K}\,\underset{\sim}{U}_{s+1}$ on the left-hand side of equation (8) is replaced by $\underset{\sim}{F}_{s+1}$, which depends on the unknown vector $\underset{\sim}{U}_{s+1}$. Thus an alternative procedure is necessary. We assume that

$$\underset{\sim}{F}_{s+1} = \underset{\sim}{F}_s + \underset{\sim}{K}_s\;\delta\underset{\sim}{U} \tag{11}$$

where $\underset{\sim}{K}_s$ is the tangential stiffness matrix evaluated from the conditions at time $t_s$ and

$$\delta\underset{\sim}{U} = \underset{\sim}{U}_{s+1} - \underset{\sim}{U}_s \tag{12}$$

Using equations (11) and (12) in the non-linear form of equation (8) and re-arranging terms

$$[\underset{\sim}{M} + \tfrac{1}{2}\,\Delta t\,\underset{\sim}{C} + \beta(\Delta t)^2\,\underset{\sim}{K}_s]\;\delta\underset{\sim}{U} = (\Delta t)^2[\beta\underset{\sim}{P}_{s+1} + (1-2\beta)\,\underset{\sim}{P}_s + \beta\underset{\sim}{P}_{s-1}] - (\Delta t)^2\,[(1-\beta)\underset{\sim}{F}_s+\underset{\sim}{F}_{s-1}]$$

$$+ [\underset{\sim}{M}-\tfrac{1}{2}\,\Delta t\,\underset{\sim}{C}]\;[\underset{\sim}{U}_s-\underset{\sim}{U}_{s-1}] \tag{13}$$

The solution of equation (13) gives $\delta\underset{\sim}{U}$ and an approximation for $\underset{\sim}{U}_{s+1}$ is obtained from equation (12). After obtaining the corresponding velocity and acceleration vectors, $\dot{\underset{\sim}{U}}_{s+1}$ and $\ddot{\underset{\sim}{U}}_{s+1}$, from equations (6) it is necessary for accuracy and stability to determine the error vector $\underset{\sim}{\varepsilon}_{s+1}$ from the equilibrium equation.

$$\underset{\sim}{\varepsilon}_{s+1} = \underset{\sim}{P}_{s+1} - \underset{\sim}{F}_{s+1} - \underset{\sim}{M}\,\ddot{\underset{\sim}{U}}_{s+1} - \underset{\sim}{C}\,\dot{\underset{\sim}{U}}_{s+1}$$

Using $\quad [\underset{\sim}{M} + \tfrac{1}{2}\,\Delta t\,\underset{\sim}{C} + \beta(\Delta t)^2\,\underset{\sim}{K}_s]\;\Delta\underset{\sim}{U} = (\Delta t)^2\,\underset{\sim}{\varepsilon}_{s+1}$

a new iterated value of $\underset{\sim}{U}_{s+1}$ is formed by adding $\Delta\underset{\sim}{U}$ to the original value. This iterative procedure is repeated until $\Delta\underset{\sim}{U}$ is sufficiently small; then the computation advances to the next time step.

Comparing the above approaches, the computational advantages of the central difference method are apparent.

Owen [9] describes the application to non-linear problems of some major implicit algorithms, but concludes that explicit methods are considered to be more economical. The small time steps that are required to follow path-dependent stress histories may be of the same order as those imposed by stability considerations. Thus, as in linear wave propagation problems, where accuracy requires very small time steps, the disadvantage of the explicit formulation vanishes, but the computational advantages remain. Comparative studies and surveys of earlier work are given in References [10] and [11].

## 5. Partitioning

If an explicit scheme is applied to a soil-structure interaction problem, the maximum time step to preserve stability is controlled by the maximum frequency for the idealization of the relatively stiff structure. A considerably larger critical time step would be predicted for an idealization of the flexible soil alone, but in a conventional approach to the complete system the lower critical time step must be used. As a major contribution to computational efficiency recent work has developed mesh partitioning, in which different integration schemes, or at least different time steps, are used for the various partitions. References [4] and [12] contain surveys. Explicit-explicit partitions with different time steps, explicit-implicit and implicit-implicit partitions have been developed. (Explicit-explicit partitions with the same time step, but with different computer programs for the two media, have applications in interaction problems). Explicit-implicit partitions are useful for soil-structure interaction problems; implicit integration is used only for the structural mesh, making possible the use of consistent mass matrices. The advantage over an implicit formulation for the complete system is the resulting smaller bandwidth of the implicit part of the mesh, and the consequential diminished computational effort, particularly for non-linear problems.

Emphasis has been on recent advances in the determination of response. However, this should not be interpreted as implying that advances are not occurring in other areas of structural vibration. For instance, reduction of vibration is of continuing interest. When calculations predict, or more likely measurements show, that unacceptable vibration levels will occur in components, there are various possible remedies. First, the source of excitation should be studied and its magnitude reduced if possible. Materials with higher damping, which could be achieved by the introduction of visco-elastic layers, or of high-damping joints or connections between sub-structures, will reduce response levels. If a particular component, such as a sensitive instrument, may be subjected to excessive vibrations, it can be isolated from the source of vibrations by a resilient mounting. Similarly, if excessive vibrations are being transmitted from a machine to its surroundings, vibration isolation of the machine will be effective. If other methods of vibration reduction are unsuitable or impossible, the addition of a properly designed absorber system - in effect a damped single degree-of-freedom system - can achieve large reductions in vibration levels when the initial dynamic magnification factor is large. Reference 13 gives a survey of vibration reduction; the author [14] has described how to determine absorber parameters in order to provide the optimum reduction of response of vibrating systems.

## References

1. Park, K.C., "An improved stiffly stable method for direct integration of non-linear structural dynamic equations", Journal of Applied Mechanics, 42, 1975, 464-470.

2. Hilber, H.M., T.J.R. Hughes and R.L. Taylor, "Improved numerical dissipation for time integration algorithms in structural dynamics", Earthquake Engineering and Structural Dynamics, 5, 1977, 283-292.

3. Hilber, H.M. and T.J.R. Hughes, "Collocation, dissipation and 'overshoot' for time integration schemes in structural dynamics", Earthquake Engineering and Structural Dynamics, 6, 1978, 99-117.

4. Key, S.W., "Transient response by time integration: a review of implicit and explicit operators", Chapter 3 of 'Advanced Structural Dynamics', (Editor J. Donéa), Applied Science, London, 1980.

5. Krieg, R.D., "Unconditional stability in numerical time integration methods", Journal of Applied Mechanics, 40, 1973, 417-421.

6. Newmark, N.M., "A method of computation for structural dynamics", Journal of the Engineering Mechanics Division, ASCE, 85, No. Em3, 1959, 67-94.

7. Goudreau, G.L. and R.L. Taylor, "Evaluation of numerical integration methods in elasto-dynamics", Computer Methods in Applied Mechanics and Engineering, 2, 1972, 69-97.

8. Bathe, K.J. and E.L. Wilson, "Stability and accuracy analysis of direct integration methods", Earthquake Engineering and Structural Dynamics, 1, 1973, 283-291.

9. Owen, D.R.J., "Implicit finite element methods for the dynamic transient analysis of solids with particular reference to non-linear situations", Chapter 5 of 'Advanced Structural Dynamics', (Editor J. Donéa), Applied Science, London, 1980.

10. Stricklin, J.A. and W.E. Haisler, "Formulations and solution procedures for nonlinear structural analysis", Computers and Structures, 7, 1977, 125-136.

11. Adeli, H., J.M. Gere and W. Weaver Jr., "Algorithms for non-linear structural dynamics", Journal of the Structures Division, ASCE, 104, 1978, 263-280.

12. Belytschko, T., "Explicit time integration of structure-mechanical systems", Chapter 4 of 'Advanced Structural Dynamics', (Editor J. Donéa), Applied Science, London, 1980.

13. Snowdon, J.C., and E.E. Ungar (editors) 'Isolation of mechanical vibration, impact and noise', ASME AMD Vol. 1, 1973.

14. Warburton, G.B., "Optimum absorber parameters for various combinations of response and excitation parameters", Earthquake Engineering and Structural Dynamics, 10, 1982, 381-401.

CHAPTER 13

# FLUID STRUCTURE INTERACTION PROBLEMS

by

C.A. Brebbia

## 1. Introduction

In this chapter we will review those fluid-structure interaction problems which due to their dynamic effects are important in the design of off-shore structures. These effects can then be introduced into the equations of motion of the system, whose solution gives the response (deterministic or probabilistic) of the structure.

In the past most off-shore structures were built in relatively shallow waters and had fundamental periods of less than one second. For these periods the sea spectrum does not present a significant energy content and the structures could be designed using equivalent static loads. In recent years off-shore structures have started to be built in deep waters and their fundamental periods have increased considerably. These periods may be in the order of 4 to 8 seconds for structures in between 100 to 200 m. of depth. For these structures the equivalent static approach is not sufficient and a fully dynamic analysis is required. This analysis is necessary because their dynamic loads effects can be magnified by the response of the structure. A dynamic analysis can also provide an idea of fatigue damage.

Two different methods of dynamic analysis are currently used for off-shore structures. The first is the design wave approach and consists of assuming a wave of a given period and height which represents the maximum wave occurring for certain environmental conditions. The second approach is to work with the wave energy spectrum using probabilistic theory, which allows to obtain maximum results for stresses, displacements, etc. of the structure within a known confidence level. The latter approach is to be preferred since the design wave does not generally give the maximum response of the structure, which depends on how the structure reacts to the loading (e.g. a wave of the same or smaller height as the design wave but of longer period could produce a larger response if its period is nearest to the resonance one than the period of the design wave). One will generally expect the response from a spectral analysis to be larger than the one obtained by the design wave, as the former will take into consideration all the periods.

Different environmental phenomenon contribute to the dynamic loading of off-shore structures, the most important of which is the wave loading. Wind blowing on the superstructure only represents a 5% to 10% of the total side loading. Currents due

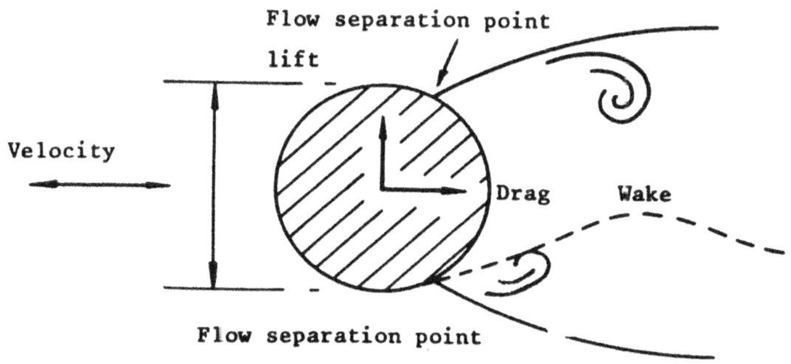


Figure 1 Vortex Shedding

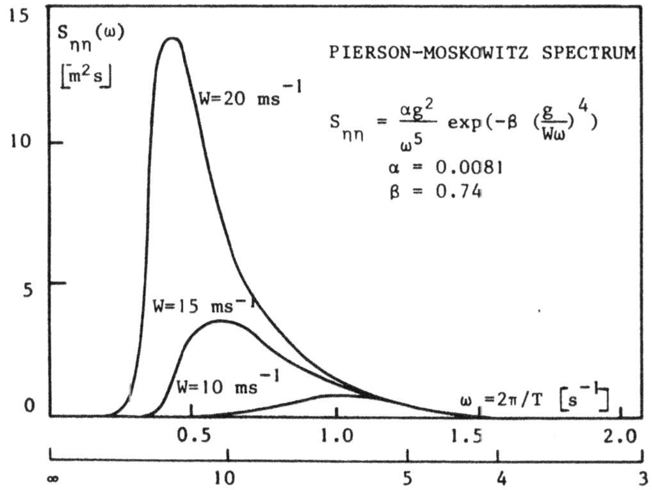


Figure 2 Wave Spectrum

to tides, winds, etc. are important because they produce significant drag forces and vortex shedding in structural members. Prediction of currents due to surges imply the simultaneous study of tidal, storm surges and currents due to wind and may be attempted using a numerical model which can correlate those different phenomena. In addition to these effects the designer should also consider the effect of water temperature variation on the structure and the foundation problems, including scouring. Consideration of the latter problems is beyond the scope of this chapter.

Let us first discuss wave effects which produce the major loading forces in the system and can be introduced in the analysis using the design wave or the wave spectrum approach. In both cases the basic information about the waves is height and period which have to be converted into forces by applying in the first place, a wave theory. In this way velocities and accelerations are obtained from the wave height and period data. Wave theories range from the simplest linear theory to rather complex non-linear ones. The differences to be expected by using linear or non-linear theories are reported to be small [11], except for waves near breaking point. As the spectral analysis also requires a linear theory, in what follows we will use this. Once the velocities and accelerations are expressed in function of the wave height we can compute the forces produced by them on the members. Disregarding for the moment lift forces produced by the phenomenon of vortex shedding, those forces are of two types, drag and inertia. For members for which the significant length (e.g. such as diameter in a cylinder) to wave length ratio is small, one can use a Morrison's type formula to obtain the forces. This formula gives inertia and drag forces without considering any modification in the shape of the wave and is generally accepted to be valid for diameter to wave length ratio of less than 0.2. Structures of large diameter such as some of the gravity structures being built, are outside this range. In this case we have to take into consideration the change in the shape of the waves due to diffraction effects. For these structures the drag components are very much less important than the inertia ones, which tend to be large. It is the inertia effects the ones that the diffraction theory which is valid for inviscid fluids (i.e. drag), is able to calculate.

Viscous effects around the member produce in addition to drag, a shedding of vortices in the wake. These vortices occur alternatively in one or the other side of the wake behind the obstruction (Figure 1). The succession of them is called a vortex street and has a frequency which is given by a dimensionless number called Strouhal number. If the shedding frequency of the vortices is similar to the frequency of the structure lift forces which are several times larger than the drag forces and also oscillatory, will occur. In many cases the natural frequencies of the structure are above the wave spectrum significant frequencies but within the shedding frequency of the vortices and resonance occurs. The designer ought to be aware of this phenomenon as several failures of this type have been registered.

When using the probabilistic approach, it is essential to determine carefully the wave spectrum. The form of this spectrum is of primary importance for the response and experiments should be carried out to find the spectra for different environmental conditions. Unfortunately in most cases the designer will not have full experimental results. If this is so, one can use an empirical developed expression for the wave spectrum. One of the most commonly used is the Pierson and Moskowitz [19] spectrum, which gives the following expression for the wave elevation,

$$S_{\eta\eta}(\omega) = \frac{\alpha g^2}{\omega^5} \exp\left[ -\beta \left(\frac{g}{\omega W}\right)^4 \right] \qquad (1)$$

where g is the acceleration due to gravity, ω the frequency and W the wind speed at 20 metres above the sea surface. α and β are dimensionless constants. They can be taken as $\alpha = 8.1 \times 10^{-3}$ and $\beta = 0.74$ for the North Sea [14]. In Figure 2, equation (1) has been plotted for wind speeds of 10, 15 and 30 metres per second. Sometimes the designer is given the significant wave height, $H_s$, and the mean wave period, $T_m$. The α and β constants can then be obtained using the following relationships presented by Scott [20],

$$H_s = 2\frac{W^2}{g}\sqrt{\frac{\alpha}{\beta}}$$

$$T_m = 2\pi \frac{W}{g}\left(\frac{1}{\beta\pi}\right)^{1/4} \qquad (2)$$

In what follows we will discuss in more detail the environmental effects mentioned, and how they contribute to the loading of the structure. The governing equations of a multi-degree of freedom system are written in terms of mass, inertia and stiffness matrices plus any forces acting on the system, i.e.

$$\underset{\sim}{M}\,\ddot{\underset{\sim}{U}} + \underset{\sim}{C}\,\dot{\underset{\sim}{U}} + \underset{\sim}{K}\,\underset{\sim}{U} = \underset{\sim}{F}(t) \qquad (3)$$

where $\underset{\sim}{U}$ are the structure displacements and interaction effects will be included in the $\underset{\sim}{F}(t)$ term.

## 2. The Mechanics of Drag, Inertia and Lift

We will now study how drag, inertia and lift forces are created by looking at the case of a circular cylinder in steady and unsteady flow. The case of the cylinder is interesting because of its many structural applications and also is the best documented structural section.

Consider the cylinder in a uniform flow for which vortex shedding has not yet occurred. Typical streamlines around the section are shown in Figure 3, though the figure only shows a low Reynolds number which are the best documented ones; high Re numbers will give similar results. The shear ($\tau$) and pressure stresses (p) produced by the fluid are distributed around the body as shown in Figure 4 [1]. The components of these stresses in the direction of flow produce a force on the cylinder. Note that the hydrostatic pressure does not produce any net force on the cylinder. It is important to take it into account when designing the cylinder itself in order to prevent buckling.

The momentum equations for the two dimensional incompressible viscous fluid surrounding the cylinder are,

$$\frac{D\dot{v}_1}{Dt} = b_1 - \frac{1}{\rho}\frac{\partial p}{\partial x_1} + \mu \nabla^2 \dot{v}_1$$

$$\frac{D\dot{v}_2}{Dt} = b_2 - \frac{1}{\rho}\frac{\partial p}{\partial x_2} + \mu \nabla^2 \dot{v}_2 \qquad (4a)$$

where $\dot{v}_1$, $\dot{v}_2$ are the velocities of the fluid, $b_1$, $b_2$ the body forces, p the pressure and $\mu$ the dynamic viscosity. $\frac{D}{Dt}$ is the particle derivative and is equal to a local time derivative part plus a convective part, i.e.

$$\frac{D}{Dt} = \frac{\partial}{\partial t} + \dot{v}_1 \frac{\partial}{\partial x_1} + \dot{v}_2 \frac{\partial}{\partial x_2}$$

This derivative is used when working in an Eulerian system of coordinates, for which the variables are referred to fixed cartesian axis. In solid mechanics instead, the variables are normally referred to a Lagrangian system for which accelerations are only function of the local time derivative, i.e. no convective part is required.

In addition to equations (4a), the velocities have to satisfy the incompressibility condition,

$$\frac{\partial \dot{v}_1}{\partial x_1} + \frac{\partial \dot{v}_2}{\partial x_2} = 0 \qquad (4b)$$

The stresses are usually given by the following relationship,

$$\sigma_{11} = -p + 2\mu \frac{\partial \dot{v}_1}{\partial x_1}, \qquad \sigma_{22} = -p + 2\mu \frac{\partial \dot{v}_2}{\partial x_2}$$

$$\tau_{12} = \mu \left( \frac{\partial \dot{v}_1}{\partial x_2} + \frac{\partial \dot{v}_2}{\partial x_1} \right)$$

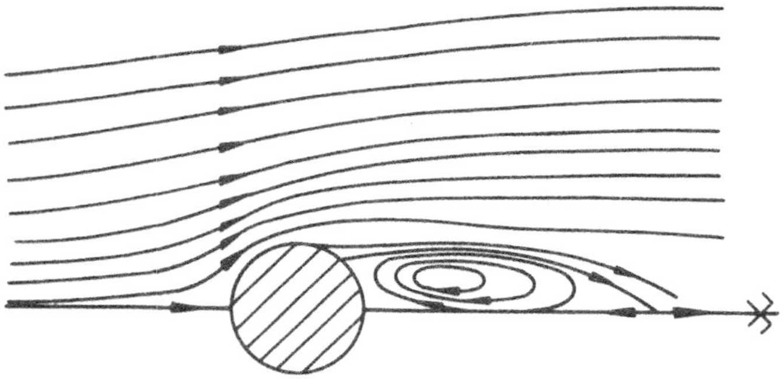

Figure 3 Streamlines (Re = 40)

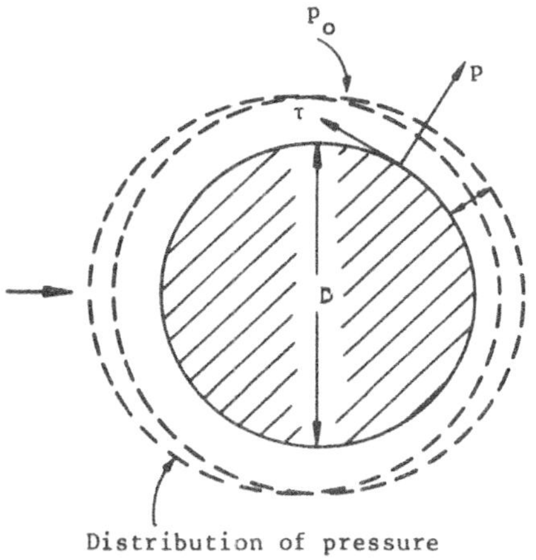


Figure 4 Pressure Distribution over a Cylinder.

The numerical solution of equations (4a) and (4b) has been satisfactorily achieved for a reduced number of cases, all of them at low Reynolds numbers (Reynolds number is $Re = \dot{v}d/\nu$, where $\nu$ is the kinematic viscosity) and using numerical techniques the distribution of shear and pressure stresses around the obstruction are determined at any time free of experimental errors. However, even if such solutions were possible for high Re numbers, those distributions would be too complex to be used in the practice. Hence the designer usually represents their effects as a drag force and in addition, if the fluid or the obstruction accelerates, as an inertia force. The inertia component for high Re number is mainly due to the potential behaviour of the flow (i.e. as if there was no viscosity) the drag component is due to the viscous components (if the viscosity is zero, the drag disappears). Drag forces are function of the velocity field around the obstruction and inertia forces of the accelerations. For steady flow inertia will disappear.

We will soon see an expression for these forces valid for slender member. In order to compute them we start by defining the velocities and accelerations in function of the wave height and period, using the following wave theory.

Linear Wave Theory

Consider a wave incident on a cylinder (Figure 5) and defined by the wave height $H = 2a$ and the wave length $\lambda$, $D$ is the diameter of the member and $d$ the depth, $c$ is the velocity or celerity of the wave.

We will review the linear solution for these periodic waves which has the advantage that can be applied in spectral super-position analysis. The equations governing the problem can be deduced from equations (4a) and (4b) after neglecting viscosity, i.e.

$$\frac{D\dot{v}_1}{Dt} = b_1 - \frac{1}{\rho}\frac{\partial p}{\partial x_1}$$

$$\frac{D\dot{v}_2}{Dt} = b_2 - \frac{1}{\rho}\frac{\partial p}{\partial x_2} \tag{5}$$

$$\frac{\partial \dot{v}_1}{\partial x_1} + \frac{\partial \dot{v}_2}{\partial x_2} = 0$$

Note that we are again considering the two dimensional case for simplicity. The final results can easily be generalized for three dimensional problems.

Assume that the only body force acting on the system is gravity. Hence,

$$b_1 = 0, \quad b_2 = -g$$

TABLE I

INERTIA CONSTANTS (Ref. 7)

| Cross section | | $C_M$ Added Mass per unit length |
|---|---|---|
| 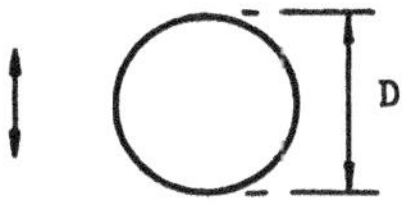 Circle | | $\frac{\rho\pi D^2}{4}$ |
| 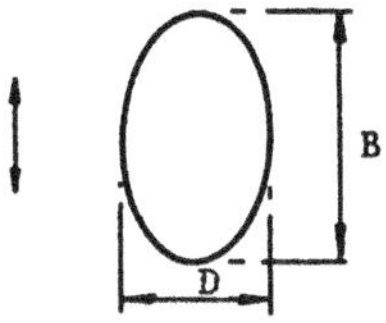 Ellipse | | $\frac{\rho\pi D^2}{4}$ |
| 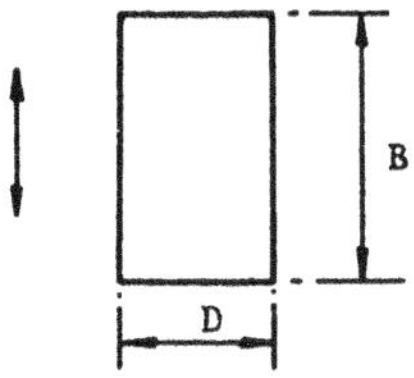 Rectangle | D/B = 1 | 1.51 × $\rho\pi D^2/4$ |
| | D/B = 1/2 | 1.7 " |
| | D/B = 1/5 | 1.98 " |
| | D/B = 1/10 | 2.23 " |
| | D/B = 2 | 1.36 " |
| | D/B = 5 | 1.21 " |
| | D/B = 10 | 1.14 " |
| 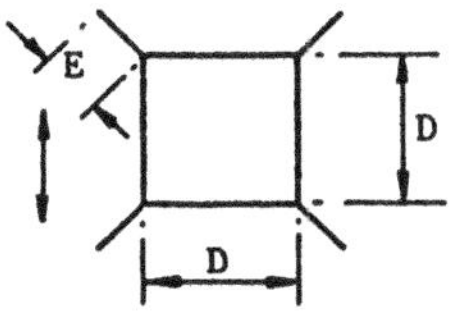 | E/D = 0.05 | 1.61 × $\rho\pi D^2/4$ |
| | E/D = 0.1 | 1.72 " |
| | E/D = 0.25 | 2.19 " |
| 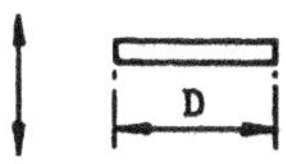 | | $\frac{\rho\pi D^2}{4}$ |

↕ Direction of Motion

Taking into account that the flow is irrotational and substituting $\dot{v}_1$, $\dot{v}_2$ in function of a velocity potential ($\dot{v}_1 = \frac{\partial \phi}{\partial x_1}$ , $\dot{v}_2 = \frac{\partial \phi}{\partial x_2}$ ), the equations of motions will give the following condition

$$\frac{\partial \phi}{\partial t} + \frac{1}{2} (\dot{v}_1^2 + \dot{v}_2^2) + \frac{P}{\rho} - g(d - x_2) = 0 \tag{6}$$

Small waves theory assumes that all motions are small which allows to neglect the velocity squared components. Hence equation (6) becomes,

$$\frac{\partial \phi}{\partial t} + \frac{P}{\rho} - g(d - x_2) = 0 \tag{7}$$

The differential equation to be solved is the continuity condition which can now be written in function of $\phi$ as,

$$\nabla^2 \phi = \frac{\partial^2 \phi}{\partial x_1^2} + \frac{\partial^2 \phi}{\partial x_2^2} = 0 \tag{8}$$

where $\nabla^2$ is the Laplacian operator.

The boundary conditions to be satisfied are,

$$\dot{v}_2 = \frac{\partial \phi}{\partial x_2} = 0 \qquad \text{at } x_2 = 0 \text{ (bottom)} \tag{8a}$$

and from (7) we obtain on the surface,

$$\eta = -\frac{1}{g} \left[\frac{\partial \phi}{\partial t}\right]_{x_2 = \eta + d} \qquad -\frac{1}{g} \left[\frac{\partial \phi}{\partial t}\right]_{x_2 = d} \tag{8b}$$

where $\eta$ is the wave elevation.

The solution of this system which can be seen in detail in reference [4] is periodic in time and space, and can be written for a right going wave as the one depicted in Figure 5.

Potential $\qquad \phi = a \frac{g}{\omega} \frac{\cosh \kappa x_2}{\sinh \kappa d} \sin(\kappa x_1 - \omega t)$

Wave amplitude $\qquad \eta = a \cos(\kappa x_1 - \omega t)$ at the surface $x_2 = d$ $\qquad$ (9)

Velocity in $x_2$ direction $\dot{v}_2 = \kappa \frac{ag}{\omega} \frac{\sinh \kappa x_2}{\sinh \kappa d} \sin(\kappa x_1 - \omega t)$

Velocity in $x_1$ direction $\dot{v}_1 = \kappa \frac{ag}{\omega} \frac{\cosh \kappa x_2}{\sinh \kappa d} \cos(\kappa x_1 - \omega t)$

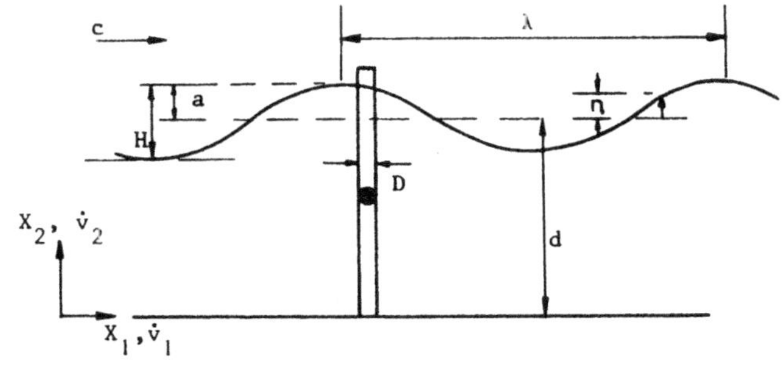


Figure 5 Small Amplitudes Waves

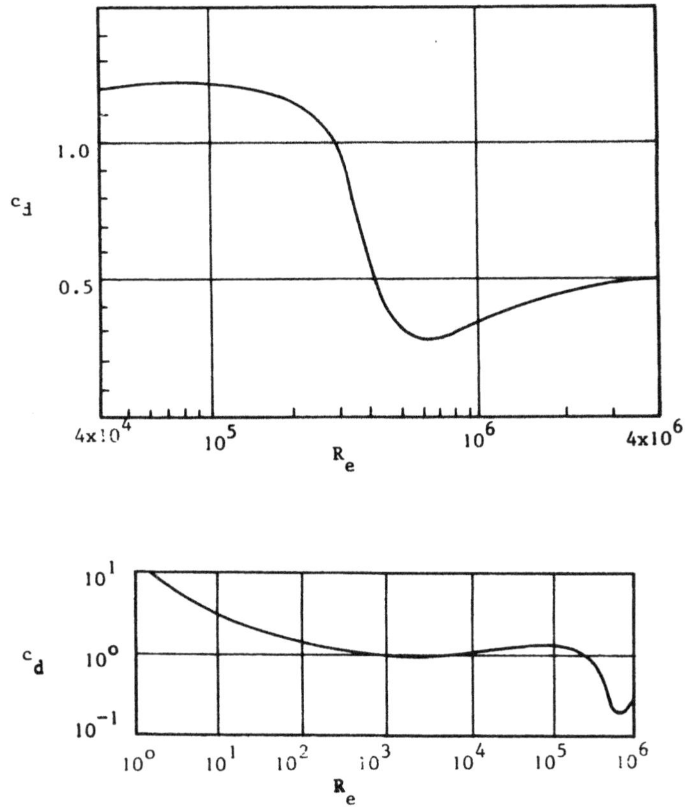


Figure 6 Drag Coefficients for a Circular Cylinder

where $\omega$ is the angular frequency of the wave, $\omega = 2\pi/T$ and $\kappa$ is the wave number, $\kappa = 2\pi/\lambda$. (See Figure 5).

From the condition that $\dot{v}_2 \simeq \frac{\partial \eta}{\partial t}$ at $x_2 = d$ and $\eta$ as given by equation (8b), one can write,

$$\frac{1}{g}\frac{\partial^2 \phi}{\partial t^2} = -\frac{\partial \phi}{\partial x_2} \quad \text{at } x_2 = d \tag{10}$$

Substituting the first of equations (9) into (10) we obtain,

$$\omega^2 = g\ \kappa\ \tanh(\kappa d) \tag{11}$$

which relates the wave number to the depth and the angular frequency of the wave. The celerity of the wave is,

$$c^2 = \frac{\omega^2}{\kappa^2} = \frac{g}{\kappa}\tanh(\kappa d)$$

If the waves are short, i.e. the depth divided by the length $\lambda$, is greater or equal to ½, one has that $\kappa d > \pi$. Hence $\tanh \kappa d \simeq 1$ and equation (11) becomes,

$$\omega = \sqrt{g\kappa} \tag{12}$$

Having defined the relationships (9) we can relate the wave amplitude spectrum $S_{\eta\eta}$ to the horizontal water-particle velocity and accelerations spectra. This gives,

$$S_{\dot{v}}(\omega) = \left[\frac{\kappa g}{\omega}\right]^2 \frac{\cosh^2 \kappa x_2}{\sinh^2 \kappa d}\ S_{\eta\eta}(\omega) \tag{13}$$

$$S_{\ddot{v}}(\omega) = \left[\frac{\kappa g}{\omega}\right]^4 \frac{\cosh^2 \kappa x_2}{\sinh^2 \kappa d}\ S_{\eta\eta}(\omega)$$

where $\dot{v}$ and $\ddot{v}$ are the horizontal water particle velocity and acceleration respectively

Drag and Inertia Forces for Slender Members

If the characteristic dimension D (e.g. diameter) of the structural member is small compared with the wave length, the incident wave can be considered undisturbed by the member for all practical purposes. This approximation is valid for $D/\lambda < 0.20$ ratios. The drag and inertia forces per unit length for the case of rigid slender members in unsteady flow fields are given by,

$$p(x,t) = \underbrace{C_I\ \ddot{v}(x,t)}_{\text{inertia}} + \underbrace{C_D\ \dot{v}(x,t)\ |\ \dot{v}(x,t)\ |}_{\text{drag}} \tag{14}$$

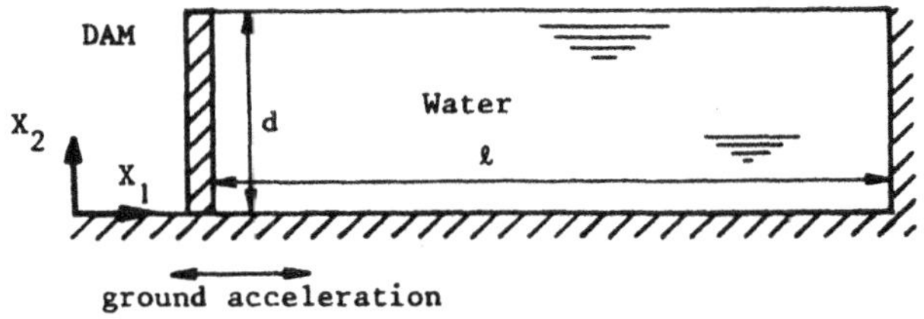


(a) Dam moving horizontally in a reservoir

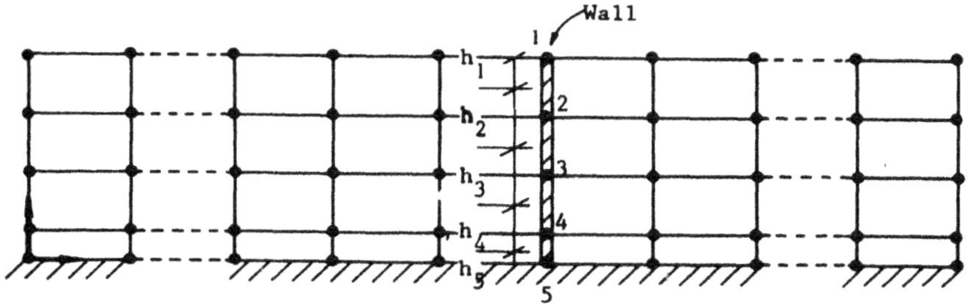


b) Wall moving inside the water

Figure 7 Fluid Structure Systems

$C_I$ is a constant due to inertia and consisting of two terms, one due to the 'hydrodynamic' mass contribution and the other to the variation of the pressure gradient within the accelerating fluid

$$C_I = C_M + C_A \approx c_m \frac{\rho\pi D^2}{4} + \rho A \tag{14a}$$

$c_m$ : inertia coefficient for the section, A is the cross sectional area.

The first term in equation (14a) can be explained by considering a body moving within a perfect fluid at rest. The acceleration of the body produces a series of pressure stresses at the interface between solid and fluid. Once the stresses are integrated we can find a force acting in the opposite direction to the body movement and proportional to the acceleration. This force is called added or hydrodynamic mass force and is written as,

$$P_H = - c_m \frac{\rho\pi D^2}{4} \ddot{u}(x,t)$$

Consider now instead the case of having the cylinder rigid and the fluid being accelerated. For this case, and in addition to the inertia or added mass forces which depend only in the relative motion between fluid and structure, we have a force due to the variation of the pressure gradient within the accelerating fluid. This force is always independent of the structure acceleration.

In order to compute this force one can analyse how the case of a uniformly accelerating fluid can be transformed to the case of a stationary fluid but accelerating cylinder. For the case of the accelerating fluid we can select a Newtonian moving frame, such that the velocity of the fluid far from the body is always zero. This is done by assuming that the frame is moving with a $\ddot{v}(x,t)$ acceleration. Hence all the fluid in the new frame has a fictitious uniform body force $- \ddot{v}(x,t)$ per unit mass. The system can now be transformed to a similar case as the one accelerating cylinder - stationary water. The cylinder will have a $- \ddot{v}$ acceleration with respect to the 'stationary' fluid which has now a body force term due to the change of reference frame. Note that the additional contribution will be $\rho A\ddot{v}$, where A is the cross sectional area of the member and is analogous to the body forces in a member due to gravity action (i.e. buoyance forces). Hence the inertia force on the body per unit length can be written

$$P_I = C_I \ddot{v} = C_M \ddot{v} + C_A \ddot{v} = (c_m \frac{\rho\pi D^2}{4} \ddot{v} + \rho A\ddot{v})$$

In Table I inertia coefficients for some sections are presented in relation to the coefficient for an infinite cylinder in an infinite medium, $c_m = 1$. Drag coefficients are more difficult to find in the references but those for a circular cylinder are shown in Figure 6 in function of Re numbers. The relationship between $c_d$ and $C_D$ coefficients is given by

$$C_D = \tfrac{1}{2} c_d \rho D \dot{v}^2$$

If the member is flexible and moving with $\dot{u}$, $\ddot{u}$ velocity and acceleration, equation (14) can be generalized to take interaction into account as,

$$p = C_M (\ddot{v} - \ddot{u}) + C_A \ddot{v} + C_D (\dot{v} - \dot{u}) |\dot{v} - \dot{u}| \quad (15)$$

Note that this formula is different from the one used in references [5] and [6] where the term $C_A\ddot{v}$ has been left out. We think that formula (15) is correct. One can also write (15) as

$$p = C_M\ddot{r} + C_D\dot{r} |\dot{r}| + C_A\ddot{v} \quad (16)$$

where $\dot{r} = \dot{v} - \dot{u}$ and $\ddot{r} - \ddot{v} - \ddot{u}$.

The drag term in equation (16) is non-linear and a linearization is now necessary to solve the problem. If we assume that a linearized damping coefficient $\bar{C}$ exists [6], we can write,

$$\begin{aligned} p &= C_M\ddot{r} + \bar{C}\dot{r} + C_A\ddot{v} + C_D\dot{r} |\dot{r}| - \bar{C}\dot{r} \\ &= C_M\ddot{r} + \bar{C}\dot{r} + C_A\ddot{v} + E \end{aligned}$$

E can be interpreted as an error function that we try to make zero. We can minimize E using the least square technique.

$$< \frac{\partial E^2}{\partial \bar{C}} > = - 2 \; <(C_D\dot{r} \; |\dot{r}| - \bar{C}r)\dot{r} > = 0 \quad (17)$$

$< \;\; >$ denotes time average. From (17) we obtain

$$\bar{C} = C_D \frac{<\dot{r}^2 \; |\dot{r}|>}{< \dot{r}^2 >} \quad (18)$$

For a Gaussian process with a zero mean one has [5]

$$\begin{aligned} < \dot{r}^2 > &= \sigma_{\dot{r}}^2 \\ < |\dot{r}| > &= \sqrt{\frac{8}{\pi}} \; \sigma_{\dot{r}} \end{aligned} \quad (19)$$

$$< \dot{r}^2 \, |\dot{r}| > \; = \sqrt{\frac{8}{\pi}} \; \sigma_{\dot{r}}^{\,3} \tag{19}$$

Hence we obtain,

$$\bar{C} = C_D \sqrt{\frac{8}{\pi}} \; \sigma_{\dot{r}} \tag{20}$$

This coefficient can be obtained for each of the degrees of freedom of a system and put into a diagonal matrix form. Note that to obtain (20) we need to know the distribution of $\dot{r}$ which is unknown at the beginning of the process. We can start the problem with the $\sigma_{\dot{v}}$ corresponding to the velocities for a rigid pile and use a cyclic procedure to obtain $\sigma_{\dot{r}}$ . Using this procedure convergence of the solution can be assured [5]. It is important to note that the use of the initial $\sigma_{\dot{v}}$ in the calculations without any cyclic improvement to obtain $\sigma_{\dot{r}}$ , can give place to errors in off-shore frame structures for which the drag effects are not negligible when compared against the inertia ones. The inertial drag effects ratio for these structures will also depend on the wind velocities (which determines the wave spectrum). The inertial effects for them may dominate at lower wind velocities but at higher velocities, the drag contributes significantly to the total hydrodynamic forces acting on the structures.

Drag forces coefficients calculations are not yet as accurate as those for the inertia forces, which can be reasonably well computed as potential effects using numerical methods.

Computation of Inertia Coefficients using Potential Theory

Let us now deduce the mass coefficients for the inertia forces acting on an arbitrary structure. These coefficients are then arranged in a matrix which is sometimes called hydrodynamic mass matrix.

The first study of this topic is due to Westergaard [23] who determined the hydrodynamic forces on dams during earthquakes. He made two main hypotheses, a) incompressibility of the water and b) that there were no surface waves. The reservoir was considered to be infinite. Westergaard solved the problem by using an analytical solution for the pressure distribution over a rigid wall subjected to harmonic ground motion. The hydrodynamic forces produced are opposite in direction to the ground motion and are proportional to the acceleration. Hence they can be represented as an equivalent mass added to the mass of the dam.

More recently there have been a series of experimental and numerical studies to determine equivalent masses for dams.

The equations governing the problem for a two dimensional case such as the one in Figure 7a) can be obtained by introducing the following assumptions in equations (4a).

a) The convective terms can be neglected for this type of motion.

b) The fluid is inviscid, as well as incompressible.

c) The only body force acting on the system is gravity.

Hence we have the two momentum equations,

$$\rho \frac{\partial \dot{v}_1}{\partial t} = - \frac{\partial p}{\partial x_1}$$
$$\rho \frac{\partial \dot{v}_2}{\partial t} = - \frac{\partial p}{\partial x_2} - g \qquad (21)$$

The continuity equation (4b) can be derivated with respect to time, which gives,

$$\frac{\partial^2 \dot{v}_1}{\partial t \partial x_1} + \frac{\partial^2 \dot{v}_2}{\partial t \partial x_2} = 0 \qquad (22)$$

Substituting (21) into (22) we obtain the following Laplacian

$$\nabla^2 P = \frac{\partial^2 P}{\partial x_1{}^2} + \frac{\partial^2 P}{\partial x_2{}^2} = 0 \qquad (23)$$

where P is the total pressure and equal to,

$$P = p + \rho\, g(d - x_2) \qquad (24)$$

$\rho g x_2$ represents the hydrostatic pressure and p the pressure due to motion.

The boundary conditions corresponding to equation (23) are,

i) We assume that there is no separation between fluid and structure. Hence on the face of the structure in contact with the fluid we have,

$$\frac{\partial P}{\partial n} = - \rho\, \ddot{u}_n \qquad (25)$$

where $\ddot{u}_n$ is the normal component of acceleration at the interface.

ii) The surface waves are assumed to be of small amplitude and such that their effect on pressure can be neglected. Thus,

$$P = 0 \quad \text{on free surface} \qquad (26)$$

iii) At the bottom of the reservoir and sides other than the structure face the boundary condition is

$$\frac{\partial P}{\partial n} = 0 \quad (27)$$

Numerical solution

The numerical solution of the problem indicated in Figure 7 can be attempted using finite difference, finite elements or boundary elements (References [4] and [5]). Boundary elements are nowadays usually preferred because they do not require the discretization of the internal domain as is the case with finite elements (see Figure 7b)).

Once the numerical method has been chosen one can study the motion of each of the points 1 to 5 on the solid (see Figure 7b)) by assuming a unit acceleration for each of the $h_i$ segments ($i = 1,5$). Then we determine the pressures at any of the points of the grid from the solution of Laplace's equation with boundary conditions (25) to (27). This can be done by using a finite differences, finite elements or boundary elements code.

The forces at nodes 1 to 5 due to the pressures obtained after applying the unit accelerations can be interpreted as the coefficients of a $\underset{\sim}{M}_H$ square matrix such that,

$$\underset{\sim}{F}_H = \underset{\sim}{M}_H \ddot{\underset{\sim}{U}} \quad (28)$$

where $\ddot{\underset{\sim}{U}}$ is a vector of accelerations and $\underset{\sim}{F}_n$ is a vector of nodal forces, both for nodes 1 to 5.

In principle the hydrodynamic mass matrix can be obtained for any type of structures. In the practice the limitations are the computer time and storage necessary for the solution of complex structures. When using a numerical technique one generally approximates the infinite domain by a finite one, although this does not need to be always the case for boundary elements. This requires accepting that at certain distance from the body we can assume that a solid boundary with boundary conditions of the type $\frac{\partial \phi}{\partial n} = 0$ exists. As an indication of the necessary length for this finite domain we quote that results for the problems shown in Figure 7a) and for a ratio $\ell/d = 4$ are in fair agreement with those obtained for $\ell/d = \infty$. Hence a ratio $\ell/d = 4$ in the numerical solution may be taken to represent $\ell/d \to \infty$.

A note of caution should be given about the theoretical inertia coefficients deduced by using potential theory. These coefficients are not strictly valid as the viscous effects are not taken into account. Specially the wake behind the obstruction may

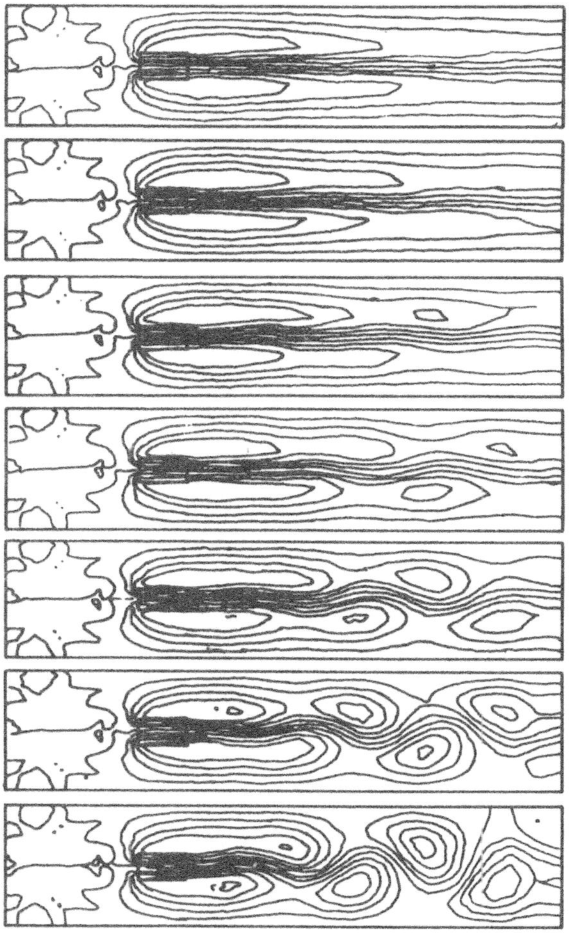

29.5 seconds

29.8

30.1

30.4

30.7

31.0

31.3

Figure 8 Vortex street development for Re = 100 showing stationary streamlines (stationary streamlines are obtained by sub tracting to the streamlines those corresponding to flow without the above motion).

significantly alter the results. The only practical solution in many cases is to use experimental coefficients when available.

Wave Diffraction

If the size of the body is such that $D/\lambda > 0.2$ one can not any longer assume that the incident waves are undisturbed. The body will now diffract the waves which will change in shape. The inertia forces cannot be obtained using the previous potential theory, but it is necessary to apply diffraction theories which take into account the scattering of the wave energy due to the body shape.

The diffraction analyses still involve the assumption of irrotationality, i.e. the fluid is inviscid, which means that drag forces or vortex shedding cannot in principle occur. This is not as bad a supposition as it may seem because drag forces can be neglected for large bodies, i.e. those with a ratio $D/H > 0.2$. This is due to the fact that for these structures the drag effects are negligible by comparison with the inertia ones.

For the diffraction analysis one assumes that,

i) The fluid is irrotational and incompressible.

ii) The motions are small.

iii) At infinity the wave follows an undisturbed harmonic motion but its shape is disturbed in the vicinity of the body.

The problem can be expressed in terms of $\phi(x_i t)$ the velocity potential by equations (7) and (8), that is,

$$\frac{\partial\phi}{\partial t} + \frac{P}{\rho} - g(d - x_2) = 0 \tag{29}$$

and

$$\nabla^2\phi = 0 \tag{30}$$

For the solution of (30) one can propose a harmonic function with a frequency such that,

$$\phi = \Phi\, e^{i\omega t} \tag{31}$$

which gives

$$\nabla^2\phi = 0 \tag{32}$$

The boundary conditions are,

i) At the free surface one has (see equation (8a))

$$\eta = -\frac{1}{g}\frac{\partial \phi}{\partial t} \quad \text{and} \quad \dot{v}_2 = \frac{\partial \eta}{\partial t} \quad \text{at} \quad x_2 = d$$

Hence $\dot{v}_2 = -\frac{1}{g}\frac{\partial^2 \phi}{\partial t^2}$ , which gives in terms of $\Phi$,

$$\frac{\partial \Phi}{\partial x_2} = \frac{\omega^2}{g} \tag{33}$$

ii) At the bottom, wall or body surface one has

$$\frac{\partial \Phi}{\partial n} = 0 \tag{34}$$

Far away from the body the incident waves are assumed to be described by the linear wave theory already seen. The solution of the above system of equations can be attempted in different ways. Boreel [2] has presented results obtained by superimposing the known solution for the incident wave to a solution obtained from a series of sources distributed over the body. The potentials of the latter are described using Green's functions selected in such a way that at infinity, they satisfy the radiation condition corresponding to a radially outgoing progressive wave. Once the potentials are known the pressure distribution can be found through equations (29). Nowadays many authors use boundary elements to solve this type of problem (see reference [5] ).

Vortex Shedding

We have already mentioned that there is a region of low pressure behind the body, which account for most of the drag force. Up to certain velocity this wake region is symmetric and does not produce any lift forces. When the velocity of the fluid increases, asymmetric eddies can be produced behind the obstruction which are alternatively shed. For a few cases (low Re numbers) this phenomenon has been numerically computed and the shedding frequencies have been found. For steady flow the frequency of these vortices and hence the frequency of the lift forces, is defined by the Strouhal number ($S = Df/\dot{v}$ , where f is the frequency, D the diameter or representative length and $\dot{v}$ the velocity of the fluid).

In Figure 8 the vortices produced behind a rectangular obstruction for a Re = 100 are shown. This figure shows numerical results obtained using finite elements and the method of solution is discussed in detail in reference [22]. The Strouhal number calculated from these results is approximately 0.12, which is in agreement with other known results and similar to those plotted in Figure 9 for a circular cylinder.

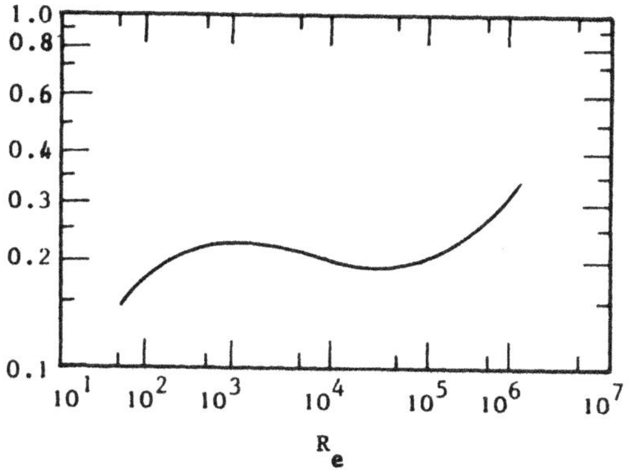


Figure 9 Strouhal Number for Vortex Shedding behind a Circular Cylinder

Once vortex shedding occurs one has to take the lift forces into consideration. They are usually defined in a similar way as the drag forces, i.e.

$$p_L = c_L \rho D \frac{\dot{v}}{2} \quad \text{per unit length} \tag{35}$$

$c_L$ will vary according with the shape of the body.

The flexibility of the body is also important for calculating the lift forces. Harleman in reference [13] mentions that experiments carried out on flexible cylinders have produced lift forces of up to four times the corresponding drag forces for vortex shedding frequencies near the cylinder frequencies. If the natural frequencies of the cylinder were instead between 10% to 60% of the vortex shedding frequencies the lift forces were of the order of the drag forces. The increase for vortex shedding frequencies near the cylinder frequencies is apparently due to the eddies and cylinder systems becoming coupled, i.e. the cylinder triggers the shedding of the vortices.

The designer has also to be aware that the system can start shedding vortices at a range of frequencies in the vicinity of Strouhal number. For instance vortices can start to be shed at the natural frequency of the structure even if the frequency is not in full agreement with the Strouhal frequency [24].

<u>In Conclusion</u>

The ratio between drag and inertia forces depends on the dimensions of the structural component and the wave properties. The drag forces are important for values of $D/H < 0.2$, otherwise the inertia forces are of much more importance. Drag forces also decrease with depth more rapidly than the inertia forces. In general inertia forces are decisive in off-shore structures, specially if the diameter of the members is large.

<u>Example</u>

Consider a circular wooden pile which has been driven into a river bed. The length of the pile and depth of the river is 20 metres, the diameter 0.40 m. Compute the drag forces in the pile assuming the velocity of the water $\dot{v}$ is constant and determine the value of $\dot{v}$ for which vortex shedding occurs. The following values apply,

- $E$ : modulus of elasticity of wood $= 10^{10}\ \mathrm{Nm^{-2}}$
- $\rho$ : density of timber $= 0.8 \times 10^3\ \mathrm{K\ m^{-3}}$
- $\rho_w$ : density of water $= 10^3\ \mathrm{K\ m^{-3}}$
- $I$ : moment of inertia $= \pi R^4/4$ ($R$ : radius $= D/2$).

The drag forces can be computed using the following formulae, ($c_D = 1$ from Fig. 6).

$$P_D = \tfrac{1}{2}\rho_w D c_D \dot{v} |\dot{v}| =$$

$$= 2 \cdot 10^2 \dot{v} |\dot{v}| \qquad (a)$$

The fundamental frequency of the pile can be obtained using Rayleigh's quotient,

$$\omega_n^2 = \frac{EI \int_0^\ell \left[\frac{\partial^2 f}{\partial x^2}\right]^2 dx}{\rho A \int_0^\ell f^2 dx + C_M \int_0^\ell f^2 dx} \qquad (b)$$

where $C_M = c_m \cdot \frac{D^2 \rho_w \pi}{4}$ and $c_m = 1$ from Table I. This term in $C_M$ is due to the hydrodynamic mass.

In order to solve (b) let us assume that the shape of the pile can be given as $f = \bar{x}^2$, where $\bar{x} = x/\ell$. Hence,

$$\omega_n^2 = \frac{\frac{4EI}{\ell^3}}{\frac{\rho A \ell}{5} + c_m \frac{D^2 \rho_w \pi}{4} \frac{\ell}{5}} \simeq 10 \text{ sec}^{-1} \qquad (c)$$

Thus, $\omega_n$ : circular frequency $\simeq 3.16 \text{ sec}^{-1}$.

$f_n$ : frequency $\simeq 0.5 \text{ sec}^{-1}$.

The Strouhal number for the cylinder at high Re numbers is

$$S = \frac{f_s D}{\dot{v}} \simeq 0.2$$

$$f_s = \frac{0.2\,\dot{v}}{D} \qquad (d)$$

Note that in order for $f_s$ to equal $f_n$ we need a velocity of

$$\dot{v} = \frac{f_n D}{0.2} = 1 \text{ m/sec.} \qquad (e)$$

There is a danger that the pile will break due to vortex shedding. The calculation is approximate due to the modal shape we have assumed. It can be improved by taking a better shape or by dividing the pile into a number of beam elements.

3. Total Hydrodynamic Forces

For a given location $x_i$ we can write the total hydrodynamic forces as,

$$P_i = C_M \ddot{r}_i + \bar{C} r_i + C_A \ddot{v}_i \qquad (36)$$

where $\bar{C} = C_D \sqrt{\frac{8}{\pi}} \sigma_{\dot{r}_i}$ and as a first approximation we take $\sigma_{\dot{r}_i} \approx \sigma_{\dot{v}_i}$ in order to linearize the problem. Equation (36) can also be written as,

$$P_i = C_M (\ddot{v}_i - \ddot{u}_i) + \bar{C} (\dot{v}_i - \dot{u}_i) + C_A \ddot{v}_i \qquad (37)$$

The term $-C_M \ddot{u}_i$ gives origin to the hydrodynamic matrix and $-\bar{C} \dot{u}_i$ is the hydrodynamic drag force. Both terms are passed to the left hand side of the equilibrium equations of the system (equations (3)). On the right hand side we are left with $(C_M + C_A)\ddot{v}_i + \bar{C} \dot{v}_i$ terms.

Knowing (37) we can now calculate the $\underset{\sim}{F}$ vector of equation (3). This will be seen in Chapter 17.

4. Final Remarks

There is a need for further information about many different aspects of fluid-structure interaction problems. First the designer ought to have more data concerning wave spectra and in particular directional wave spectra. Prediction of the worst possible situations should also be attempted by correlating wind, waves, currents, etc. with probabilistic criterion.

Another research area is to investigate how valid are the water velocity and acceleration fields obtained from the wave elevation observations by using waves theories.

The scatter and scarcity of drag and lift coefficients indicate that more fundamental research (experimental and theoretical) is required in these fields.

A better understanding of the behaviour of the foundations is essential. Small relaxation of the support conditions of an off-shore structure (e.g. considering them as elastically supported) can greatly alter the frequencies of the structure and produce a different response.

References & Bibliography

1. Batchelor, G.G. *An Introduction to Fluid Dynamics*, Cambridge University Press, 1970.

2. Boreel, L.J. *Wave Action on Large Off-Shore Structures*, Conference on Off-shore Structures at the Institution of Civil Engineers, London, October 1974.

3. Borgman, L.E. The Spectral Density for Ocean Wave Forces, *Coastal Engineering Conference, A.S.C.E.* Santa Barbara, October 1965.

4. Brebbia, C.A. and S. Walker *Dynamic Analysis of Offshore Structures.* Butterworths, London, 1979.

5. Brebbia, C.A., J.T.Telles and L. Wrobel *The Boundary Element Technique - Theory and Applications in Engineering.* Springer-Verlag, Berlin - N.Y., 1983.

6. British Ship Research Association *A critical solution of the data on Wave Force Coefficients.* OSFLAG Project 10, Report No. W278, 2 Vols, 1976.

7. Chakrabarti, S.K., Wolbert, A.L. and Tam, W.A. *Wave forces on Vertical Circular Cylinders.* Proc. ASCE, Water and Coastal Engineering Division, 102, May, 1976.

8. Charlton, P. *Fluid-Structure Interaction with Particular Reference to the dynamic Behaviour of Framed Structures Subjected to Random Waves.* M.Sc. Thesis, Department of Civil Engineering, University of Southampton, 1974.

9. Connor, J.J. and Brebbia, C.A. *Finite Element Techniques for Fluid Flow.* Butterworths, London, 1976.

10. Heaf, N.J., R.K. Henrywood and L.R. Wootton, *Assessment of the Accuracy of the Design Process for Off-Shore Structures*, Conference on Off-Shore Structures at the Institution of Civil Engineers, London, October 1974.

11. Hogben, N. *Wave Loads on Structures.* Proc. Behaviour of Offshore Structures Conf. Trondheim, 1976.

12. Hogben, N. *Fluid Loading of Offshore Structures, a state of art appraisal: Wave Loads.* R. Inst. Naval Arch., 1974.

13. Ippen, A.T. (editor) *Estuary and Coastline Hydrodynamics*, McGraw-Hill, 1966.

14. Malhotra, A.K. and J. Penzien, Response of Off-Shore Structures to Random Wave Forces, *J. of Struct. Div., Proc. of A.S.C.E.*, October 1970.

15. Malhotra, A.K. and J. Penzien, Non-deterministic Analysis of Off-Shore Structures, *J. of Eng. Mech. Div., Proc. of A.S.C.E.*, December 1970.

16. McCormick, M.E. *General Engineering Wave Problems*, J. Wiley, New York, 1973.

17. McDowell, D.M. and P. Holmes, *General Research Problems in the Design of Off-Shore Structures*, Conference on Off-Shore Structures at the Institution of Civil Engineers, London, October 1974.

18. Morrison, J. et al, The Forces exerted by Surface Waves on Piles, *J. Petroleum Technology, Am. Inst. M. Eng.*, 189, 1950, pp. 149-154.

19. Pierson, W.J. and K. Moskowitz, A proposed spectral form for fully developed wind seas based on the Similarity Theory of S.A. Kitrigorodskii, *Journal of Geophysical Research*, Vol. 69, No.24, 1964, pp. 5181-5190.

20. Scott, J.R., A Sea Spectrum for Model Tests and Long-term Ship Prediction, *Journal of Ship Research*, Vol. 9, No. 3, December, 1965.

21. Shaw, T.L. *Wave Induced Loading on Dynamic Structures*, Conf. on Off-Shore Structures at the Institution of Civil Engineers, London, October 1974.

22. Smith, S. and C.A. Brebbia, Finite Element Solution of Navier Stokes Equations for Transient Two Dimensional Incompressible Flow, *J. of Comp. Physics*, Vol. 17, No. 3, March 1975.

23. Westergaard, H.M. Water Pressure on Dams during Earthquake *Trans. Amer. Soc. Civ. Engrs.*, Vol. 98, 1933, pp 418-433.

24. Wootton, L.R., M.H. Warner, R.N. Sainsbury and D.H. Cooper, *Oscillation of Piles in Marine Structures*, CIRIA Technical Note 40, April 1972.

# CHAPTER 14

## INTRODUCTION TO RANDOM VIBRATIONS

by

H. Tottenham

When we consider the effects of winds or earthquakes on buildings or the passage of vehicles over bridge decks we cannot say precisely how the load will vary with time. We can, however, frequently give a *statistical* description of the loads in such a form that it is possible to derive a *statistical* description of the response of the structure. We are, then, concerned with

a) the probabilistic description of the loads,

b) the effects of these loads on the structure, and

c) the meaning of the probabilistic description of the response.

Before considering these topics, however, we will establish some of the basic ideas of *random processes*.

### 1. Random Processes

Let us consider the time-wise variation of some factor $x_1(t)$ such as is shown in Figure 1. We call this a continuous random sequence if the factor $x_1(t)$ varies continuously with t and if it is not possible to predict precisely what will happen at time $t + \delta t$ from the record of events up to time t. We call a collection of such sequences a process and write this as

$$\{ x(t) = x_1(t), \quad x_2(t), \ldots x_n(t) \}$$

Naturally we will consider only collections of records of similar factors, e.g. pressure due to wind on a particular part of similar buildings in similar locations. Under these circumstances we would expect the individual sequences to have some properties in common. Thus we would expect the mean values of the pressures, averaged over a relatively long time to be equal to one another, i.e.

$$\frac{1}{T}\int_0^T x_1(t)dt = \frac{1}{T}\int_0^T x_2(t)dt = \ldots = \frac{1}{T}\int_0^T x_n(t)dt \tag{1}$$

If, over a fairly long time interval, we measured the proportion of the total time that the value of $x_1(t)$ did not exceed some value, a, say, we would expect that over any other suitably long time the value of $x_1(t)$ would be less than a for the same proportion of the time. We call this proportion of the time the "probability" that $x_1(t) < a$, and denote this by $p_{x_1}(a)$. The suffix $x_1$ denotes the particular sequence being considered. By taking different values of a we could construct a graph showing how $p_{x_1}(a)$ varied with a, such a graph we call the *probability distributed function;* it will be of a form similar to that shown in Figure 2. Obviously $p_{x_1}(-\infty) = 0$ and $p_{x_1}(\infty) = 1$, and the curve is always ascending.

The probability that the value of $x_1(t)$ lies between a and $(a+\delta a)$ is

$$p_{x_1}(a+\delta a) - p_{x_1}(a) = \delta p_{x_1}(a)$$

We now define a new function $P_{x_1}(a)$ such that

$$P_{x_1}(a)\delta a = \delta p_{x_1}(a)$$

and hence

$$P_{x_1}(a) = \frac{dp_{x_1}(a)}{da} \qquad (2)$$

We call this function the *probability density* of the sequence $x_1(t)$. We will also have

$$P_{x_1}(a) = \int_{-\infty}^{\infty} P_{x_1}(b)db \qquad (3)$$

and note that $P_{x_1}(a)$ is always positive.

A process x(t) is said to be *stationary* if the probability distribution functions $p_{x_n}(a)$ of each of the sequences $x_n(t)$ do not depend upon the actual instant chosen as the origin of time, t = 0, i.e.

$$p_{x_n}(t) = p_{x_2}(t_1) \qquad (n = 1,2, \ldots ) \qquad (4)$$

where $t_1 = t+t^*$ and $t^*$ can take any value.

The process is said to be *ergodic* if the probability distribution function for the process is the same as that for each of the individual sequences,

$$p_x(a) = p_{x_1}(a) = \ldots p_{x_n}(a) \qquad (5)$$

We will be considering only stationary ergodic random processes.

We now introduce the notation $< y >$ to denote the time averaged value of y, taken over a sufficiently long time period:

$$< y > = \frac{1}{T} \int^{T} y(t)dt \quad (T \text{ large}) \qquad (6)$$

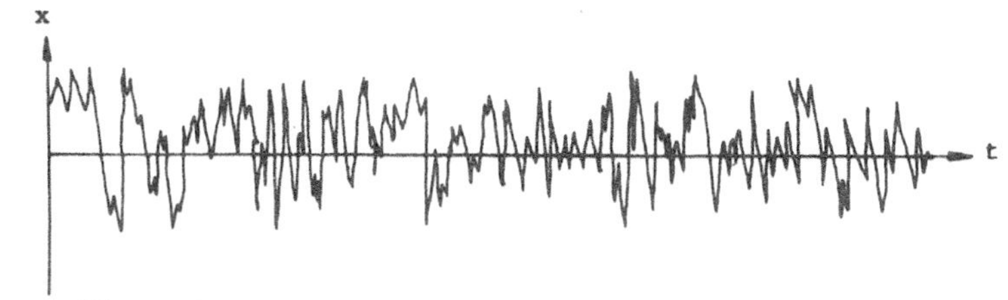


Figure 1

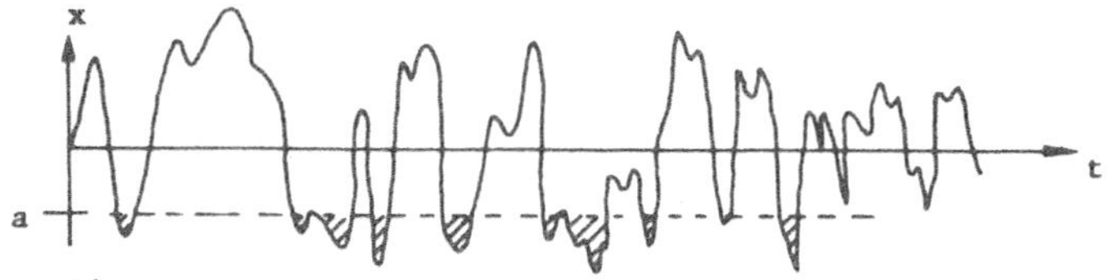


Figure 2a

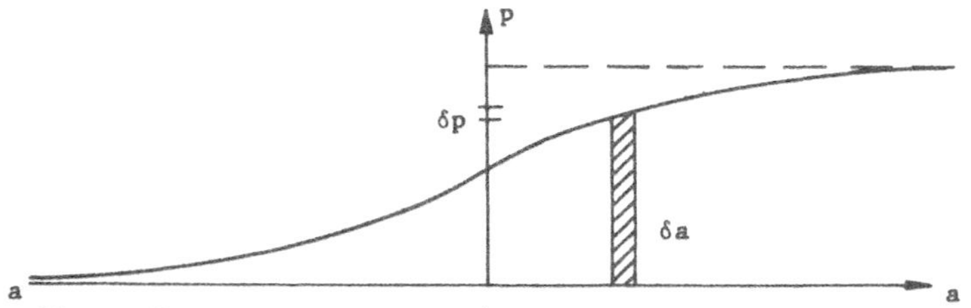


Figure 2b

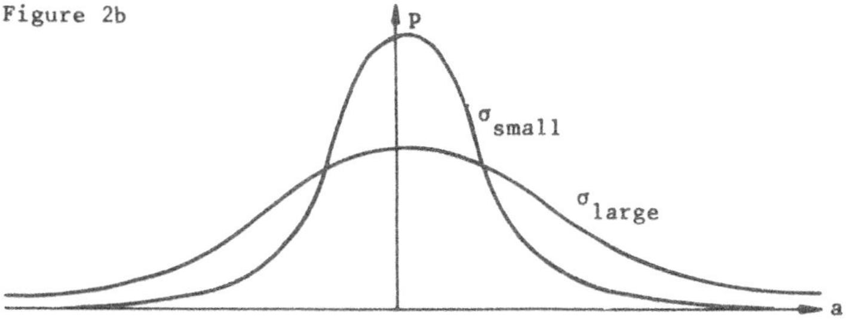


Figure 3

The time averaged value of y can also be found from the probability density function. Since the proportion of time that y(t) lies between a and (a+δa) is $P_y(a)\delta a$ the time average value is

$$< y > = \int_{-\infty}^{\infty} y\, P_y(a)da \tag{7}$$

This integral is the first moment of area of the probability density curve about the origin, y = 0, and is thus equal to $A\bar{y}$, where A is the area of the curve and $\bar{y}$ is the coordinate of the centroid of the area under the curve. Since A = 1 we have $< y > = \bar{y}$. We will consider y(t) as being the sum of a constant value $\bar{\mathbf{y}}$ plus a fluctuating component where time average is zero. The constant part is a 'static' event and does not concern use here. We consider therefore events for which $< y > = 0$.

A useful means of describing the probability density curve is by means of its mean square value or *variance*, $\sigma^2$. We will see that this enables us to evaluate the probability of an event.

$$\sigma^2 = < y^2 > = \frac{1}{T}\int_{0}^{T} y^2(t)dt$$

$$= \int_{-\infty}^{\infty} y^2\, P_y(y)dy \tag{8}$$

Thus the variance is the second moment of area of the probability density curve. Since the area of this curve is always unity the value of $\sigma^2$ gives a measure of the "peakiness" of the curve. A low value of $\sigma^2$ means a low value of second moment and hence the area must be concentrated near the origin, whilst a large value of $\sigma^2$ means a high value of second moment and hence there will be a significant part of the area not near the origin (Figure 3). A particular form of the probability density function is the GAUSSIAN distribution:

$$P_y(y) = \frac{1}{\sqrt{2\pi}\sigma}\int_{-\infty}^{y} e^{-x^2/2\sigma^2}\, dx \tag{9a}$$

giving

$$P_y(y) = \frac{1}{\sqrt{2\pi}\sigma}\; e^{-y^2/2\sigma^2} \tag{9b}$$

We can normalize these by putting $\bar{y} = y/\sigma$, to get

$$P_y(\bar{y}) = \int_{-\infty}^{\bar{y}} e^{-x^2/2}\, dx$$

Since this quantity contains only the variable $\bar{y}$ we can tabulate it and $p(\bar{y})$, tables of this distribution are given in most standard text books on statistics. For the tables we would find, for example that $p(3) = 0.9985$, and hence the probability that $y > 3\sigma$ is $1 - 0.9985 = 0.0015$ or 0.15%.

A very useful property of the Gaussian distribution is that if a process has a Gaussian distribution so also will any process which is linearly related to it. Since we are dealing with linear elastic structures we can thus say that if the loading has a Gaussian distribution so also will the response. There are other distributions which are, for example, skew or have a definite cut-off point, but we will assume that in the range of probability in which we are interested our process may be considered to have a Gaussian distribution.

Before leaving the general properties of a random process we should define the *auto-correlation* functions, which we denote by $R(\tau)$. This is defined as

$$R(\tau) = \langle y(t)\, y(t+\tau) \rangle \tag{10}$$

That is, we find the time average of the product two events which occur at a distance apart in time $\tau$. The value of $R(\tau)$ at $\tau = 0$ is obviously the mean square value $\langle y^2 \rangle$ and as $\tau$ gets longer we would expect positive and negative products to cancel one another out, and hence $R(\tau)$ to decrease. The more rapid the decrease in $R(\tau)$ the more sudden and disconnected the process.

## 2. Spectral Density Function

A periodic function can be represented by a Fourier series:

$$y(t) = a_o + \sum_{n=1}^{\infty} a_n \cos n\omega t + \sum_{n=1}^{\infty} b_n \sin n\omega t \tag{11}$$

where $\omega = 2\pi/T$ and T is the period of the function. We could write the above using complex notation as

$$y(t) = \sum_{n=-\infty}^{\infty} c_n\, e^{in\omega t}$$

where

$$c_n = \tfrac{1}{2}(a_n - ib_n) = \frac{1}{T} \int_{-T/2}^{T/2} y(t)\, e^{-in\omega t}\, dt \tag{12}$$

Substituting this back into (11) gives

$$y(t) = \sum_{n=-\infty}^{\infty} \{ (\frac{\omega}{2\pi} \int_{-T/2}^{T/2} y(t)\, e^{-in\omega t} dt)\, e^{in\omega t} \}$$

$$= \omega \sum F(n\omega) \quad \text{say.} \tag{13}$$

Let us now plot a "histogram" of $F(n\omega)$ using values $\omega$, $2\omega$, ... $n\omega$ along the base, the ordinate of the histogram from $i\omega$ to $(i+1)\omega$ being $F(i\omega)$, as in Figure 4, we can see that since the width of each block is $\omega$ the total area of the histogram is in fact $y(t)$ , i.e. the area = $\sum \omega \times F(n\omega)$. We now let $\omega$ become very small so that the interval becomes $\delta\omega$ in place of $\omega$. The succession of points $\omega$, $2\omega$, $3\omega$ etc. become the succession of points $\delta\omega$, $2\delta\omega$, $3\delta\omega$ ... or in other words the continuous variable $\omega$ as $\delta\omega \to 0$. If then we let $\delta\omega \to 0$ we get an integral in place of the summation, and hence

$$y(t) = \int_{-\infty}^{\infty} \frac{1}{2\pi} \left\{ \int_{-T/2}^{T/2} y(t) e^{-i\omega t}\, dt \right\} e^{i\omega t}\, d\omega$$

But as the circular frequency $\omega$ in the original expression becomes very small the period T becomes very large and the limits on the inner integral become $\pm\infty$ . For convenience we replace the circular frequency $\omega$ by the frequency f, and since $\omega = 2\pi f$, $d\omega = 2\pi df$, and hence

$$y(t) = \int_{-\infty}^{\infty} \left\{ \int_{-\infty}^{\infty} y(t)\, e^{-i2\pi ft} dt \right\} e^{i2\pi ft} df \tag{14}$$

The inner integral will be a function of f and not of t and we denote it by A(if) and hence get

$$y(t) = \int_{-\infty}^{\infty} \bar{Y}\, e^{i2\pi f}\, df \tag{15a}$$

and

$$\bar{Y} = \int_{-\infty}^{\infty} y(t)\, e^{-i2\pi ft} dt \tag{15b}$$

We say that $\bar{Y}$ is the Fourier transform of $y(t)$ and the two quantities $y(t)$ and $\bar{Y}$ form a *Fourier Transform pair*. We will see that Fourier transform pairs play an important role in the analysis of the response of elastic structures to random loads. We now look to the mean square value of a process $x(t)$, we have

$$< x^2 > = \lim_{T\to\infty} \frac{1}{T} \int_{-T/2}^{T/2} x^2(t) dt$$

$$= \lim_{T\to\infty} \frac{1}{T} \int_{-T/2}^{T/2} x(t).x(t) dt$$

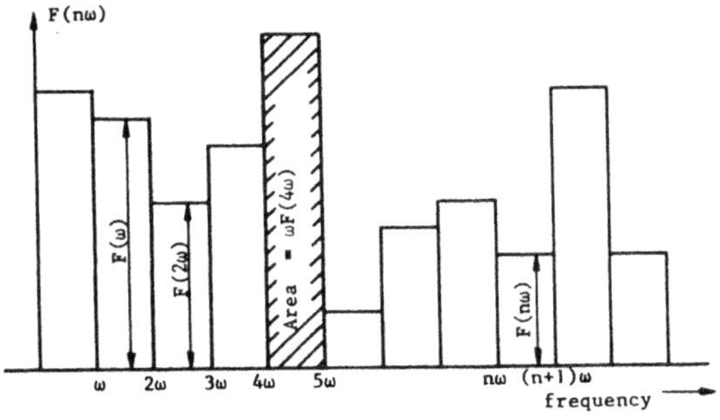


Figure 4

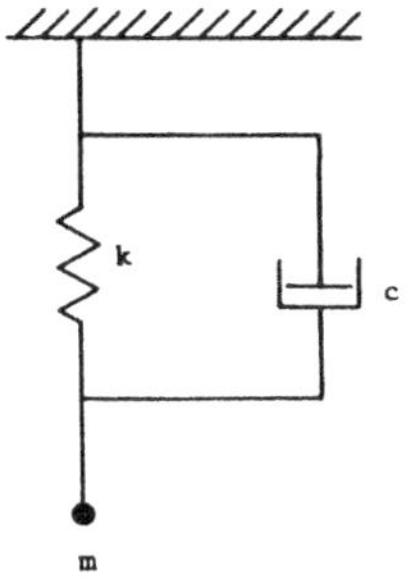


Figure 5

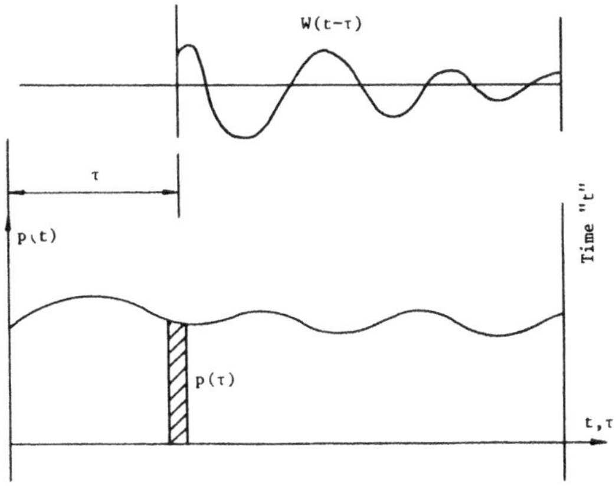


Figure 6

Expressing one of the factors x(t) from the transform pair this gives

$$< x^2 > = \lim_{T\to\infty} \frac{1}{T} \int_{-T/2}^{T/2} x(t) \int_{-\infty}^{\infty} \bar{X}\, e^{i2\pi ft} df.dt$$

Changing the order of integration this gives

$$< x^2 > = \int_{-\infty}^{\infty} \bar{X} \left(\lim_{T\to\infty} \frac{1}{T} \int_{-T/2}^{T/2} x(t)\, e^{i2\pi ft} dt\right) df$$

When $T \to \infty$ the inner integral is the Fourier transform of x(t) with -i in place of i, that is it is the complex conjugate of $\bar{X}$ which we denote by $\hat{X}$. We cannot evaluate this quantity as it stands as we will be dividing an infinite quantity by another infinite quantity, T. We can however take $x_T(t)$ as being zero for $t < -T/2$ and for $t > T/2$ and equal to x(t) in between these values, then putting

$$\bar{X} = \int_{-\infty}^{\infty} x_T(t)\, e^{-i2\pi ft}\, dt$$

we get

$$< x^2 > = \lim_{T\to\infty} \frac{1}{T} \int_{-\infty}^{\infty} \bar{X} \,.\, \hat{X}\, df$$

$$= \lim_{T\to\infty} \frac{2}{T} \int_{0}^{\infty} |\bar{X}(\omega)|^2 df$$

$$= \int_{0}^{\infty} S(f)\, df \qquad (16)$$

where

$$S(f) = \lim_{T\to\infty} \frac{2}{T} |\bar{X}(\omega)|^2 \qquad (17)$$

We call the quantity S(f) the spectral density function of the process x(t). The spectral density function gives a measure of the amount of $<x^2>$ is associated with the frequency f. It is equivalent to a continuous variation of coefficient in the Fourier series taking the whole of time as the basic period.

## 3. The Weiner-Khinchin Relationship

Let us now look at the autocorrelation function of the random process x(t),

$$R(\tau) = < x(t)\; x(t+\tau)>$$

We redefine this as the limit as $T\to\infty$ of the autocorrelation function of the modified process $x_T(t)$,

$$R(\tau) = \lim_{T\to\infty} R_T(\tau) = \lim_{T\to\infty} < x_T(t)\ x_T(t+\tau) >$$

e can form this over longer time intervals until a steady value is reached. orming the Fourier transform pair of x(t+τ) we have

$$\bar{X} = \int_{-\infty}^{\infty} x(t+\tau)\ e^{-i2\pi ft}dt$$

$$x(t+\tau) = \int_{-\infty}^{\infty} \bar{X}\ e^{i2\pi f(t+\tau)}df$$

since the time taken as t = 0 is immaterial for a stationary process. We now have

$$R(\tau) = \lim_{T\to\infty} \frac{1}{2T} \int_{-T}^{T} x_T(t) \int_{-\infty}^{\infty} \bar{X}\ e^{i2\pi f(t+\tau)}\ df\ dt$$

$$= \frac{1}{2} \int_{-T}^{T} \lim_{T\to\infty} \frac{1}{T}\, x_T(t) \int_{-\infty}^{\infty} \lim_{T\to\infty} \bar{X}\ e^{i2\pi ft} e^{i2\pi f\tau}\ df.dt$$

We now change the order of integration to get

$$R(\tau) = \frac{1}{2} \int_{-\infty}^{\infty} \lim_{T\to\infty} \bar{X}\ (\lim_{T\to\infty} \int_{-T}^{T} \frac{1}{T}\, x_T(t)\ e^{i2\pi ft}dt)\ e^{i2\pi ft}dt) e^{i2\pi f\tau}df$$

$$= \frac{1}{2} \int_{-\infty}^{\infty} \lim_{T\to\infty} \bar{X}\ .\ \hat{X}\ e^{i2\pi f\tau}\ df$$

$$= \frac{1}{2} \int_{-\infty}^{\infty} S_x(f)\ e^{i2\pi f\tau}df \qquad (18)$$

Thus $2R(\tau)$ is the Fourier transform of the spectral density function of the process x, its pair is

$$S_x(f) = 2 \int_{-\infty}^{\infty} R(\tau)\ e^{-i2\pi f\tau}d\tau \qquad (19)$$

That is the spectral density function and twice the autocorrelation function form a Fourier transform pair. Given either f of these we can form the other. Equations (18) and (19) are known as the WEINER-KHINCHIN relationships.

Since S (f) is an even function of f, i.e. $S_{.}(f) = S_{..}(-f)$, we have

$$R(\tau) = \int_0^{\infty} S_x(f) \cos 2\pi f\tau \, df$$
$$S_x(f) = 4 \int_0 R(\tau) \cos 2\pi f\tau \, d\tau \qquad (20)$$

4. Response of a Simple Spring System to Random Load

Consider a simple spring loaded by a force, P(t), as shown in Figure 5. The equation of motion is

$$m\ddot{x} + c\dot{x} + kx = P(t)$$

where dots denote derivatives with respect to time. If P(t) is of the form

$$P(t) = P e^{i\omega t}$$

we can find x(t) as $X e^{i\omega t}$

$$X = \frac{P}{(-\omega^2 m + i\omega c + k)}$$
$$= \frac{P}{k} \left\{ \frac{1}{\{(1 - \frac{\omega}{\omega_1})^2)^2 + 4(\frac{\omega}{\omega_1})^2 \gamma^2\}^{1/2}} \right\}$$

where

$$\omega_1^2 = \frac{k}{m} \qquad \gamma = \frac{c}{2\sqrt{mk}}$$

That is $\omega_1$ is the *natural frequency* of the undamped system and $\gamma$ is the *critical damping ratio.*

The factor P/k represents the "static" deflection of the system and the remainder shows the influence of the dynamic nature of P, the part is called the admittance function of the system and is denoted by $|H(\omega)|$. We put

$$\alpha(i\omega) = \frac{1}{k - m\omega^2 + i\omega c}$$

or

$$\alpha(if) = \frac{1}{k - 4\pi^2 mf^2 + 2i\pi fc} \qquad (21)$$

We call this the receptance of the system

$$|\alpha(if)| = \frac{|H(\omega)|}{k}$$

The response of the system can also be found by means of the impulse response function. If we give the system, which is initially at rest an impulsive load I the subsequent motion will be defined by

$$m\ddot{x} + c\dot{x} + kx = 0$$

with initial conditions $x = 0$, $\dot{x} = I/m$ at $t = 0$. The solution is

$$x = \frac{I}{m\bar{\omega}} e^{-\beta t} \sin \omega_d t = I\,W(t) \tag{22}$$

$$\omega_d^{\,2} = \frac{k}{m} \cdot \left(\frac{c}{2m}\right)^2 \qquad \beta = \frac{c}{2m} = \gamma\omega_1$$

$$= \omega_1^{\,2}(1-\gamma^2)$$

A load $P(t)$ can be considered as a series of impulses $I = P(t)\delta t$ and the response at time t will be the sum of the responses of all the impulses up to the time t.

The response at time t to the impulse $P(\tau)\delta\tau$ applied at time $\tau$ is

$$x(\tau) = W(t-\tau)\, P(\tau)\delta\tau$$

and hence the total response for the impulses up to time t is

$$x(t) = \int_{-\infty}^{t} W(t-\tau)\, P(\tau)d\tau \tag{23}$$

This is known as DUHAMEL's integral or the *convolution* integral. By changing the dummy variable to $\tau' = t-\tau$, and then since $\tau'$ is a dummy variable replacing it by $\tau$ we get

$$x(t) = \int_0^{\infty} W(\tau)\, P(t-\tau)d\tau \tag{24}$$

If now we put $P = P\,e^{i\omega t} = P\,e^{i2\pi ft}$ we get

$$x(t) = \int_0^{\infty} W(\tau)\, P\, e^{i2\pi f(t-\tau)} d\tau$$

$$= P\, e^{\,i2\pi ft} \int_0^{\infty} W(\tau)\, e^{-i2\pi f\tau} d\tau$$

But we know that $x(t) = \alpha(if)\, P\, e^{i2\pi ft}$ and hence

$$\alpha(if) = \int_0^{\infty} W(\tau)\, e^{-i2\pi f\tau} d\tau \tag{25}$$

This is the Fourier transform of $W(\tau)$ if we assume that $W(\tau) = 0$ for $\tau < 0$. We thus see that $W(\tau)$ and $\alpha(if)$ form a Fourier transform pair.

We now let the load be a random load which has spectral density function $S_p(t)$ and autocorrelation function $R_p(\tau)$. We want to find the spectral density function $S_x(f)$ and the autocorrelation function $R_x(\tau)$ of the response x(t).

Considering first the autocorrelation function $R_x(\tau)$

$$R_x(\tau) = < x(t)\ x(t+\tau)>$$

We have

$$x(t) = \int_0^\infty W(\tau_1)\ P(t-\tau_1)\ d\tau_1$$

$$x(t+\tau) = \int_0^\infty W(\tau_2)\ P(t+\tau-\tau_2)\ d\tau_2$$

The two dummy variables $\tau_1$, $\tau_2$ and the variable $\tau$ must be kept distinct as we must integrate the product of x(t) and $x(t+\tau)$ over all values of $\tau_1$ and $\tau_2$ and still have $\tau$ as the variable in $R_x(\tau)$.

$$R_x(\tau) = \int_0^\infty W(\tau_1).\int_0^\infty W(\tau_2) < P(t-\tau_1)\ P(t+\tau-\tau_2)> d\tau_2.d\tau_1$$

since the time averaging is over t. For a stationary process the origin of time is immaterial and we can take our time from $t-\tau_1$ giving $t-\tau_1$ in place of $\tau$, thus we find

$$\int_0^\infty W(\tau_1) \int_0^\infty W(\tau_2) < P(t)\ P(t+\tau_1-\tau_2+\tau)> d\tau_2.d\tau_1$$

$$= \int_0^\infty W(\tau_1) \int_0^\infty W(\tau_2)\ R_p(\tau_1-\tau_2+\tau)\ d\tau_2\ d\tau_1 \qquad (26)$$

From this it is, in principle, possible to evaluate $R_x(\tau)$ but such an evaluation would be very tedious. If we go on now to look at the spectral density function $S_x(f)$ we will see that a much simpler method can be used.

Since

$$S_x(f) = 2\int_\infty^\infty R_x(\tau)\ e^{-i2\pi f\tau} d\tau$$

we have

$$S_x(f) = 2\int_{-\infty}^\infty \{\int_0^\infty W(\tau_1) \int_0^\infty W(\tau_2)\ R_p(\tau_1-\tau_2+\tau)d\tau_2 d\tau_1\}\ e^{-i2\pi f\tau} d\tau$$

Changing the order of integration we can write this as

$$2\int^\infty W(\tau_1) \int_0^\infty W(\tau_2)\ \{\int_{-\infty}^\infty R_p(\tau_1-\tau_2+\tau)\ e^{-i2\pi f\tau} d\tau\}\ d\tau_2 d\tau_1$$

We now multiply and divide by $e^{i2\pi f\tau_1}$ and $e^{-i2\pi f\tau_2}$ to get

$$2\int_0^\infty W(\tau_1)e^{i2\pi f\tau_1}\int_0^\infty W(\tau_2)e^{-i2\pi f\tau_2}\int_0^\infty R_p(\tau_1-\tau_2+\tau)e^{-i2\pi f(\tau_1-\tau_2+\tau)}$$

$$d(\tau_1-\tau_2+\tau).\ d\tau_2 d\tau_1$$

In this we have used the fact that in the innermost integral $\tau_1$ and $\tau_2$ are treated as constants so that $d\tau = d(\tau_1-\tau_2+\tau)$. Each of the variables $\tau_1,\tau_2$ and $(\tau_1-\tau_2+\tau) = \tau^*$ now separate out to give

$$S_x(f) = 2\int_0^\infty W(\tau_1)e^{i2\pi f\tau_1}\,d\tau_1.\int_0^\infty W(\tau_2)e^{-i2\pi f\tau_2}\,d\tau_2\int_{-\infty}^\infty R_p(\tau^*)e^{-i2\pi f\tau^*}\,d\tau^* =$$

$$= 2\alpha(if)^*.\ \alpha(if)\ .\ \tfrac{1}{2}S_p(f) = \qquad (27)$$

$$= |\alpha(if)|^2\ S_p(f)$$

We thus find that there is a very simple relationship between the spectral density functions of the applied load and the response. The autocorrelation function of the response can then be found, if required from $S_x(f)$.

For our simple one mass system

$$\alpha(if)^2 = \frac{1}{(k-4\pi^2mf^2)^2 + 4\pi^2f^2c^2} \qquad (28)$$

and hence

$$S_x(f) = \frac{S_p(f)}{(k-4\pi^2mf^2)^2 + 4\pi^2f^2c^2} \qquad (29)$$

If the load has Gaussian properties so also will the response and hence

$$< x^2 > = \int_0^\infty S_x(f)\,df$$

$$= \int_0^\infty \frac{S_p(f)}{(k-4\pi^2mf^2)^2 + 4\pi^2fc^2}\,df \qquad (30)$$

If we consider the spectral density of the applied load as "white" or uniform over the range of frequencies when $|\alpha(if)|$ is large we have

$$< x^2 > \;=\; S_p \int_0^\infty \frac{df}{(k-4\pi^2 mf^2)^2 + 4\pi^2 fc^2} \;=\; \frac{S_p}{4ck}$$

If the damping is hysteretic, $c = i\gamma k$, and

$$< x^2 > \;=\; \frac{S_p}{4\,k\,\sqrt{km}}$$

For small damping, such as usually occurs, then two results are equal.

Since the response curve is usually very "peaky" we can generally take the spectral density $S_p(t)$ as constant near the peak values of $|\alpha(if)|^2$.

5. More Complex Systems with a Single Random Load

In the previous section the receptance $\alpha(if)$ and the impulse function $W(\tau)$ were given for a simple single spring system. Using classical methods we can evaluate these for more complex systems. The relationship between $R_p$, $S_p$, $R_x$ and $S_x$ did not involve the form of the structure but merely the fact that there was a single load. We can thus use the same expressions for systems with several degrees of freedom provided that there is only one applied load which varies in a random manner.

Bibliography: General Reading

1. Robson, J.D. *An Introduction to Random Vibration*, Edinburgh, 1963.

2. Crandall, S.H. (Editor) *Random Vibration*, Wiley, 1958.

3. Thomson, W.T. *Vibration Theory and Application*, Chap.10, London, 1971.

4. Hurty, W.C. and Rubinstein, M.F. *Dynamics of Structures*, Prentice Hall, 1964.

5. Clough, R.W. and Penzien, J. *Dynamics of Structures*, McGraw-Hill Book Company, 1975.

6. Newland, D.E. *An Introduction to Random Vibration and Spectral Analysis*, Longman, 1975.

# CHAPTER 15

# EARTHQUAKE RESPONSE OF STRUCTURES

by

H. Tottenham

## 1. INTRODUCTION

The design of structures in earthquake regions has been based upon the assumption of a lateral load proportional to the mass of the building. For example in many areas the structure is designed to carry a "horizontal gravity" force of 0.2 g. Such designs are based upon the assumption that the period of natural vibration of the structure is large compared with the typical period of the ground motion. For many modern structures this is not so and although such simple design rules may be "conservative" they in fact lead to structures where the strength is not properly distributed. That is to say for the same expenditure a safer structure could be built.

Earthquakes are unpredictable both as to when they will occur and the exact form the ground movement will take. However experience has shown that in various areas the ground motions, or accelerations, have definite statistical properties. Such properties are common to both major and minor seismic movements. It is now becoming usual in the design of important structures in seismic regions to investigate the response of a proposed structure to an expected earthquake in two ways. The first response calculation is to check the motion of the structure to the onset of the motion, such a calculation is performed by integrating the equations of motion in time. The second response calculation is to estimate the probable maximum response during the life of the building. It is in this calculation that the theory of random vibrations is used.

Although the theory of non-stationary random processes is being developed as yet the stage has not been reached where it can be used in design. The best we can do is to treat the ground motion as random during the whole of the earthquake. This implies that the auto-correlation function of the ground acceleration must reduce effectively to zero in a time less that the duration of any ground movement. This is in fact true for most ground tremors.

The data available for use in earthquake analysis of structures takes one of the following forms

i) the ground acceleration-time curve

ii) the ground acceleration response spectrum

iii) the ground velocity response spectrum

iv) the ground acceleration spectrum

v) the ground velocity spectrum

The first of these is the quantity that is actually recorded, the instruments at various sites being used to measure the component of the ground accelerations in two directions.

The second form in which data may be presented, also known as the Displacement Response Spectrum, is a plot of the maximum displacement that a single degree of freedom system, whose natural frequency is $\omega$ and damping coefficient $\eta$, would suffer when the support is given the recorded ground movement, this displacement being plotted against $\omega$. The third form is of a similar nature but uses velocities rather than displacements of the system.

The fourth form, the ground acceleration spectrum is the one usually employed in random vibration analysis of seismic response. Many researchers have studied the spectral density function of various major earthquake records and proposed formulae which may be used to give the expected spectral density functions of ground motions in the area. The most frequently used is of the form proposed by Houssner, Tajimi, Jennings and others, this can be written as

$$\omega\, S_{\ddot{x}}(\omega) = \frac{4}{\pi}\,\frac{b\,\lambda}{(1+4b^2)}\,\frac{(1+4b^2\lambda^2)}{(1-\lambda^2)^2+4b^2\lambda^2}\cdot\alpha_g^{\,2} \tag{1}$$

where $\lambda = \omega/\omega_g$, $\omega_g$ being a ground dominant frequency, and b is a constant depending on the ground surface layer. Typically $\omega_g \approx 15$ to 25 radians/sec and b varies from 0.3 for soft ground to 0.7 for fairly hard ground. $\alpha_g$ is the earthquake intensity factor, generally expressed in terms of g. It will be seen that this form of spectral density functions can be considered as describing the response of the surface to a base rock movement of the "white noise" type, the surface layer acting as a single degree of freedom system whose natural frequency is $\omega_g$ and whose damping coefficient is b.

## 2. BEAM ANALYSIS

We will now consider the structural response to such ground movements and take as a particular case in the analytical work a cantilever structure.

We let the ground movement be denoted by y and the elastic displacement in the structure be denoted by w. The total movement of any point is thus $w + y$. The equation of motion of an elementary part of the beam is now

$$EI \frac{\partial^4 w}{\partial x^4} + c \frac{\partial w}{\partial t} + m \frac{\partial^2 w}{\partial t^2} = - m \frac{\partial^2 y}{\partial t^2} \qquad (2)$$

where the symbols have their usual meaning. The free vibration of the structure can be written as

$$w(x,t) = \sum_r w_r(x)\, q_r(t) \qquad (3)$$

when $w_r(x)$ is the $r^{th}$ mode shape defined by

$$EI \frac{d^4 w_r}{dx^4} - m\, \omega_r^2 w = 0 \qquad (4)$$

and the appropriate boundary conditions. These modes are "weighted orthogonal" in that

$$\int_o^L w_r(x)\, w_s(x)\, m(x)\, dx = 0, \quad r \neq s$$
$$= M_r, \quad r = s \qquad (5)$$

We take a similar form for the response of the structure and make the assumption that the damping is proportional to the mass, so that we can re-write the equation of motion as

$$\sum m\, w_r(x)\ddot{q}_r + \sum \eta_r\, m\, \omega_r\, w_r(x) q_r + \sum m\, \omega_r^2\, w_r(x) q_r = - m\ddot{y} \qquad (6)$$

Multiplying by $w_r(x)$ and integrating over the length of the beam reduces the above to

$$M_r\, \ddot{q}_r + C_r\, \omega_r\, \dot{q}_r + \omega_r^2\, M_r\, q_r = - F_r(t) \qquad (7)$$

where:

$$M_r = \int_o^L m\, w_r^2\, dx \quad \text{is the } r^{th} \text{ generalized mass}$$

$$C_r = \int_o^L c\, w_r^2\, dx \quad \text{is the } r^{th} \text{ generalized damping constant} \qquad (8)$$

$$F_r(t) = \int_o^L m\ddot{y}\, w_r(x)\, dx \text{ is the } r^{th} \text{ generalized force.}$$

We put $C_r = 2\, \eta_r\, M_r$ where

$$\ddot{q}_r + 2\eta_r\, \omega_r\, \dot{q}_r + \omega_r^2\, q_r = - \frac{F_r(t)}{M_r} \qquad (9)$$

If we take a point load $P_o e^{i\omega t}$ at some point $x_1$ then

$$F_r(t) = w_r\, (x_1)\, P_o e^{i\omega t}$$

and the function $q_r$ will be

$$q_r = \frac{w_r(x_1)}{M_r \omega_r^2} \left\{ \left[ \frac{1}{(1-\lambda_r^2) + 2i\eta_r\lambda_r} \right] \right\} P_o e^{i\omega t} \tag{10}$$

where

$$\lambda_r = \omega/\omega_r$$

For all the modes

$$w(x,t) = \sum w_r(x)\, q_r(t)$$

$$= \left[ \sum \frac{w_r(x)\; w_r(x_1)}{M_r\; \omega_r^2} \; \frac{1}{1-\lambda_r^2+2i\eta_r\lambda_r} \right] P_o e^{i\omega t} \tag{11}$$

The receptance at x due to loads at $x_1$ is thus

$$\alpha(x,x_1,\omega) = \sum_r \frac{w_r(x)\; w_r(x_1)}{M_r\; \omega_r^2} \; \frac{1}{1-\lambda^2+2i\eta_r\lambda_r} \tag{12}$$

3. SPECTRAL DENSITY OF RESPONSE

For many loads at $x_i$, $x_j$ ... one can write a spectral density function such as,

$$S_w(x,\omega) = \sum \sum \alpha(x,x_i,\omega)\; \overline{\alpha(x,x_j,\omega)}\; S_p(x_i,x_j,\omega) \tag{13}$$

where $S_p(x_i,x_j,)$ is the cross spectral density of the loads at $x_i$ and $x_j$.

If all the loads have the same timewise variation, i.e.

$$P(x,t) = P(x)\, f(t) \tag{14}$$

we can find the cross correlation $R(x_i,x_j,\omega)$ since

$$\begin{aligned} R(x_i,x_j,\tau) &= \langle P(x_i)f(t)\; P(x_j)f(t-\tau) \rangle \\ &= P(x_i)\; P(x_j) \langle f(t)\; f(t-\tau) \rangle \\ &= P(x_i)\; P(x_j)\; R_f \end{aligned} \tag{15}$$

It follows then that the cross spectral density with p related to the spectral density of the variation f by

$$S_p(x_i,x_j,\omega) = P(x_i)\; P(x_j)\; S_f \tag{16}$$

If now we put

$$P(x_i) = p(x_i)\ \delta x \tag{17}$$

and

$$f(t) = -m\ddot{y}$$

we find

$$\begin{aligned} S_w(x,\omega) &= \int_o^L\int_o^L \alpha(x,x_i,\omega)\ \overline{\alpha(x,x_j,\omega)}\ m(x_i)\ m(x_j)\ S_{\ddot{y}}\ dx_i\ dx_j \\ &= \left\{\int_o^L \alpha(x,x_i,\omega)\ m(x_i)\ dx_i\right\}\left\{\int_o^L \overline{\alpha(x,x_j,\omega)}\ m(x_j)\ dx_j\right\}\ S_{\ddot{y}} \end{aligned} \tag{18}$$

and since $x_i$ and $x_j$ are both "dummy" variables we can write this as

$$S_w(x,\omega) = \left|\int_o^L (x,x_i,\omega)\ M(x_i)dx_i\right|^2 S_{\ddot{y}} \tag{19}$$

Using the value we had earlier for $(x,x_i,\omega)$ this gives

$$S_w(x,\omega) = \left|\int_o^L \sum_r \frac{w_r(x)w_r(x_1)m(x_1)}{M_r\omega_r^2(1-\lambda^2+2i\eta_r\lambda_r)}\ dx_i\right|^2 S_{\ddot{y}} \tag{20}$$

$$= \left|\sum_r \frac{w_r(x)\ P_r}{\omega_r^2\ M_r}\cdot\frac{1}{(1-\lambda_r^2+2\eta_r\lambda_r)}\right|^2 S_{\ddot{y}} \tag{21}$$

where $P_r = \int_o^L m(x)\ w_r(x)\ dx$ is the $r^{th}$ generalized load.

The mean square response $<w^2>$ is given by

$$\begin{aligned} <w^2(x)> &= \int_o^\infty S_w(x,\omega)\ d\omega \\ &= \int_o^\infty \left|\sum_r \frac{w_r(x)\ P_r}{\omega_r{}^2\ M_r}\ \frac{1}{(1-\lambda_r^2+2i\eta_r\lambda_r)}\right|^2 S_{\ddot{y}}\ d\omega \end{aligned} \tag{22}$$

When, as is usually the case with cantilever structures, the natural frequencies $\omega_r$ are well separated and the damping is small we can neglect the cross product terms in the above and write

$$<w^2(x)> = \int^\infty \left|\sum_r \frac{w_r^2(x)\ P_r^2}{\omega_r^4\ M_r}\ \frac{1}{(1-\lambda_r^2)^2+2\eta_r^2\lambda_r^2}\right| S_{\ddot{y}}\ d\omega \tag{23}$$

Also when the damping is small the frequency dependent term $1/|(1-\lambda_r^2)^2+2\eta_r^2\lambda_r^2|$ is very peaky, and if $S_{\ddot{y}}$ only varies slowly with we can consider that for any term in the summation $S_{\ddot{y}}$ is constant at the value $S_{\ddot{y}}(\omega_r)$ we can then carry out the integrations to give

$$< w^2(x)> \quad = \quad \frac{\pi}{4} \sum \frac{w_r^2(x)\ P_r^2}{\omega_r^4\ M_r^2} \cdot \frac{1}{\eta_r} \cdot S_{\ddot{y}}\ (\omega_r) \tag{24}$$

It will generally be found that one term dominates - almost always the first - so we can put

$$< w^2(x)> \quad \approx \quad \frac{\pi}{4} \ \frac{w_1^2(x)\ P_1^2}{M_1^2\ \omega_1^4} \ \frac{1}{\eta_1}\ S_{\ddot{y}}(\omega_1) \tag{25}$$

The root mean square value is the "standard deviation" of the response

$$\sigma_w \quad = \sqrt{< w\ (x)\ >} \quad = \quad \frac{w_1(x)P_1}{\omega_1^2\ M_1} \ \sqrt{\frac{\pi\ S_y(\omega_1)}{4\ \eta_1}} \tag{26}$$

This enables us to estimate the maximum value of w(x).

In design it is usual to allow a ten per cent increase in $\sigma$ to allow for the higher modes and take the estimated maximum response as 3.5 times the result so obtained, i.e.

$$w(x)_{max} \quad = \quad 2.85\ \sigma_w \tag{27}$$

For some structures the modes have near equal frequencies and the above simplifications cannot be used. It is then necessary to use numerical integration, including all the modes that make a significant contribution. Simpsons rule integration can be used if the interval in $\omega$ is small, of the order of $0.02\ \omega_1$.

In the above we have seen how the displacements may be estimated. We can use exactly the same method to estimate the magnitudes of the internal forces and bending moments. Associated with each mode shape then will be a bending moment and shear force distribution which we can denote by $M_r$ and $Q_r$. These will vary in the time dimension in exactly the same manner as does $w_r(x)$. The mean squared value will thus be given by formulae analogous to those for the mean square displacement but with $M_r(x)$ and $Q_r(x)$ substituted in place of $w_r(x)$.

For a uniform cantilever the first mode is given by

$$w_1(x) \quad = \quad \frac{(\sinh \lambda L + \sin \lambda L)(\cosh \lambda x - \cos \lambda x) - (\cosh \lambda L + \cos \lambda L)(\sinh \lambda x - \sin \lambda x)}{2(\tanh \lambda L - \tan \lambda L)} \times w_{tip} \tag{28}$$

where $\lambda L = 1.8751$. The shear force $Q_1(x)$ is given by

$$Q_1(x) = EI \frac{d^3 w_1(x)}{dx^3} \tag{29}$$

$$= EI\lambda^3 \left\{ \frac{(\sinh \lambda L+\sin \lambda L)(\sinh \lambda x-\sin \lambda x)-(\cosh \lambda L+\cos \lambda L)(\cosh \lambda x+\cos \lambda x)}{2(\tanh \lambda L - \tan \lambda L)} \right\} w_{tip}$$

At the base, x = 0 this reduces to

$$Q_o = - EI\lambda^3 \left\{ \frac{\cosh \lambda L + \cos \lambda L}{\tanh \lambda L - \tan \lambda L} \right\} w_{tip} \tag{30}$$

and using $\lambda^4 = m\omega^2/EI$, $\lambda L = 1.8751$, this gives

$$Q_o = 0.3914\ m\omega^2 L\ w_{tip} \tag{31}$$

Example: Consider a cooling tower shell with the following characteristics.

| | | |
|---|---|---|
| Dimensions | Height | 100.78 m |
| | Throat height | 82.19 m |
| | Base radius | 41.71 m |
| | Throat radius | 25.60 m |
| | Thickness | 0.127 m |
| | Density | 2400 $Kg/m^3$ |
| | Young's Modulus | $2.11 \times 10^5 Kg/m^2$ |
| | Poisson's Ratio | 0.15 |

| | 1st Mode | 2nd Mode | 3rd Mode |
|---|---|---|---|
| Natural Frequency | 20.76 | 42.72 | 65.99 |
| Generalized mass ($\times 10^9$) | 1.494 | 2.469 | 5.250 |
| Generalized load ($\times 10^9$) | 2.318 | -1.923 | 1.564 |

Taking internal damping as 0.05 and the ground characteristics b = 0.65, $\omega_g = 15.7$, $\alpha = 0.2$ g, we find peak response (at 3.5 × standard deviation)

| | |
|---|---|
| 1st mode only | 67.21 mm |
| 1st & 2nd modes | 67.34 mm |
| 1st, 2nd & 3rd modes | 67.34 mm |

## Bibliography

### General Reading


1. Robson, J.D. *An Introduction to Random Vibration*, Edinburgh, 1963.

2. Crandall, S.H. (Editor) *Random Vibration*, Wiley, 1958.

3. Thomson, W.T. *Vibration Theory and Application*, Chap. 10, London, 1971.


### Earthquake Response


4. Housner, G.W. *Earthquake Engineering*, Edited by Wiezel, R.L. Chapters 4 and 5. Prentice Hall, 1970.

5. Tajimi, H. A statistical method for determining the maximum response of a structure during an earthquake. *Proc. 2nd World Conf. on Earthquake Engineering*, Japan, 1960.

6. Rosenblueth, E. Probabilistic design to resist earthquakes. *Proc. A.S.C.E.*, Vol. 93, No.EM1, Feb. 1967.

7. Newmark, N.M. and Rosenblueth, E. *Fundamentals of Earthquake Engineering.* Prentice Hall, 1970.

8. Earthquake Regulations, World List. *International Association for Earthquake Engineering*, 1970.

9. Okamoto, S. *Introduction to Earthquake Engineering.* Univ. of Tokyo, 1974.

10. Seed, H.B., Ugas, C. and Lysmer, J. Site dependent spectra for earthquake resistant design. *EERC Report 74-12*, 1974. Earthquake Engineering Research Centre, Berkeley, California.

11. Lomnitz, C. and Rosenblueth, E. *Seismic risk and engineering decisions.* Elsevier, 1976.

12. Romstad, K.M., Bruce, J. and Hutchinson, J.R. Site dependent earthquake motions. *Journal of the Geotechnical Engineering Division, ASCE*, November 1978, pp. 1389-1400.

## CHAPTER 16

## RESPONSE OF STRUCTURES TO WIND LOADING

by

H. Tottenham

### 1. INTRODUCTION

Our knowledge of the general structure of the wind and the loads it imposes on buildings is still somewhat limited. However, a few general comments may be made.

On the large scale we can consider the wind as the steady movement of air from region to region, the speed of the wind depending on the distribution of atmospheric pressure over the surface of the world. In the lower levels of the atmosphere this flow is retarded by friction with the ground, the extent to which this retardation affects the air flow depends upon the surface roughness. We can assume that the mean wind speed will be about zero at the actual ground surface, rapidly increasing with height in the lower regions then more slowly increasing with height until a fairly constant speed is reached. This constant speed is known as the Gradient Velocity and the height at which it occurs is called the Gradient Height. The gradient height depends upon the ground roughness and is about 500 m in very rough terrain, for example built up city centres, 400 m in ordinary landscape with houses and trees, and about 300 m in flat open country. The basic wind velocity profile can be written in the form

$$\left(\frac{v}{v_g}\right)^2 = \left(\frac{h}{h_g}\right)^{\alpha} \qquad (1)$$

where $v_g$ and $h_g$ are the gradient velocity and height, $\alpha$ is a constant depending upon terrain. For the three types mentioned above Davenport gives $\alpha$ the values 0.4, 0.28 and 0.16 respectively.

This large scale pattern is not uniform in time. The timewise variation can be best illustrated by the spectral density of the velocity made up from very long time records. Van der Hoven gives the results shown in figure 1. There is a long flat part in this curve between about 0.5 cycles per hour and 10 cycles per hour. This gap is used to separate, in meteorological terms, the large scale on "weather map" wind speeds from those due to guests or turbulence. Wind variation with a period of less than, say, 0.1 hour we consider as gusts.

The mean wind speed is calculated on an hourly basis. From the records it is possible to estimate the likely maximum wind velocity in any length of time. This maximum (for the required length of time) is used as a steady wind and its effect on the structure assessed using a static pressure.

To the steady wind velocity we must add the effects of turbulence. It is this aspect of the wind velocity which may call for dynamical analysis of the structure.

Although turbulence decreases with height on the large scale measurements indicate that it is practically constant up to the height of all but the very tallest of buildings. According to Davenport we can assume that the standard deviation of the wind velocity is about 0.1 of the gradient velocity. This includes all directional components of the turbulent wind: the vertical component is small and the component in the direction of the mean wind speed predominates. In terms of the standard 10 m level wind velocity $V_{10}$ we would have

$$\sigma_v = k\, V_{10} \tag{2}$$

where k = 0.58, 0.32, 0.18 according to the ground being rough, normal or smooth as described earlier.

The spectrum of the turbulent air has been investigated by many research workers and various formulae have been proposed. The two most frequently quoted are due to Davenport and Harris. Thus Davenport has suggested

$$S_v(\omega) = \frac{kL^2\omega}{\pi^2\left[1 + \left(\frac{\omega L}{2\pi V_{10}}\right)\right]^{4/3}} \qquad (0 < \omega < \infty) \tag{3}$$

when k is the roughness parameter, varying from 0.0015 to 0.15 according to the terrain.

$$\sigma_v^2 = 6k\, V_{10}^2 \tag{4}$$

The formula suggested by Harris does not have zero value when $\omega = 0$ and has a similar form to the above:

$$S_v(\omega) = \frac{2kLV_{10}}{\left[2 + \left(\frac{\omega L}{2\pi V_{10}}\right)^2\right]^{5/6}} \tag{5}$$

with

$$\sigma_v^2 = \frac{40}{6}\, k\, V_{10}^2$$

The pressure on a building is closely proportional to the square of the velocity and Handa proposes a spectrum of the squared velocity of the form

$$S_{v^2}(\omega) = \frac{\sigma_v^{\,2}\, a}{\pi(\omega^2 + a^2)} \tag{6}$$

where

$$a = 2\sqrt{3}\,\pi\,\frac{V_{10}}{L}$$

and

$$\sigma_{v^2} = (\frac{40}{6}\,k\,V_{10})^2 \tag{7}$$

This form of the power spectrum is the most useful for design, not only because it deals with velocity squared which can be converted directly to pressure, i.e. it can be used as the pressure spectrum, but also because it has a simple form leading to the autocorrelation function

$$R_{v^2} = \sigma_v^2 \;.\; e^{-a|\tau|} \tag{8}$$

In order to procede with a dynamical analysis using available wind data it is necessary to make a set of assumptions.

a) the wind speed variations and the pressure variations may be treated as a random process

b) the wind pressure is normal to the surface; drag could be included but there is no data on the effect of turbulence on drag

c) the pressure field is homogeneous. That is to say the mean squared wind pressure and the spectrum is the same for all points on the surface. This in turn implies that the dimensions of the structure are small compared with those of a "gust". This is in fact not true for larger structures.

d) a suitable spectral density function and cross correlation function is known.

In practice there is not enough information on the cross correlation of wind velocities at points separated horizontally, cross correlation of wind velocity in the vertical direction although not adequately established has been investigated by many workers. At present there are several experiments which should help to produce the necessary data.

Shimizu has proposed, for rotational shells such as wind shields and cooling towers, etc. cross spectrum in the form

$$S_p(x_1,x_2,\omega) = S_p(\omega)\; e^{-A_1|\bar{z}|}\; e^{-A_2|\bar{\theta}|} \tag{9}$$

where $x_1$, $x_2$ are points on the shell surface, $\bar{z}$ and $\bar{\theta}$ are dimensionless separation distances $z_2 - z_1$ and $\theta_2 - \theta_1$, and the quantities $A_1$ and $A_2$ are parameters which will vary according to the structure. Using results of measurements of pressure and wind velocities on cooling towers in Germany and Japan he suggests

$$S_p(\omega) = \frac{2C_p}{\pi} \frac{a}{\omega^2+a^2} \tag{10}$$

$$S_p(x_1,x_2,\omega) = \frac{2C_p}{\pi} \frac{a}{\omega^2+a^2} e^{-3.77 |\bar{z}|} . e^{-0.94 |\bar{\theta}|}$$

$C_p$ being the mean squared pressure, and a, the exponential decay factor in the autocorrelation function has a value of approximately 0.3.

## 2. RESPONSE OF SHELLS

The equations of equilibrium of a thin elastic shell can be written in the form

$$\begin{aligned} L_1(u,v,w) &= p_u \\ L_2(u,v,w) &= p_v \\ L_3(u,v,w) &= p_w \end{aligned} \tag{11}$$

Here u, v, w are displacements in two directions on the shell surface and the normal direction to the surface, $p_u$, $p_v$, $p_w$ are the loads applied in these directions and $L_1$, $L_2$, $L_3$ are some differential expressions. We can use these equations to derive the equations of motion

$$\begin{aligned} L_1(u,v,w) + \rho h\,\ddot{u} + c\,\dot{u} &= p_u \\ L_2(u,v,w) + \rho h\,\ddot{v} + c\,\dot{v} &= p_v \\ L_3(u,v,w) + \rho h\,\ddot{w} + c\,\dot{w} &= p_w \end{aligned} \tag{12}$$

where h is the thickness of the shell and $\rho$ is its density. For free vibrations

$$\begin{aligned} L_1(u,v,w) - \rho h\ \omega^2\ u &= 0 \\ L_2(u,v,w) - \rho h\ \omega^2\ v &= 0 \\ L_3(u,v,w) - \rho h\ \omega^2\ w &= 0 \end{aligned} \tag{13}$$

Let the natural frequencies be $\omega_r$ and the corresponding mode shapes $u_r$, $v_r$, $w_r$, so that, for example

$$L_1(u_r,v_r,w_r) - \rho h\,\omega_r^2\ u_r = 0 \tag{14}$$

Now let the response be written in the form

$$u = \sum u_r q_r \qquad v = \sum v_r q_r \qquad w = \sum w_r q_r \tag{15}$$

when $q_r$ is the time dependent function defining the amplitude of the mode at any instant. Substituting these expressions in the equation of motion and using the relationships such as (13) above we get

$$\begin{aligned}
\sum \rho h\, u_r \ddot{q}_r + \sum c\, u_r \dot{q}_r + \sum \rho h\, \omega_r^2 u_r q_r &= p_u \\
\sum \rho h\, v_r \ddot{q}_r + \sum c\, v_r \dot{q}_r + \sum \rho h\, \omega_r^2 v_r q_r &= p_v \\
\sum \rho h\, w_r \ddot{q}_r + \sum c\, w_r \dot{q}_r + \sum \rho h\, \omega_r^2 w_r q_r &= p_w
\end{aligned} \tag{16}$$

The orthogonality of the normal modes has the form

$$\begin{aligned}
\int \rho h(u_r u_s + v_r v_s + w_r w_s)\, dA &= 0 \qquad r \neq s \\
&= M_r \qquad r = s
\end{aligned} \tag{17}$$

the integration being over the whole surface of the shell.

If then we multiply the first of the above equations by $u_r$, the second by $v_r$, the third by $w_r$, add them together and integrate over the shell surface we will obtain

$$\ddot{q}_r + 2\eta_r \omega_r \dot{q}_r + \omega_r^2 q_r = \frac{F_r}{M_r} \tag{18}$$

Here again we have assumed that the damping constant is proportional to $\rho h$, and

$$F_r = \int (u_r p_u + v_r p_v + w_r p_w)\, dA \tag{19}$$

We have thus separated the modes and can carry on the analysis as before.

Particularizing for rotational shells, taking u in the meridional direction, v in the circumferential direction and w in the outward normal direction, and denoting the angle between the normal and the axis of rotation by $\phi$, using z to denote the coordinate along this axis we have

$$dA = \frac{r}{\sin\theta}\, d\theta\, dz \tag{20}$$

The various modes will have terms such as $\cos\theta$, $\cos 2\theta$, ..., and for each of these there will be a series of longitudinal modes, we therefore write

$$u_r = u^{mn} = u_{mn}(z) \cos n\theta$$
$$v_r = v^{mn} = v_{mn}(z) \sin n\theta \qquad (21)$$
$$w_r = w^{mn} = w_{mn}(z) \cos n\theta$$

In wind analysis we can take $p_u = p_v = 0$, since the wind will apply only normal pressure to the shell surface. Very many modes may contribute to the overall response.

A horizontal displacement y of the shell gives

$$u = - \cos\theta \cos\phi \, u,$$
$$v = \sin\theta \, y \qquad (22)$$
$$w = \cos\theta \sin\phi \, y$$

so that the loads in the equation of motion have the form $p_u = - \rho h \cos\theta \cos\phi \, \ddot{y}$ etc. This means that all the generalized loads with $n \neq 1$ will be zero, and hence only modes with n = 1 will be excited by the ground movement.

Example 1

Considering the same hyperbolic cooling tower, and wind factors $c_p = (100 \text{ kg/m}^2)^2$, a = 0.3, and using material damping 0.02 we find

| Mode n | m | Natural Frequency | $w^2$ (cm$^2$) |
|---|---|---|---|
| 1 | 1 | 20.76 | 0.058 |
| | 2 | 42.72 | 0.002 |
| | 3 | 65.99 | 0.000 |
| 2 | 1 | 11.12 | 0.358 |
| | 2 | 23.24 | 0.022 |
| | 3 | 43.73 | 0.002 |
| 3 | 1 | 8.64 | 0.088 |
| | 2 | 12.60 | 0.079 |
| | 3 | 27.32 | 0.006 |
| 4 | 1 | 7.45 | 0.317 |
| | 2 | 9.12 | 0.039 |
| | 3 | 17.52 | 0.026 |
| 5 | 1 | 6.53 | 0.110 |
| | 2 | 9.03 | 0.031 |
| | 3 | 12.92 | 0.053 |
| 6 | 1 | 7.24 | 0.000 |
| | 2 | 8.36 | 0.022 |
| | 3 | 12.67 | 0.003 |
| 7 | 1 | 8.21 | 0.031 |
| | 2 | 9.57 | 0.002 |
| | 3 | 12.16 | 0.026 |
| | | $\Sigma$ | 1.276 |

$\sigma = 1.13$ cm.

Bibliography

General Reading

1. Robson, J.D. *An Introduction to Random Vibration*, Edinburgh, 1963.

2. Crandall, S.H. (Editor) *Random Vibration*, Wiley, 1958.

3. Thomson, W.T. *Vibration Theory and Application*, Chap. 10, London, 1971.

4. Davenport, W.B. *Probability and random processes*, McGraw-Hill Book Co. 1970.

Wind Response

5. Van Der Hoven, I. "Power spectrum of horizontal wind speed in the frequency range from 0.0007 to 900 cycles per hour", *Journal of Meteorology*, 1957, vol. 14, pp. 160-164.

6. Davenport, A.G. The spectrum of horizontal gustiness near the ground in high winds, *J. Royal Meteorological Soc.*, 87(1961), pp. 194-211.

7. Davenport, A.G. The application of statistical concepts to the wind loading of structures. *Proc. I.C.E.*, Vol. 19, 1961.

8. *International Conference on Wind Effects on Buildings and Structures*, Teddington, 1963.

9. Davenport, A.G. "The buffeting of large superficial structures by atmospheric turbulence", *Annals of the New York Academy of Sciences*, Volume 116, pp 135-160, June 1964.

10. Symposium 16 *Wind Effects on Buildings and Structures*, H.M.S.O., London, 1965.

11. Harris, R.I. On the spectrum and auto correlation function of gustiness in high winds, *ERA Report 5273*, October 1968.

12. *Modern Design of Wind-Sensitive Structures*, C.I.R.I.A., London, 1971.

13. Kaimal, J.C. Spectral characteristics of surface-layer turbulence. *J. Royal Meteorological Soc.* 98 (1972) pp. 563-589.

14. Macdonald, A.J. *Wind Loading on Buildings*, Applied Science, London, 1975.

15. Houghton, E.L. and Carruthers, N.B. *Wind Forces on Buildings and Structures*, Arnold, London, 1976.

16. Saul, W.E. and Jayachandran, P. and Peyrot, A.H. Response to stochastic wind of n-degree tall buildings. *Journal Struct.Div. ASCE*, May 1976.

17. Simiu, E. and Scanlon, R.U. *Wind Effects on Structures*, Wiley, New York, 1978.

18. "Wind Engineering in the Eighties" *Proc. of the C.I.R.I.A. Conference* held on 12/13 November 1980.

19. Lawson, T.V. *Wind Effects on Building*, Vols. 1 and 2, Applied Science, London, 1981.

20. Abu Sitta and Gould, J. *Dynamic Response of Structures to Wind and Earthquake Loading*, Pentech Press, London, 1980.

CHAPTER 17

# RANDOM RESPONSE ANALYSIS OF OFF-SHORE STRUCTURES

by

C.A. Brebbia

## 1. Introduction

In this chapter the dynamic response of off-shore structures in deep water under random waves forces will be analysed. The structures will be considered to be such that they do not substantially change the motion of the waves, and that this motion can be represented by linear gravity theory. In this theory horizontal velocities and accelerations are linearly related to the wave heights, which are represented by a spectrum and assumed to be a zero-mean Gaussian process. Once the velocities and accelerations are known we can compute the forces on the structure. During this process drag forces coefficients are linearized in the form described in chapter 13. To obtain a better estimate of the drag, the designer could correct the linearized values during a second cycle and run the problem again. This is not important however in many practical cases where the inertia forces tend to be very much larger than the drag components.

The forces acting on the system can be defined in terms of spectrum density functions. The product of these by the transfer functions gives the spectral density of the response. This process is explained in detail for a one degree of freedom and then generalized for multi-degree of freedom systems, under the following assumptions,

a) The damping modal coefficients are uncoupled.
b) The lower natural frequencies which are taken to be the only significant ones, are well separated.
c) Damping is low.

Under these assumptions the now uncoupled system of equations is thought to be such that the cross spectral density terms do not contribute to the response of the system.

Finally the spectral density response of the system can be used to find the variance of displacements, stresses or other parameters of interest. Once the variance is known the probability of any such parameters to exceed a given value can be calculated.

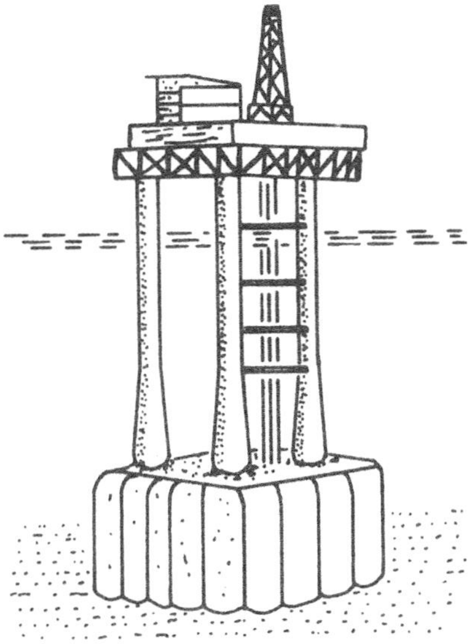

Figure 1 Typical Oil Rig Structure

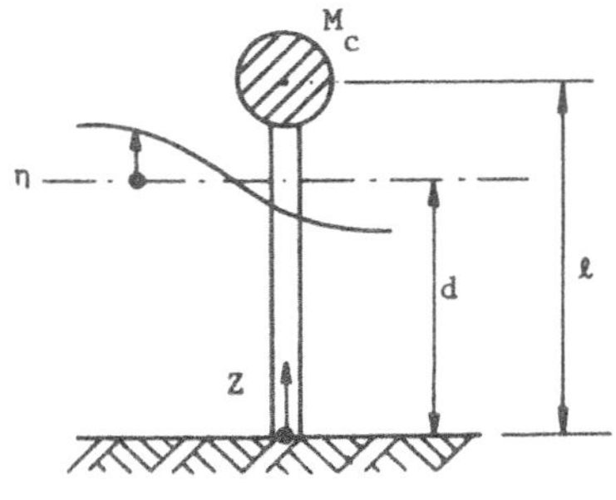


Figure 2 One Degree of Freedom Idealization

## 2. One Degree of Freedom System

Consider now that the structure can be idealized as a one degree of freedom system. The complexity of a typical off-shore rig (Figure 1) is such that this idealization as depicted in Figure 2 may be suspected. However the response of these structures tends to occur predominantly in the first mode, which indicates that a one degree of freedom idealization may be useful as a preliminary design tool. Later on the analysis will be extended to multi-degree of freedom systems.

For the one degree of freedom system one can compute equivalent values which can be written as,

$EI$ $[Nm^2]$, equivalent stiffness of the column

$A_c$ $[m^2]$ , equivalent area of the column.

In addition we define,

$V_w$ $[m^3/m]$ , volume of water displaced per unit length

$M_c$ $[K]$ , mass of the platform

$\rho$ $[K/m^3]$ , density of the water.

We can also calculate the following drag and inertia coefficients,

$$C_D = \frac{c_d \rho D}{2} \quad [K\ m^{-2}]$$

$$C_M = c_m \rho V_w \quad [K\ m^{-1}] \tag{1}$$

$$C_A = \rho V_w \quad [K\ m^{-1}]$$

where $c_d$ is a drag coefficient (1.0 for cylinders) and $c_m$ is inertia coefficient (also 1.0 for cylinder).

The equilibrium equation of the system can be written as,

$$M \ddot{u} + C \dot{u} + K u = F(t) \tag{2}$$

where u is the displacement at the top of the structure ($x = \ell$). The term C includes the structural and hydrodynamic damping. M is obtained by addition of the mass of the column, the mass of the platform and the hydrodynamic mass. Note that the $C_A$ term does not enter into M because $C_A$ only affects the water particles accelerations.

If the shape of the column is assumed to be $f(\bar{z})$, where $\bar{z} = z/\ell$, the M term in equation (2) is

$$M\ [K] = \ell\, \rho_c\, A_c \int_0^1 [f(\bar{z})]^2\, d\bar{z} + C_M \ell \int_0^{d/\ell} [f(\bar{z})]^2\, d\bar{z} + M_c \qquad (3)$$

The inertia term for the column is

$$K\ [Nm^2] = \frac{EI}{\ell^3} \int_0^{\ell} \left(\frac{\partial^2 f(\bar{z})}{\partial \bar{z}^2}\right)^2 d\bar{z} \qquad (4)$$

If we now write the deflection of the column as,

$$u = q\, e^{i\omega t}$$

we have

$$\ddot{u} = -\,\omega^2\, \ddot{q}\, e^{i\omega t} \qquad (5)$$

The natural frequency of the system is given by the solution of,

$$(K - \omega^2 M) q = 0 \qquad (6)$$

$$\therefore \quad \omega_n^2 = \frac{K}{M}, \qquad \omega_n = \sqrt{\frac{K}{M}} \qquad (7)$$

Equation of motion (2) can now be written,

$$M \ddot{u} + C \dot{u} + \omega_n^2 M u = F(t) \qquad (8)$$

In order to write the C term (where $C = C_s + C_h$) in its usual form $C = 2M \gamma \omega_n$, let us consider its two components, the structural $C_s$ and hydrodynamic $C_h$ damping terms. The component $\bar{C}$ of the last term was written (see chapter 13 formula (20)) as

$$\bar{C} = C_D \sqrt{\frac{8}{\pi}}\ \sigma_{\dot{v}} \qquad (9)$$

where $C_D$ is given in equations (1). For a column the hydrodynamic damping can be written,

$$C_h\, \dot{u} = \{ C_D \sqrt{\frac{8}{\pi}} \int_0^1 \sigma_{\dot{v}}\, [f(\bar{z})]^2\, d\bar{z} \}\, \dot{u} \qquad (10)$$

We can compute the velocity variance $\sigma_{\dot{v}}$ which is a function of the velocity spectrum as

$$\sigma_{\dot{v}}^2 = \int_0^{\infty} S_{\dot{v}\dot{v}}\, (\omega)\, d\omega =$$

$$= \int_0^{\infty} \omega^2 \frac{\cosh^2 \kappa z}{\sinh^2 \kappa d} S_{\eta\eta}(\omega)\, d\omega \qquad (11)$$

$$\sigma_{\dot{v}} = \sqrt{\sigma_{\dot{v}}^2} \quad m^2/sec.$$

Note that for deep waters $\kappa = \omega^2/g$.

Once the $C_h$ term has been computed we can write

$$\gamma = \gamma_s + \frac{C_h}{2M\,\omega_n} \qquad (12)$$

$\gamma_s$ is the structural damping.

The equation of motion (8) can now be written as,

$$M\,\ddot{u} + 2M\,\omega_n\,\gamma\,\dot{u} + \omega_n^2\,M\,u = F(t) \qquad (13)$$

Let us now compute the F(t) term. The hydrodynamic forces due to the water velocity were given by (Chapter 13),

$$P = (C_M + C_A)\,\ddot{v} + C_D \sqrt{\frac{8}{\pi}}\,\sigma_{\dot{v}}\,\dot{v} =$$

$$= (C_M + C_A)\,\omega^2 \frac{\cosh \kappa z}{\sinh \kappa d}\,\eta + C_D \sqrt{\frac{8}{\pi}}\,\sigma_{\dot{v}}\,\omega \frac{\cosh \kappa z}{\sinh \kappa d}\,\eta' \qquad (14)$$

where $\eta$ is the wave height.

The generalized force for the column is now given by

$$F = \int_0^d p\, f(\bar{z})\, dz$$

or,

$$F(t) = \eta\,(C_M + C_A) \frac{\omega^2}{\sinh \kappa d} \int_0^d \cosh \kappa z\, f(\bar{z})\, dz$$

$$+ \eta^r\, C_D \sqrt{\frac{8}{\pi}} \frac{\omega}{\sinh \kappa d} \int_0^d \cosh \kappa z\, \sigma_{\dot{v}}\, f(\bar{z})\, dz \qquad (15)$$

where $\eta' = a \sin(\kappa x_1 - \omega t)$.

Taking the Fourier transform of equation (8) we can write,

$$M(-\omega^2 + 2i\gamma\,\omega\,\omega_n + \omega_n^2)\,\bar{u}(\omega) = \bar{F}(\omega) \qquad (16)$$

where $\bar{u}(\omega)$ is the Fourier transform of u(t) and $\bar{F}(\omega)$ the transform of F(t).

We can write (16) as,

$$\alpha(\omega)\ \bar{F}(\omega) = \bar{u}(\omega) \tag{17}$$

where

$$\alpha(\omega) = [(-\omega^2 + 2i\gamma\omega\omega_n + \omega_n{}^2)\ M]^{-1}$$

The complex conjugate of (17) is obtained by taking the complex conjugate of the Fourier transform (that is with $-\omega$ instead of $\omega$) on equation (8). This gives,

$$\alpha^*(\omega)\ \hat{F}(\omega) = \hat{u}(\omega) \tag{18}$$

where

$$\alpha^*(\omega) = [(-\omega^2 - 2i\gamma\ \omega\ \omega_n + \omega_n{}^2)\ M]^{-1}$$

We multiply each side of (17) by its complex conjugate (18) and divide by the period T multiplied by $\pi$, which can be written,

$$\alpha(\omega)\ \frac{\bar{F}(\omega)\ \hat{F}(\omega)}{\pi T}\ \alpha^*(\omega) = \frac{\bar{u}(\omega)\ \hat{u}(\omega)}{\pi T} \tag{19}$$

When $T \to \infty$ we can write formulae (19) in terms of the spectral densities as,

$$|\alpha(\omega)|^2\ S_{FF}(\omega) = S_{uu}(\omega) \tag{20}$$

where,

$$|\alpha(\omega)|^2 = \alpha(\omega)\ \alpha^*(\omega) = \frac{1}{M^2\omega_n{}^4\left[\left[1 - (\frac{\omega}{\omega_n})^2\right]^2 + \left[2\gamma\ \frac{\omega}{\omega_n}\right]^2\right]}$$

$S_{FF}$ and $S_{uu}$ are the power spectral densities of responseand excitation respectively.

The above transformation for the forces gives,

$$S_{FF}(\omega) = S_{\eta\eta}(\omega)\ \{\frac{(C_M+C_A)^2\omega^4}{\sinh^2\ \kappa d}\ (\int_0^d \cosh \kappa z\ f(\bar{z})dz)^2 + $$
$$+ c_D{}^2\ \frac{8}{\pi}\ \frac{\omega^2}{\sinh^2 \kappa d}\ (\int_0^d \cosh \kappa z\ \sigma_{\dot{v}}(z)\ f(\bar{z})dz)^2\} \tag{21}$$

Hence we can now find the spectral density for the generalized coordinate u using formula (20). Once $S_{uu}(\omega)$ is known we can calculate the variance of u by integrating,

$$\sigma_u{}^2 = \int_0^\infty S_{uu}(\omega)\ d\omega. \tag{22}$$

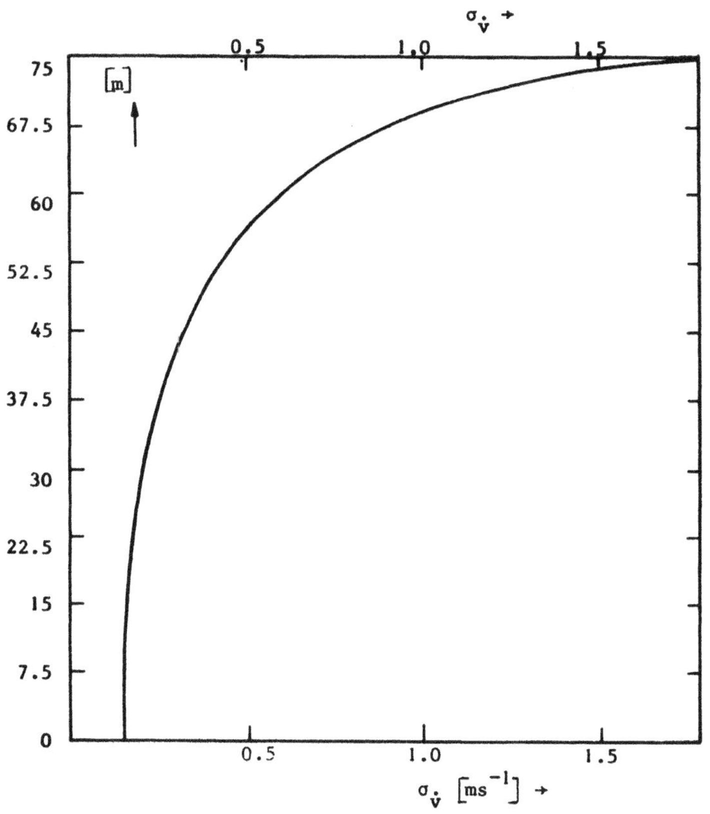


Figure 3 Variation of $\sigma_{\dot{v}}$ with depth

$$C_M = \sum_i (c_m \rho V_w)_i = 78000 \text{ K/m} \qquad \text{(b)}$$

$$C_A = \sum_i (\rho V_w)_i = 78000 \text{ K/m} .$$

$\sum_i$ indicates summation over all the structural elements, D is the diameter of each cylindrical member ($c_m = c_d = 1$ in all the cases). The wave spectrum used is the one given by Pierson and Moskowitz (equation (1) of chapter 13) for a wind velocity $W = 200 \text{ ms}^{-1}$, and is plotted in Figure 4a.

The deflected shape of the structure will be approximated by $f(\bar{z}) = \bar{z}^2$, where $\bar{z} = z/\ell$. Hence the mass of the system can be written,

$$M = \ell \rho_c A_c \int_0^1 \bar{z}^4 \, d\bar{z} + C_M \ell \int_0^{3/4} \bar{z}^4 \, d\bar{z} + M_C = 3860 \times 10^3 \text{ K} \qquad \text{(c)}$$

and the stiffness is,

$$K = \frac{EI}{\ell^3} \int_0^1 \left[\frac{\partial^2 f}{\bar{z}}\right]^2 d\bar{z} = 9000 \times 10^3 \text{ N/m} \qquad \text{(d)}$$

We can now find the natural frequency of the system, $\omega_n$,

$$\omega_n^2 = \frac{K}{M} = \frac{9000 \times 10^3}{3860 \times 10^3} = 2.334$$

$$\omega_n = 1.54 \qquad \text{(e)}$$

In order to calculate the damping constant $\gamma$ we compute the hydrodynamic damping constant $C_h$ from formula (11), having first found the variance $\sigma_{\dot{v}}$ at different heights using Pierson-Moskowitz spectrum for a wind speed $20 \text{ ms}^{-1}$. The variation of $\sigma_{\dot{v}}$ is shown in Figure 3 and was obtained integrating numerically equation (11).

Now we can compute $C_h \dot{u}$ using equation (10), which gives,

$$C_h \dot{u} = 113000 \dot{u} \qquad \text{(f)}$$

The damping constant can now be calculated

$$\gamma = \gamma_s + \frac{113000}{118900} \times 10^{-2} \simeq \qquad \text{(g)}$$

$$\simeq \gamma_s + 0.01$$

For the structural damping we take $\gamma_s = 0.05$. Hence $\gamma = 0.06$ for this case.

Next we evaluate the force spectrum given by equation (21) using numerical integration. They are shown in Figure 4a where the part corresponding to drag and inertia can be seen.

The transfer function for a one degree of freedom system can be computed using formulae (20), and the results are plotted in Figure 4b. Next we can multiply the transfer functions values by the spectral density $S_{FF}$ to obtain the response spectrum

$$S_{uu} \;=\; |\alpha(\omega)|^2 \, S_{FF} \;=\; |\alpha'(\omega)|^2 \frac{1}{M^2\omega_n^{\,4}} S_{FF} \qquad \text{(h)}$$

The response spectrum is shown in Figure 4b. Integrating this spectrum the variance of the generalized displacement can be obtained,

$$\sigma_u^{\,2} \;=\; \int_0^\infty S_{uu}(\omega)\, d\omega \qquad \text{(i)}$$

It gives,

$$\sigma_u^{\,2} \;=\; 0.092 \text{ m}^2$$

$$\sigma_u \;=\; 0.304 \text{ m.}$$

The probability of the u value being within $\pm\, 3\sigma_u = \pm\, 0.912$ m is 99.7%.

## 3. Multi-degree of Freedom System

For a multi-degree of freedom system we have structural stiffness damping mass and force matrices of the type

$$\underset{\sim}{K}\,\underset{\sim}{U} + \underset{\sim}{C}\,\dot{\underset{\sim}{U}} + \underset{\sim}{M}\,\ddot{\underset{\sim}{U}} \;=\; \underset{\sim}{F} \qquad (24)$$

where the matrix $\underset{\sim}{C}$ is proportional to $\underset{\sim}{K}$ or $\underset{\sim}{M}$ or linear combination of both.

These matrices can be formed by considering the individual element matrices, i.e.

$$\underset{\sim}{k}\,\underset{\sim}{U}^n + \underset{\sim}{c}\,\dot{\underset{\sim}{U}}^n + \underset{\sim}{m}\,\ddot{\underset{\sim}{U}}^n \;=\; \underset{\sim}{f} \qquad (25)$$

Matrix $\underset{\sim}{k}$, $\underset{\sim}{m}$ and $\underset{\sim}{f}$ are given by,

$$\underset{\sim}{k} \;=\; \int \underset{\sim}{B}^T\,\underset{\sim}{D}\,\underset{\sim}{B}\; d(\text{vol})$$

$$\underset{\sim}{m} \qquad \int \underset{\sim}{G}^T\,\rho\,\underset{\sim}{G}\; d(\text{vol})$$

$$\underset{\sim}{f} \qquad \int_{S_\sigma} \underset{\sim}{G}^T\,\underset{\sim}{p}\; dS$$

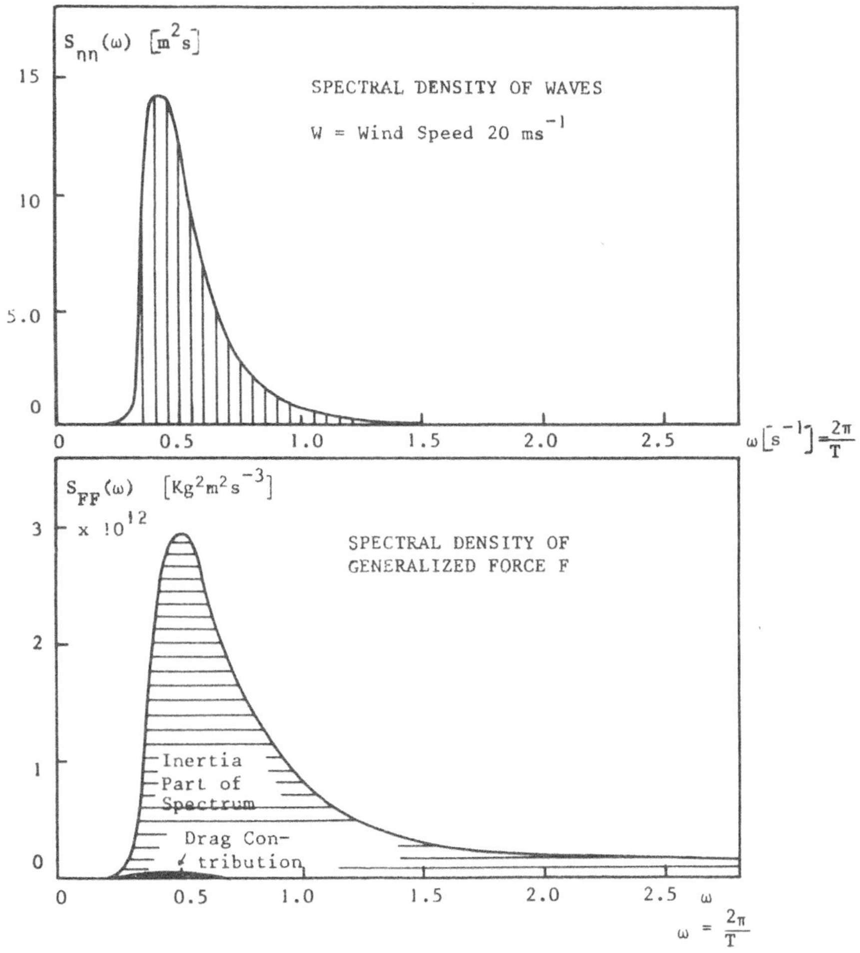


**Figure 4a** Spectral Density of Waves and Generalized Force.

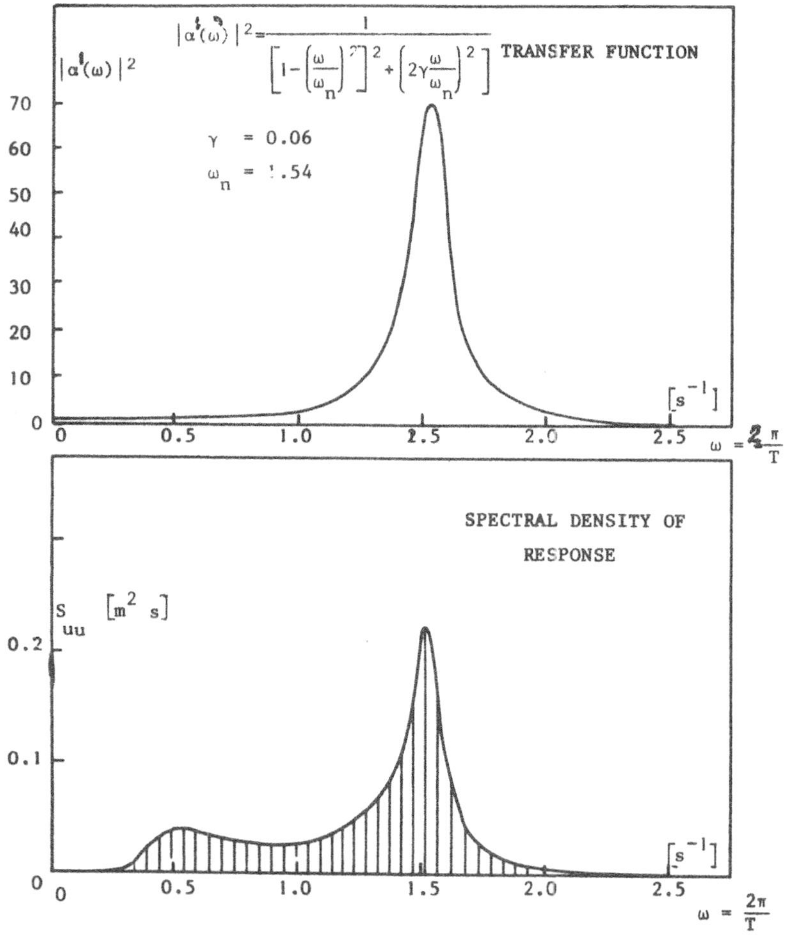


Fig. 4b Transfer Functions and Response Spectrum

If we are working with a Gaussian distribution process with zero mean, the probability of the value of u being within a $\pm \lambda\sigma_u$ value is given in the following table.

| $\lambda$ | Probability of $-\lambda\sigma \le u \le \lambda\sigma$ | Probability of $\lvert u \rvert > \lambda\sigma$ |
|---|---|---|
| 1 | 68.3% | 31.7% |
| 2 | 95.4% | 4.6% |
| 3 | 99.7% | 0.3% |

It is usual to take a value of $\lambda = 3$ in structural applications such as the one discussed here.

In addition to $\sigma_u$ we can calculate the variance of any other quantities such as stresses. Assume for instance that the bending moment at the base is related to q by a B function such that

$$M = B\,u$$

The spectral density of this function is

$$S_{MM}(\omega) = B^2\, S_{uu}$$

and

$$\sigma_{MM}^2 = \int_0^\infty S_{MM}(\omega)d\omega = B^2 \int_0^\infty S_{uu}\, d\omega \qquad (23)$$

Example

Assume that we have an off-shore structure that can be made equivalent to the one degree of freedom structure shown in Figure 2 which has the following characteristics,

$$\begin{aligned} d &= 75\text{ m}, \quad \ell = 100\text{ m} \\ M_C &= 2 \times 10^6 \text{K}, \quad EI = 2250 \times 10^9\ \text{Nm}^2 \\ A_C &= 28^2\text{m} \text{ (cross sectional area of concrete)} \\ V_W &= 78\ \text{m}^3/\text{m (total volume of water displaced per unit length)} \\ \rho &= 10^3\ \text{K/m}^3 \text{ (density of water)} \\ \rho_C &= 2.5 \times 10^3\ \text{K/m}^3 \text{ (density of concrete).} \end{aligned} \qquad (a)$$

The drag and inertia coefficients for the equivalent column are,

$$C_D = \sum_i \left[\frac{c_d\ \rho\ D}{2}\right]_i = 5000\ \text{K/m}^2$$

We substitute the p force vector by the vector of hydrodynamic forces, whose components have been discussed in chapter 13. This vector can be written as,

$$\underset{\sim}{p} = \underset{\sim}{p}_M(\ddot{\underset{\sim}{v}} - \ddot{\underset{\sim}{u}}) + \underset{\sim}{p}_D(\dot{\underset{\sim}{v}} - \dot{\underset{\sim}{u}}) + \underset{\sim}{p}_A \ddot{\underset{\sim}{v}} \tag{26}$$

For instance for a beam element perpendicular to the flow and with three degrees of freedom $u_1$ $u_2$ $\theta_1$ this vector becomes,

$$\begin{Bmatrix} p_1 \\ p_2 \end{Bmatrix} = \begin{bmatrix} C_M & 0 & 0 \\ 0 & 0 & 0 \end{bmatrix} \left\{ \begin{Bmatrix} \ddot{v}_1 \\ \ddot{v}_2 \\ \ddot{v}_\theta \end{Bmatrix} - \begin{Bmatrix} \ddot{u}_1 \\ \ddot{u}_2 \\ \ddot{u}_\theta \end{Bmatrix} \right\} + \tag{27}$$

$$+ \begin{bmatrix} \bar{C} & 0 & 0 \\ 0 & 0 & 0 \end{bmatrix} \left\{ \begin{Bmatrix} \dot{v}_1 \\ \dot{v}_2 \\ \dot{v}_\theta \end{Bmatrix} - \begin{Bmatrix} \dot{u}_1 \\ \dot{u}_2 \\ \dot{u}_\theta \end{Bmatrix} \right\} + \begin{bmatrix} C_A & 0 & 0 \\ 0 & 0 & 0 \end{bmatrix} \begin{Bmatrix} \ddot{v}_1 \\ \ddot{v}_2 \\ \ddot{v}_\theta \end{Bmatrix}$$

where $p_1$ is the component in $x_1$ direction and $p_2$ in $x_2$. For an inclined element one has to decompose the forces in $p_1 p_2$ directions according with the angle of inclination.

The element velocities and accelerations can be written in terms of interpolation functions $\underset{\sim}{G}$ such that

$$\dot{\underset{\sim}{u}} = \underset{\sim}{G}\,\dot{\underset{\sim}{U}}_e \quad , \qquad \ddot{\underset{\sim}{u}} = \underset{\sim}{G}\,\ddot{\underset{\sim}{U}}_e \tag{28}$$

We can assume the same variation applies for $\dot{\underset{\sim}{v}}$, $\ddot{\underset{\sim}{v}}$ velocities. This gives the following expression for $\underset{\sim}{f}$.

$$\underset{\sim}{f} = \underset{\sim}{m}_H \ddot{\underset{\sim}{V}}_e - \underset{\sim}{m}_H \ddot{\underset{\sim}{U}}_e + \underset{\sim}{c}_D \dot{\underset{\sim}{V}}_e - \underset{\sim}{c}_D \dot{\underset{\sim}{U}}_e + \underset{\sim}{m}_A \ddot{\underset{\sim}{V}}_e \tag{29}$$

where,

$$\underset{\sim}{m}_A = \int \underset{\sim}{G}^T \underset{\sim}{p}_A \, dS \; : \quad \text{added mass}$$

$$\underset{\sim}{m}_H = \int \underset{\sim}{G}^T \underset{\sim}{p}_M \underset{\sim}{G} \, dS \; : \quad \text{hydrodynamic mass}$$

$$\underset{\sim}{c}_D = \int \underset{\sim}{G}^T \underset{\sim}{p}_D \underset{\sim}{G} \, dS \; : \quad \text{hydrodynamic damping}$$

Note that the $\underset{\sim}{p}_D$ terms are function of $\sigma_{\dot{v}}$ . To represent the variation of $\sigma_{\dot{v}}$ we can assume that this is a linear function over the element. Hence $\underset{\sim}{p}_D$ is also a linear function of the spatial coordinates.

Equation (25) can now be written as,

$$(\underset{\sim}{m}+\underset{\sim}{m}_H)\ddot{\underset{\sim}{U}}_e + (\underset{\sim}{c}+\underset{\sim}{c}_D)\dot{\underset{\sim}{U}}_e + \underset{\sim}{k}\,\underset{\sim}{U}_e = (\underset{\sim}{m}_H+\underset{\sim}{m}_A)\ddot{\underset{\sim}{V}}_e + \underset{\sim}{c}_D\,\dot{\underset{\sim}{V}}_e \tag{30}$$

In order to avoid proliferation of notation this equation will be written in the same form as (25), i.e.

$$\underset{\sim}{m}\,\ddot{\underset{\sim}{U}}_e + \underset{\sim}{c}\,\dot{\underset{\sim}{U}}_e + \underset{\sim}{k}\,\underset{\sim}{U}_e = \underset{\sim}{f}(\ddot{\underset{\sim}{V}}_e, \dot{\underset{\sim}{V}}_e) \tag{31}$$

For the complete structure we can write

$$\sum_{n_e} (\underset{\sim}{m}\,\ddot{\underset{\sim}{U}}_e + \underset{\sim}{c}\,\dot{\underset{\sim}{U}}_e + \underset{\sim}{k}\,\underset{\sim}{U}_e) = \sum_{n_e} \underset{\sim}{f}$$

or more simply,

$$\underset{\sim}{M}\,\ddot{\underset{\sim}{U}} + \underset{\sim}{C}\,\dot{\underset{\sim}{U}} + \underset{\sim}{K}\,\underset{\sim}{U} = \underset{\sim}{F} \tag{32}$$

If we take the Fourier transform and its complex conjugate of this equation we obtain

$$(-\omega^2\,\underset{\sim}{M} + i\,\omega\,\underset{\sim}{C} + \underset{\sim}{K})\bar{\underset{\sim}{U}} = \bar{\underset{\sim}{F}}$$

and (33)

$$(-\omega^2\,\underset{\sim}{M} - i\,\omega\,\underset{\sim}{C} + \underset{\sim}{K})\bar{\underset{\sim}{U}} = \bar{\underset{\sim}{F}}$$

Equations (33) can be written,

$$\underset{\sim}{H}(\omega)\bar{\underset{\sim}{U}} = \bar{\underset{\sim}{F}}$$
$$\underset{\sim}{H}^*(\omega)\hat{\underset{\sim}{U}} = \hat{\underset{\sim}{F}} \tag{34}$$

We can now multiply each side of these equations between themselves and divide by $\pi$ and the period T. This gives,

$$\underset{\sim}{H}(\omega)\,\frac{1}{\pi T}\,\{\bar{\underset{\sim}{U}}\,\underset{\sim}{U}^T\}\,\underset{\sim}{H}^*(\omega) = \frac{1}{\pi T}\,\{\bar{\underset{\sim}{F}}\,\underset{\sim}{F}^T\} \tag{35}$$

When $T \to \infty$ we can write (35) in terms of spectral densities,

$$\underset{\sim}{H}(\omega)\,\underset{\sim}{S}_u\,\underset{\sim}{H}^*(\omega) = \underset{\sim}{S}_F \tag{36}$$

where the matrix $\underset{\sim}{S}_u$ is the spectral density of the response and $\underset{\sim}{S}_F$ the spectral density of the applied forces. They are both square matrices.

In principle we could solve equations (36) for a series of $\omega$ values and in this way find $S_u$ density for any given point. The variance $\sigma_u^2$ can then be obtained by integration

$$\sigma_{u_i}^{\,2} = \int_o^{\infty} S_{u_i} \, d\omega \tag{37}$$

i : point under consideration.

In order to simplify the problem and economize computer time, let us use a modal decomposition technique. Starting with equation (32) we can first find the eigenvalues and eigenvectors of the system by solving the homogeneous equation,

$$\underset{\sim}{M}\,\ddot{\underset{\sim}{U}} + \underset{\sim}{K}\,\underset{\sim}{U} = \underset{\sim}{0} \tag{38}$$

which gives

$$(\underset{\sim}{K} - \omega^2 \underset{\sim}{M})\underset{\sim}{Z} = \underset{\sim}{0}$$

The response of the structure can be written,

$$\underset{\sim}{U} = \underset{\sim}{Z}\,\underset{\sim}{q} \approx \sum_{i=1}^{s} \underset{\sim}{q}_i\,\underset{\sim}{z}_i \tag{39}$$

in which $\underset{\sim}{Z}$ is the modal matrix and q are the generalized coordinates. Normally it is only necessary to consider the first two or three generalized coordinates and modes, i.e. s = 2 or 3.

In this process we have assumed that the matrix $\underset{\sim}{C}$ is orthogonal with respect to the $\underset{\sim}{z}$ modes. This is unfortunately not so because of the $\underset{\sim}{C}_D$ contribution. Note that $\underset{\sim}{C}_D$ is a diagonal matrix whose coefficients were obtained by least square minimization of an error function.

In what follows we will first assume that the $\underset{\sim}{C}$ matrix can be orthogonalized and then try to minimize the error made in this way. Hence we can multiply and premultiply (32) by the eigenvector matrix $\underset{\sim}{Z}$, (see (39)), which gives,

$$\{\underset{\sim}{Z}^T \underset{\sim}{M}\,\underset{\sim}{Z}\}\,\ddot{\underset{\sim}{q}} + \{\underset{\sim}{Z}^T \underset{\sim}{C}\,\underset{\sim}{Z}\}\,\dot{\underset{\sim}{q}} + \{\underset{\sim}{Z}^T \underset{\sim}{K}\,\underset{\sim}{Z}\}\,\underset{\sim}{q} = \{\underset{\sim}{Z}^T \underset{\sim}{F}\} \tag{40}$$

or

$$\underset{\sim}{M}'\,\ddot{\underset{\sim}{q}} + \underset{\sim}{C}'\,\dot{\underset{\sim}{q}} + \underset{\sim}{K}'\,\underset{\sim}{q} = \underset{\sim}{F}'$$

where $\underset{\sim}{M}'$, $\underset{\sim}{C}'$ and $\underset{\sim}{K}'$ are now diagonal matrices. This system of equations can also be written as S uncoupled second order differential equations,

$$M_i'\,\ddot{q}_i + C_i'\,\dot{q}_i + K_i'\,q_i = F_i' \tag{41}$$

where i = 1, S.

The elements of the matrix C' can be obtained by minimizing the error vector $\underset{\sim}{E}$ [7] where

$$\underset{\sim}{E} = \underset{\sim}{C}_o \, \dot{\underset{\sim}{q}} - \underset{\sim}{C}' \, \dot{\underset{\sim}{q}} \qquad (42)$$

$\underset{\sim}{C}_o$ is the actual (uncoupled) damping matrix obtained after multiplying by $\underset{\sim}{Z}$ and $\underset{\sim}{C}'$ a diagonal matrix we want to obtain with the least possible error.

$$\underset{\sim}{C}_o = \underset{\sim}{Z}^T \, \underset{\sim}{C} \, \underset{\sim}{Z}$$

The mean error can be minimized by doing,

$$< \frac{\partial E^2}{\partial C_i'} > = < \left( \sum_{j=1}^{s} C_{o_{ji}} \dot{q}_j - C_i' \dot{q}_i \right) \dot{q}_i > = 0$$

$$C_i' = C_{o_{ii}} + \sum_{\substack{j=1 \\ j\neq 1}}^{s} \frac{C_{o_{ij}} <\dot{q}_i \, \dot{q}_j>}{<\dot{q}_j>}$$

For a first approximation the supposition that $C_i' = C_{o_{ii}}$ appears to be reasonable. The designer can check this hypothesis once the $\dot{q}_i$ values have been obtained.

After the system of equations has been uncoupled one can apply Fourier transforms and conjugate to obtain,

$$(-\omega^2 M_i' + i\omega C_i' + K_i')\bar{Q}_i = \bar{F}_i'$$

$$(-\omega^2 M_j' - i\omega C_j' + K_j')\hat{Q}_j = \hat{F}_j'$$

Remembering that,

$$\omega_i^2 = \frac{K_i'}{M_i'} , \qquad \omega_j^2 = \frac{K_j'}{M_j'}$$

$$C_i' = 2\gamma_i \frac{K_i'}{\omega_i} , \qquad C_j' = 2\gamma_j \frac{K_j'}{\omega_j}$$

we can define two transfer functions,

$$\alpha_i(\omega) = \{(-\omega^2 + 2i\gamma\,\omega\,\omega_i + \omega_i^2)M_i'\}^{-1}$$

$$\alpha_j^*(\omega) = \{(-\omega^2 - 2i\gamma\,\omega\,\omega_j + \omega_j^2)M_j'\}^{-1} \qquad (45)$$

Hence equations (44) can be written as,

$$\bar{Q}_i = \alpha_i \bar{F}_i' \qquad \hat{Q}_j = \alpha_i^* \hat{F}_j' \tag{46}$$

Multiplying these equations and dividing by T ($t \rightarrow \infty$) we obtain,

$$S_{Q_{ij}} = \alpha_i S_{F'_{ij}} \alpha_j^* \qquad \text{for } i,j = 1,S \tag{47}$$

This can be written in matrix form as,

$$\begin{bmatrix} S_{Q_{11}} & S_{Q_{12}} & \cdots \\ S_{Q_{21}} & S_{Q_{22}} & \cdots \\ \cdots & & \end{bmatrix} = \begin{bmatrix} \alpha_1 & & \\ & \alpha_2 & \\ & & \ddots \end{bmatrix} \begin{bmatrix} S_{F'_{11}} & S_{F'_{12}} & \cdots \\ S_{F'_{21}} & S_{F'_{22}} & \\ \cdots & & \end{bmatrix} \begin{bmatrix} \alpha_1^* & & \\ & \alpha_2^* & \\ & & \ddots \end{bmatrix}$$

$$\underset{\sim}{S}_Q = \underset{\sim}{\alpha} \, \underset{\sim}{S}_F \, \underset{\sim}{\alpha}^* \tag{48}$$

The force spectral density matrix can be found by transforming the original spectral density matrix which can be easily deduced. This gives,

$$\underset{\sim}{S}_F' = \underset{\sim}{Z}^T \underset{\sim}{S}_F \underset{\sim}{Z} \tag{49}$$

Once the generalized coordinates spectral density has been deduced we can obtain the spectral density of the displacements in the original system by a linear transformation,

$$\underset{\sim}{S}_u = \underset{\sim}{Z} \, \underset{\sim}{S}_Q \, \underset{\sim}{Z}^T \tag{50}$$

The spectral density of any other parameters related to u such as the stresses can now be obtained. Assume that for an element the stresses can be written in matrix form as

$$\underset{\sim}{M}^e = \underset{\sim}{B}^e \underset{\sim}{U}^e \tag{51}$$

where e refers to the element. The spectrum for M stresses is,

$$\underset{\sim}{S}_M^e = \underset{\sim}{B}^e \underset{\sim}{S}_u^e \underset{\sim}{B}^{e,T} \tag{52}$$

The above theory can be simplified considerably if we neglect the cross-spectral density terms. Following references [4] we will assume that terms of the type

$$\alpha_i \, S_{F'_{ij}} \, \alpha_j^* \qquad (i \neq j) \tag{53}$$

can be neglected if the lower natural frequencies are well separated and if the damping is fairly low. In this way the product $\alpha_i \ \alpha_j^*$ is very small (see Figure 5).

The diagonal terms of (48) are the only to remain, hence we can write,

$$S_{Q_{ii}} = |\alpha_i|^2 \, S_{F'_{ii}} \tag{54}$$

where i = 1,S.

Once these densities are known we can find the spectral density of the displacements $\underset{\sim}{U}$ by transforming,

$$\underset{\sim}{S}_u = \underset{\sim}{Z} \, \underset{\sim}{S}_Q \, \underset{\sim}{Z}^T \tag{55}$$

where $\underset{\sim}{S}_Q$ is now a diagonal matrix with $S_{Q_{ii}}$ elements.

Let us now write the expression for the force spectral density. We have,

$$\begin{aligned} \underset{\sim}{F}(t) &= (\underset{\sim}{M}_H + \underset{\sim}{M}_A)\ddot{\underset{\sim}{V}} + \underset{\sim}{C}_D \dot{\underset{\sim}{V}} \\ &= \underset{\sim}{M}_I \ddot{\underset{\sim}{V}} + \underset{\sim}{C}_D \dot{\underset{\sim}{V}} \end{aligned} \tag{56}$$

Water particles velocities and accelerations can be written in function of two transfer expressions which give them in terms of wave height $\eta$.

$$\begin{aligned} \ddot{\underset{\sim}{V}} &= \underset{\sim}{A}_{\ddot{V}} \, \eta(t) \\ \dot{\underset{\sim}{V}} &= \underset{\sim}{A}_{\dot{V}} \, \eta'(t) \end{aligned} \tag{57}$$

where $\underset{\sim}{A}$ are transfer function vectors with elements,

$$\begin{aligned} (A_{\ddot{V}})_i &= \omega^2 \frac{\cosh \kappa z_i}{\sinh \kappa d} \cos \kappa x_i \\ (A_{\dot{V}})_i &= \omega \frac{\cosh \kappa z_i}{\sinh \kappa d} \cos \kappa x_i \end{aligned} \tag{58}$$

z is the distance from the bottom and $\cos \kappa x_i$ takes into account the horizontal distance between nodes. The amplitude of the wave and hence the spectrum is given for $x_i = 0$ position.

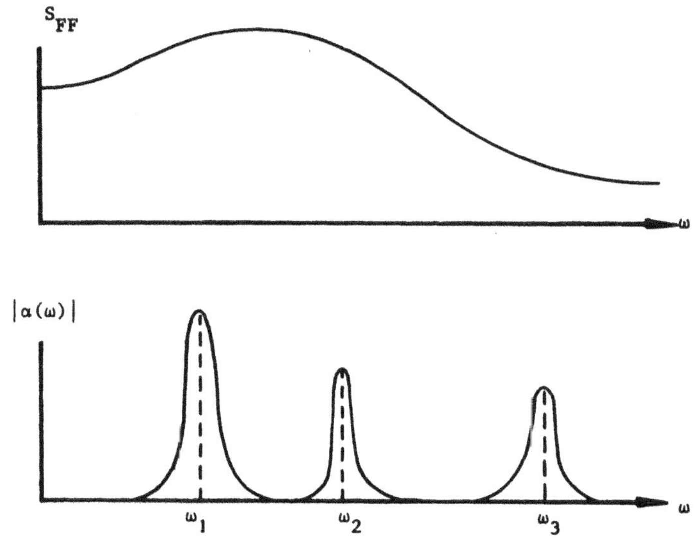


Figure 5    Power Spectral Density and Transfer Functions

The $\underset{\sim}{F}$ vector now becomes,

$$\underset{\sim}{F} = (\underset{\sim}{M}_I \underset{\sim}{A}_V \ddot{\eta} + \underset{\sim}{C}_D \underset{\sim}{A}_V \eta') \qquad (59)$$

The Fourier transform of $\underset{\sim}{F}$ and its conjugate give

$$\bar{\underset{\sim}{F}} = \bar{\underset{\sim}{A}}\,\bar{\eta} \;, \qquad \hat{\underset{\sim}{F}} = \hat{\underset{\sim}{A}}\,\hat{\eta}$$

Multiplying the F's and dividing by T we have

$$\underset{\sim}{S}_F = \underset{\sim}{A}\, S_{\eta\eta}\, \underset{\sim}{A}^T \qquad (60)$$

The force spectrum for the generalized coordinates is given by

$$\underset{\sim}{S}_F' = \underset{\sim}{Z}^T \underset{\sim}{S}_F \underset{\sim}{Z} \qquad (61)$$

This theory can be easily implemented in a computer program. In reference [3] results for some frame structures are presented. It is interesting to note that if the wave energy is at low frequencies compared with the structural frequencies, the response is practically all in the first mode. This seems to indicate that for these cases a one degree of freedom system may give a good approximate value.

## 4. Closing Remarks

The random response analysis described in this chapter was based on a series of assumptions necessary to linearize the problem and to facilitate its solution. In spite of this the method is a powerful design tool when compared against other experimental or analytical techniques. By contrast with the design wave approach it takes into consideration the way in which the wave energy of the sea is distributed. Reduced model studies of these structures are practically impossible due to the number and complexity of the phenomena involved. Thus, with all its limitations, random analysis appears to be the only feasible method of analysing them.

The main drawback of the theory is that, being a linear superposition technique, it does not take into account neither geometrical and material nonlinearities nor nonlinear waves. Waves which appear in high seas and those distorted due to wind are not well represented by linear waves theories.

Possible future extensions of the method of immediate interest are to consider the soil structure interaction and to analyse the fatigue in the structural joints.

Both analyses can be attempted with random vibrations, but they are beyond the scope of this chapter. Interested readers are referred to

References and Bibliography


1. Bendat, J.S. and A.G. Piersol, *Random Data: Analysis and Measurement*, Wiley, New York, 1971.

2. Brebbia, C.A. and S. Walker, *Dynamic Analysis of Offshore Structures*. Butterworths, London, 1979.

3. Charlton, P. *Fluid-Structure Interaction with Particular Reference to the Dynamic Behaviour of Framed Structures subjected to Random Waves*. M.Sc. Thesis, Department of Civil Engineering, Univeristy of Southampton, October 1974.

4. Hurty, W.C. and M.F. Rubinstein, *Dynamics of Structures*, Prentice Hall, 1964.

5. Kallaby, J. and J.B. Price, *Evaluation of fatigue considerations in the design of framed offshore structures*. OTC page 2609, Proc. Offshore Technology Conf, Houston, 1976.

6. Lin, Y.K. *Probabilistic Theory of Structural Dynamics*, McGraw-Hill, 1967.

7. Malhotra, A.K. and J. Penzien, *Nondeterministic Analysis of Off-shore Structures*, Proc. ASCE, Mech. Div., December 1970.

8. Marshall, P.W. *Dynamic and fatigue analysis using Directional Spectra*. OTC, 2537, Proc. Offshore Technology Conf, Houston, 1976.

9. Nataraja, R. and C.L. Kirk. *Dynamic Response of a Gravity Platform under Random Wave Forces*, OTC 2904, Proc. Offshore Technology Conf, Houston, 1977.

10. Robson, J.D. *Random Vibrations*, Edinburgh University Press, 1963.

11. Vughts, J.H. and F.K. Kirk , " *Probabilistic Analysis of Fixed Offshore Structures*," OTC 2608, Proc. Offshore Technology Conf. Houston, 1976.

# Lecture Notes in Engineering

Edited by C.A. Brebbia and S.A. Orszag

Vol. 1: J. C. F. Tellse,
The Boundary Element Method
Applied to Inelastic Problems
IX, 243 pages. 1983.

Vol. 2: Bernard Amadei,
Rock Anisotropy and
the Theory of Stress Measurements
XVIII, 479 pages. 1983.

Vol. 3: Computational Aspects of
Penetration Mechanics
Proceedings of the Army Research
Office Workshop on Computational
Aspects of Penetration Mechanics
held at the Ballistic Research Laboratory
at Aberdeen Proving Ground, Maryland,
27–29 April, 1982
Edited by J. Chandra and J. E. Flaherty
VII, 221 pages. 1983.

Vol. 4: W.S. Venturini
Boundary Element Method in Geomechanics
VIII, 246 pages. 1983.

Vol. 5: Madassar Manzoor
Heat Flow Through Extended
Surface Heat Exchangers
VII, 286 pages. 1984.

Vol. 6: Myron B. Allen III
Collocation Techniques for Modeling
Compositional Flows in Oil Reservoirs
VI, 210 pages. 1984.

Vol. 7: Derek B. Ingham,
Mark A. Kelmanson
Boundary Integral Equation
Analyses of Singular, Potential,
and Biharmonic Problems
IV, 173 pages. 1984.

Vol. 8: Linda M. Abriola
Multiphase Migration of Organic
Compounds in a Porous Medium
A Mathematical Model
VIII, 232 pages. 1984.

Vol. 9: Theodore V. Hromadka II
The Complex Variable Boundars
Element Method
XI, 243 pages. 1984.

Vol. 10: C. A. Brebbia, H. Tottenham,
G. B. Warburton, J. M. Wilson, R. R. Wilson
Vibrations of Engineering Structures
VI, 300 pages. 1985.

GPSR Compliance

*The European Union's (EU) General Product Safety Regulation (GPSR) is a set of rules that requires consumer products to be safe and our obligations to ensure this.*

*If you have any concerns about our products, you can contact us on ProductSafety@springernature.com*

In case Publisher is established outside the EU, the EU authorized representative is:

Springer Nature Customer Service Center GmbH
Europaplatz 3
69115 Heidelberg, Germany

**Batch number: 10723355**

Printed by Printforce, the Netherlands